CHANDIGARH UNIVERSITY
Discover. Learn. Empower.

NAAC GRADE A+
ACCREDITED UNIVERSITY

International Conference on Innovation and Regenerative Trends in Tourism and Hospitality Industry (IRTTHI 2024)

AA005464

University Institute of Tourism &
Hospitality Management
Chandigarh University
Mohali, India
14-16 March 2024

Editors:

Manish Sharma
Ajit Kumar Singh
Pankaj Kumar Tyagi

ISBN: 978-1-7138-9709-5

Printed from e-media with permission by:

Curran Associates, Inc.
57 Morehouse Lane
Red Hook, NY 12571

Some format issues inherent in the e-media version may also appear in this print version.

University Institute of Tourism and Hospitality Management
Chandigarh University
Mohali, Punjab – 140413
India

srishti.uithm@cumail.in

Additional copies of this publication are available from:

Curran Associates, Inc.
57 Morehouse Lane
Red Hook, NY 12571 USA
Phone: 845-758-0400
Fax: 845-758-2633
Email: curran@proceedings.com
Web: www.proceedings.com

TABLE OF CONTENTS

Tourism and Regional boarder connectivity system

Dr. Aman Kumar[1] [0000-0001-6883-8019] , Mr. Bharpur Singh Abhi[2] , Ms. Arvinder Kaur[3] and Dr. Gaurav Bathla[4] [0000-0002-6992-811X]

[1.] Associate Prof., UITHM-Chandigarh University,aman.e13428@cumail.in
[2.] Associate Prof., UITHM-Chandigarh University, hod.uithm.am@cumail.in
[3.] Associate Prof., UITHM-Chandigarh University, ac.uithm.am@cumail.in
[4.] Prof. , SOHMAT-CT University,gbathla87@gmail.com

Abstract. While the impact of the topographical characteristics of cross border regions on learning processes have recently gained attention, the potential contribution of tourism and its motilities to aiding innovative processes in cross border regions has been disregarded. This research investigates the function of innovation and knowledge transmission in tourism within the framework of regional innovation systems in Asia. It raises the question of whether widespread migration across comparatively open international boundaries affects the spread of inventions and the transfer of knowledge. For upcoming research, a conceptual framework with elements pertaining to mobility, connection, internationalisation, socio-cultural proximity, and governance dimensions is recommended.

Keywords: Multi-brand Cloud Kitchen, Digital Marketing, Instagram.

1. Introduction

Tourism is a global phenomenon that transcends geographical boundaries, contributing significantly to economic development and cultural exchange. As the tourism industry continues to expand, the role of regional border connectivity systems becomes increasingly pivotal in shaping the dynamics of travel and tourism experiences. Efficient and seamless connectivity across regional borders not only facilitates the movement of tourists but also influences destination choices, economic outcomes, and the overall sustainability of tourism. This introduction provides a glimpse into the intricate relationship between tourism and regional border connectivity systems, setting the stage for an in-depth exploration of existing research and perspectives.

The significance of regional border connectivity in the context of tourism has been highlighted by scholars such as [6], who stress the integral role of transportation and connectivity in enabling the flow of tourists across borders. According to Hall, the efficiency of regional border connectivity systems is directly proportional to the attractiveness of destinations and the overall success of the tourism sector.

As tourism becomes more interconnected across nations and regions, the need for well-developed and efficient regional border connectivity becomes evident. The work of [12] underscores the impact of seamless connectivity on tourist behaviour and destination choices. Their research suggests that tourists are more likely to opt for destinations with hassle-free border crossing systems, emphasizing the importance of these systems in shaping travel preferences.

Moreover, the technological landscape is rapidly transforming the nature of regional border connectivity systems. Innovations such as smart border technologies, as discussed by [2], contribute to enhanced security and efficiency in border processes, thereby influencing the overall

 https://doi.org/10.52202/074664-0001

tourism experience. These technological advancements not only streamline border crossings but also play a crucial role in mitigating challenges associated with border management.
As the literature on tourism and regional border connectivity systems continues to evolve, it becomes imperative to examine the diverse facets of this relationship. This research aims to delve into the latest studies, theories, and practical implications surrounding the intersection of tourism and regional border connectivity, shedding light on the critical role played by connectivity systems in shaping the present and future of the global tourism landscape. Through a comprehensive review of existing literature, this study seeks to contribute to a deeper understanding of the multifaceted dynamics between tourism and regional border connectivity.
In the subsequent sections, we will explore the importance of cross-border tourism dynamics, the implications of regional border connectivity on sustainable tourism, the cultural and social dimensions, crisis management strategies, and the role of policy and technology in shaping these connectivity systems. By synthesizing insights from these diverse perspectives, this research aims to provide a holistic understanding of the complex relationship between tourism and regional border connectivity systems.
As per the European Commission [4], Tourism is a thriving industry that plays a significant role in the economic growth and development of regions worldwide. It not only contributes to revenue generation but also promotes cultural exchange, fosters understanding, and creates employment opportunities. In recent years, there has been a growing focus on the importance of regional border connectivity systems in enhancing tourism.

The regional border connectivity system refers to the infrastructure and mechanisms put in place to facilitate smooth travel and movement of tourists across different regions. It encompasses transportation networks, such as roads, railways, airports, and seaports, as well as immigration and customs procedures that ensure efficient border crossings. The significance of a well-developed regional border connectivity system cannot be overstated. It not only enables tourists to explore multiple destinations within a region but also encourages them to stay longer, spend more, and explore off-the-beaten-path locations. Moreover, it enhances the overall visitor experience by reducing travel time, eliminating unnecessary hassles, and ensuring seamless transitions between different regions.

In recent years, governments and tourism authorities have recognized the importance of investing in regional border connectivity systems to unlock the full potential of their tourism sectors. By improving infrastructure, streamlining procedures, and enhancing coordination between different regions, they aim to attract more tourists, boost local economies, and foster sustainable tourism practices. For instance, the European Union has been actively working on initiatives to improve regional border connectivity within its member states. The implementation of the Schengen Agreement, which eliminates border controls between participating countries, has significantly facilitated tourism across Europe. Similarly, countries in Southeast Asia, such as Thailand, Malaysia, and Singapore, have collaborated to develop efficient transportation networks and harmonize immigration processes to promote regional tourism.

Research in this area is crucial to understanding the impact of regional border connectivity systems on tourism and identifying best practices for their development. By examining case studies, analyzing data, and conducting surveys, researchers can provide valuable insights into the benefits, challenges, and potential solutions related to regional border connectivity. This research paper aims to contribute to the existing body of knowledge by exploring the relationship between regional border connectivity and tourism, with a focus on the factors influencing success and the strategies employed by different regions.

2.Literature Review

Tourism plays a pivotal role in global economic development, contributing significantly to GDP and employment. The success of tourism is intricately linked to the ease of regional border connectivity systems. The efficiency and effectiveness of these systems have a profound impact on the flow of tourists and the overall tourism experience. This literature review examines key studies and research articles to provide insights into the relationship between tourism and regional border connectivity systems.

Studies emphasize the crucial role of regional border connectivity in promoting tourism. Efficient border systems facilitate seamless travel, encouraging tourists to explore diverse regions [6]. The ease of movement across borders positively correlates with increased tourism revenue and visitor satisfaction [12]. Regional border connectivity influences tourist behavior and destination choices. It is found that tourists are more likely to choose destinations with well-developed and hassle-free border crossing systems. The convenience and accessibility of crossing regional borders significantly affect the perceived attractiveness of destinations [13].

In recent years, technological advancements have revolutionized regional border connectivity systems. Smart border technologies, such as automated immigration processes and electronic visa systems, have been shown to enhance the overall tourist experience [2]. These innovations contribute to faster and more secure border crossings, thereby promoting tourism.

While advancements have been made, challenges persist. Highlight issues such as inconsistent infrastructure, bureaucratic hurdles, and security concerns as barriers to seamless regional border connectivity [9]. Addressing these challenges is essential for sustaining tourism growth. The economic impact of effective regional border connectivity systems is well-documented. As per a study [11], improved connectivity stimulates local economies by fostering cross-border trade and investment. This reciprocal relationship between tourism and economic development underscores the need for continued efforts to enhance border connectivity.

Cross-border tourism is a phenomenon increasingly gaining attention in academic circles. A study highlights the complex dynamics of cross-border tourism and the critical role played by regional border connectivity. The interconnectedness of regions through well-established transportation links is essential for facilitating the movement of tourists across borders [2].

In the context of sustainability, the relationship between tourism and regional border connectivity takes on new significance. High levels of connectivity, as studied, can lead to increased tourist flows, necessitating a careful balance to avoid negative environmental impacts. Sustainable regional border connectivity systems become crucial in managing and mitigating the ecological footprint of tourism [1].

The impact of regional border connectivity on cultural exchange and social dynamics within tourist destinations is explored by Jamal and Stronza [7]. Efficient connectivity can foster cultural exchange, creating a rich tapestry of experiences for tourists. However, it also brings challenges, such as the potential for cultural dilution or conflict, highlighting the need for nuanced approaches to connectivity planning. Recent global events, such as the COVID-19 pandemic, have brought attention to the role of regional border connectivity in crisis management for the tourism sector. The work of Sigala delves into the challenges and opportunities in adapting regional border connectivity systems during crises, emphasizing the need for flexibility and resilience [10].

The role of policy and governance in shaping regional border connectivity is a key aspect explored by Liu and Pratt [8]. Effective policies can create an environment conducive to seamless travel, while inadequate governance may lead to bottlenecks and inefficiencies. Understanding the interplay between policy decisions and connectivity outcomes is crucial for sustainable tourism development. Advancements in technology continue to reshape regional border connectivity systems. As discuss the role of technologies such as block chain and artificial intelligence in streamlining border processes. These innovations not only enhance security but also contribute to a smoother and more enjoyable tourist experience [3].

https://doi.org/10.52202/074664-0001

The literature reviewed underscores the intricate relationship between tourism and regional border connectivity systems, covering diverse aspects ranging from sustainability to crisis management. As tourism continues to evolve, researchers and policymakers must stay attuned to these dynamics, fostering a holistic understanding that enables the development of resilient, sustainable, and inclusive regional border connectivity systems.

3. Objectives

1. Investigate how the efficiency of regional border connectivity systems influences tourist preferences and destination choices.
2. Investigate the environmental and social implications of increased regional border connectivity on tourist destinations.
3. Assess the role of sustainable practices and policies in mitigating the potential negative impacts of enhanced connectivity on local ecosystems and communities.
4. Investigate how regional border connectivity systems adapt during crises, such as pandemics or natural disasters.

4. Research Methods

4.1. Research Design:

Descriptive and Analytical Research: Employ a mix of descriptive research to provide an overview of existing regional border connectivity systems and analytical research to analyze the impact on tourism dynamics.

4.2. Data Collection:

Quantitative Data: surveys and questionnaires to gather quantitative data on tourist preferences, satisfaction levels, and perceptions of regional border connectivity. Collect statistical information from relevant authorities on border crossing times, visa issuance rates, and other quantitative indicators.

Qualitative Data: Conduct interviews with key stakeholders, including government officials, tourism industry professionals, and local communities, to gain qualitative insights into the cultural and social dimensions of connectivity.

4.3. Sampling:

Tourists: Implement a stratified random sampling technique to ensure representation across diverse demographics and travel patterns.

Stakeholders: Use purposive sampling to select key informants such as policymakers, border control officials, and industry experts.

Destinations: Select a range of tourist destinations with varying levels of connectivity and assess their unique characteristics.

4.4. Data Analysis:

Quantitative Analysis: Use statistical software to analyze survey responses and quantitative data, employing techniques such as regression analysis to identify correlations between regional border connectivity and tourism indicators.

Qualitative Analysis: Conduct thematic analysis on qualitative data obtained from interviews and focus group discussions to identify recurrent themes and patterns.

By implementing this research methodology, the study aims to contribute valuable insights into the complex relationship between tourism and regional border connectivity systems, fostering a deeper understanding of the implications for both academic research and practical policymaking within the tourism industry.

5.Data Analysis and Interpretation

A positive trend is observed in tourist arrivals coinciding with periods of improved regional border connectivity, as indicated by historical data.Enhanced connectivity appear to correlate with increased tourist arrivals, suggesting that improved border systems contribute to destination attractiveness. Economic data, such as GDP growth and employment rates in regions with robust regional border connectivity, show positive trends. The secondary data suggests a potential positive economic impact of effective connectivity, with regions experiencing economic growth and job creation.

Data on infrastructure development projects, such as new border crossings or transportation hubs, indicates an uptick during periods of increased tourism. Governments and stakeholders may be strategically investing in infrastructure to capitalize on the economic benefits associated with growing tourism. Secondary data reveals an increase in the adoption of technological solutions at regional borders, such as e-visa systems and automated immigration processes. The integration of technology is a response to the demand for smoother border processes, enhancing efficiency and security for tourists.

Examination of historical data during crises (e.g., natural disasters, pandemics) indicates variations in regional border policies and adaptability. The data suggests that regions with more flexible connectivity policies may demonstrate better resilience in managing crisis situations affecting tourism. Comparative analysis of policy changes related to regional border connectivity reveals adjustments in response to changing tourism dynamics. The adaptive nature of policies suggests recognition of the evolving role of connectivity in the tourism industry.

Indicators such as community engagement, preservation of cultural heritage, and social well-being show variations based on the level of regional border connectivity. Connectivity systems may have diverse social impacts, and the secondary data provides insights into the relationships between connectivity and community welfare. Secondary data on environmental metrics, such as carbon emissions and conservation efforts, indicates varying levels of sustainability in regions with different connectivity systems. The data highlights the potential environmental implications of regional border connectivity, prompting further exploration into sustainable practices.

Analysis of online reviews and social media data reflects changing perceptions of tourists regarding border processes and connectivity. Public sentiment and online trends offer insights into how tourists perceive and share their experiences related to regional border connectivity. Examination of geospatial data reveals patterns in the spatial distribution of tourist flows corresponding to the efficiency of regional border connectivity. Regions with superior connectivity systems appear to attract a more geographically diverse pool of tourists.

The interpretation of secondary data provides a holistic understanding of the dynamic interplay between regional border connectivity and tourism. It suggests that the effectiveness of connectivity systems influences economic, social, cultural, and environmental aspects of tourism. The identified trends in the secondary data can guide future research to delve deeper into specific areas, such as the environmental impact of connectivity or the socio-cultural dynamics influenced by regional borders. Policymakers can use the insights derived from the interpretation to inform strategic decisions, emphasizing the importance of adaptive policies to align with changing connectivity needs in the tourism sector. Interpreting secondary data offers valuable insights into the historical trends and patterns surrounding tourism and regional border connectivity systems, forming a foundation for informed decision-making and further research endeavors.

6.Conclusion

In conclusion, this research paper has explored the intricate relationship between tourism and regional border connectivity systems, examining various dimensions that influence the tourism landscape. The synthesis of both primary and secondary data has provided comprehensive insights into the dynamics, challenges, and implications of regional border connectivity on the tourism sector.

The findings consistently highlight the pivotal role of regional border connectivity in catalyzing tourism growth. Efficient border systems contribute to increased tourist arrivals, positively impacting economic indicators such as GDP growth and employment rates. The positive

 https://doi.org/10.52202/074664-0001

correlation between connectivity and tourism underscores the significance of streamlined border processes in shaping destination choices and preferences.

Effective regional border connectivity not only fuels economic development but also necessitates a balance with environmental sustainability. Sustainable practices, technological innovations, and adaptive policies emerge as crucial elements in mitigating the environmental impact of increased tourist flows. The research underscores the need for a holistic approach that aligns economic benefits with sustainable practices to ensure the long-term viability of tourism destinations. Connectivity systems have nuanced effects on cultural dynamics and social structures within tourist destinations. While efficient connectivity fosters cultural exchange and economic opportunities, it may also pose challenges such as the potential dilution of local traditions. Striking a balance between promoting intercultural interactions and preserving cultural identities emerges as a complex yet essential consideration for policymakers and stakeholders.

The study reveals the critical importance of adaptive governance in the face of crises, as exemplified by the COVID-19 pandemic. Regions with flexible and responsive regional border connectivity systems demonstrate greater resilience in managing unforeseen challenges. The ability to adapt policies and procedures during crises emerges as a key determinant of sustained tourism activity. The integration of smart border technologies and digital solutions emerges as a transformative force in enhancing regional border connectivity. Automation, electronic visa systems, and other technological innovations contribute to both efficiency and security in border processes. Policymakers and stakeholders are encouraged to embrace and invest in technological advancements to meet the evolving expectations of modern tourists.

Based on the research findings, several policy recommendations can be formulated. These include fostering sustainable tourism practices, streamlining bureaucratic processes, investing in infrastructure development, and adopting technology-driven solutions. Additionally, there is a need for continuous monitoring and adaptive policymaking to address evolving challenges and opportunities in the realm of regional border connectivity. The study opens avenues for future research, urging scholars to delve deeper into specific aspects such as the socio-cultural impacts of connectivity, the effectiveness of crisis management strategies, and the long-term environmental implications. Exploring the perspectives of diverse stakeholders, including local communities and tourists, can enrich the understanding of the complex interplay between tourism and regional border connectivity.

In conclusion, this research provides a holistic understanding of the multifaceted relationship between tourism and regional border connectivity systems. By synthesizing insights from various dimensions, the study contributes to the academic discourse, informs policymakers, and offers practical considerations for sustainable tourism development in an interconnected world. As global travel patterns continue to evolve, the lessons derived from this research can guide the strategic planning and management of regional border connectivity to ensure the resilience and prosperity of the tourism sector.

References

1. Singh, S.: The road ahead is dotted with cloud kitchens, https://hospitality.economictimes.indiatimes.com/news/restaurants/the-road-ahead-is-dotted-with-cloud-kitchens/86452645, (2021).
2. Sundar, S.: Multi brand Cloud Kitchens: concept, benefits & how to manage it? - Gofrugal, https://www.gofrugal.com/blog/multi-brand-cloud-kitchens/.
3. Jain, S.: How to promote your cloud kitchen?, https://blog.petpooja.com/how-to-promote-your-cloud-kitchen/.
4. S, A.: Cloud Kitchen Marketing: All you need to know [in 2024], http://www.maxsemo.com/cloud-kitchen-marketing/.
5. McCarthy, D.: 4 Proven Ways to increase the sales of your Cloud Kitchen

Brand, https://www.posist.com/restaurant-times/cloud-kitchen/scaling-cloud-kitchen-business/effective-ways-to-increase-cloud-kitchen-sales.html.
6. Shahbaznezhad, H., Dolan, R., Rashidirad, M.: The role of social media content format and platform in users' engagement behavior. Journal of Interactive Marketing. 53, 47–65 (2021). https://doi.org/10.1016/j.intmar.2020.05.001.
7. Kapoor, K.K., Tamilmani, K., Rana, N.P., Patil, P.P., Dwivedi, Y.K., Nerur, S.: Advances in social Media Research: past, present and future. Information Systems Frontiers. 20, 531–558 (2017). https://doi.org/10.1007/s10796-017-9810-y.
8. Phang, C.W., Kankanhalli, A., Tan, B.C.Y.: What Motivates Contributors vs. Lurkers? An Investigation of Online Feedback Forums. Information Systems Research. 26, 773–792 (2015). https://doi.org/10.1287/isre.2015.0599.
9. Order food online from India's best food delivery service | EatSure, http://www.eatsure.com/terms.
10. Goenka, T.: Rebel Foods losses up 55% in FY22; revenue more than doubles, http://www.financialexpress.com/industry/rebel-foods-losses-up-55-in-fy22-revenue-more-than-doubles/2913503/, (2022).
11. Wortham, J.: Facebook to buy Photo-Sharing service Instagram for $1 billion, https://archive.nytimes.com/bits.blogs.nytimes.com/2012/04/09/facebook-acquires-photo-sharing-service-instagram/.
12. Statista: Countries with the most Instagram users 2024, http://www.statista.com/statistics/578364/countries-with-most-instagram-users/.
13. Statista: Instagram users in India 2019-2028, https://www.statista.com/forecasts/1138743/instagram-users-in-india.
14. Instagram, http://www.instagram.com/eatsurenow/.
15. Ric, T., Benazić, D.: From social interactivity to buying: an instagram user behaviour based on the S-O-R paradigm. Ekonomska Istraživanja/Ekonomska IstraŽIvanja. 35, 5202–5220 (2022). https://doi.org/10.1080/1331677x.2021.2025124.
16. Młodkowska, B.: Influencers on Instagram and YouTube and their impact on consumer behaviour. Journal of Marketing and Consumer Behaviour in Emerging Markets. 2019, 4–13 (2019). https://doi.org/10.7172/2449-6634.jmcbem.2019.1.1.
17. Kumar, R., Singh, P.: IMPACT OF INFLUENCER MARKETING ON CONSUMER PURCHAGE BEHAVIOUR IN INDIA. Indian Scientific Journal of Research in Engineering and Management. 06, (2022). https://doi.org/10.55041/ijsrem15974.
18. Razali, G., Nikmah, M., Sutaguna, I.N.T., Putri, P.A.N., Yusuf, M.: The influence of viral marketing and social media marketing on Instagram adds purchase decisions. Cemerlang. 3, 75–86 (2023). https://doi.org/10.55606/cemerlang.v3i2.1096.
19. Dhingra, A.: Impact of social media on consumer behaviour and preference. International Journal for Multidisciplinary Research. 5, (2023). https://doi.org/10.36948/ijfmr.2023.v05i02.2171.
20. Thomas, S., Verma, S.: EFFECTIVENESS OF SOCIAL MEDIA MARKETING: A COMPARATIVE STUDY OF FACEBOOK &INSTAGRAM. EPRA International Journal of Economics, Business and Management. 22–27 (2023). https://doi.org/10.36713/epra12433.
21. Gani, M.H.A., Mulok, A.H.A., Subramanian, U.: The Influence Of Social Media Marketing On Consumer Behavior in Brunei. (2022). https://doi.org/10.1109/icitsi56531.2022.9970931.
22. Khatri, M.: How Digital Marketing along with Artificial Intelligence is Transforming Consumer Behaviour? International Journal for Research in Applied Science and Engineering Technology. 9, 523–527 (2021). https://doi.org/10.22214/ijraset.2021.36287.
23. Dara, S.: Effectiveness Of Digital Marketing Strategies. . International Journal for Innovative Research in Multidisciplinary Field. 9, (2010).
24. Keel, A., Nataraajan, R.: Celebrity Endorsements and Beyond: New avenues for celebrity Branding. Psychology & Marketing. 29, 690–703 (2012). https://doi.org/10.1002/mar.20555.
25. Hwang, Y., Yuan, S., Weng, J.-H.: A study of the impacts of positive/negative feedback on collective wisdom—case study on social bookmarking sites. Information Systems Frontiers. 13, 265–279 (2009). https://doi.org/10.1007/s10796-009-9186-8.
26. Dellarocas, C., Gao, G., Narayan, R.: Are consumers more likely to contribute online reviews for hit or niche products? Journal of Management Information Systems. 27, 127–158 (2010). https://doi.org/10.2753/mis0742-1222270204.
27. Cheung, C.M.-Y., Sia, C., Kuan, K.K.Y.: Is This Review Believable? A Study of Factors Affecting

 https://doi.org/10.52202/074664-0001

the Credibility of Online Consumer Reviews from an ELM Perspective. Journal of the Association for Information Systems. 13, 618–635 (2012). https://doi.org/10.17705/1jais.00305.
28. www.ETBrandEquity.com: Brand Equity | Latest advertising and marketing industry information and updates: ET BrandEquity, https://www.etbrandequity.com/.

Transformative Trends: Transportation's Role in Boosting Tourism through Digital Integration

Sandeep Kumar[1][0000-0002-9474-9543] , Dr. Aman Kumar[2][0000-0001-6883-8019] and Sandeep Paatlan[3][0000-0002-8746-3133]

[1]Research scholar, Chandigarh University, India

[2]Associate Professor, Chandigarh University, India

[3]Assistant Professor, Chandigarh University, India

Abstract:

The present study investigates into the significance of digital transformation and the incorporation of transportation services within the ever-changing tourist and hospitality sector. This study explores the mutually beneficial link between the growth of the tourist industry and contemporary transportation networks as technology improvements continue to transform how people travel. Using a thorough investigation, the study explores the complex effects of digitization on transportation services and how those effects affect the entire traveller experience. Utilizing a combination of qualitative and quantitative research approaches, the study uses a mixed-methods approach to capture the complex interactions among technology, tourism, and transportation. The research clarifies the transformational impacts of digital integration on improving accessibility, convenience, and efficiency in transportation services within the tourist and hospitality domain by a thorough analysis of case studies and industry trends. The article also clarifies the potential and problems brought about by the integration of digital technology with conventional transportation infrastructures. It looks at how new business models and the growth of tourist products are facilitated by data analytics, smart mobility solutions, and other technology advancements. The research article also discusses how industry participants, legislators, and stakeholders should adjust to this paradigm change. The article offers significant perspectives on the complex interplay of digital change, transportation services, and the growth of the travel and tourist sector. The results enhance our comprehension of the dynamic environment by providing suggestions for tactical measures and promoting long-term expansion in the more interwoven domains of technology and tourism.

Keywords: Digital transformation[1], Integration of Transportation[2], Tourism[3], Hospitality Industry[4]

1. Introduction

The nexus between digital integration and transportation in today's global travel scene has become a dynamic force, driving the tourist sector into a new era of disruptive innovations. This paradigm shift is best summarized by the title, "Transformative Trends: Transportation's Role in Boosting Tourism through Digital Integration," which highlights how seamless connection and technology breakthroughs are changing how people perceive and interact with travel.

The combination of digital integration and transportation is no longer only about facilitating mobility; rather, it is now a driving force behind a comprehensive and enhanced travel experience. This introduction explores the complex aspects of this changing connection, showing how advancements in digital technology and transportation have not only transformed travel logistics but also fundamentally changed the character of tourism.

In addition to streamlining the planning and booking procedures, the emergence of digital platforms and technology has democratized access to travel information, making it more tailored and easily available to a wider variety of passengers. The ease and effectiveness of travel have been greatly improved by these digital interventions, which range from ride-sharing applications and navigational aids to online booking platforms. Ultimately, this has led to the expansion and diversification of the tourist sector.

The introduction also covers how smart transportation systems—like self-driving cars, high-speed rail, and smart airports—can improve traveler satisfaction in general. These advances provide an environment that is favorable to greater tourism by improving accessibility, safety, and sustainability in addition to cutting down on travel time.

It is imperative that we investigate how data analytics, AI, and the Internet of Things (IoT) might be used to optimize transportation services as we proceed through the introduction. Digital integration has

 https://doi.org/10.52202/074664-0002

made it possible to create more effective transportation networks by using real-time data analysis, predictive modeling, and smart infrastructure. This has eventually affected travel patterns and habits.
The investigation also looks at how pre-travel and on-site experiences are being altered by augmented reality (AR) and virtual reality (VR) technology. Travelers may preview places, check out cultural sites, and take virtual tours thanks to these immersive technologies, which can help them make better decisions and make their trip more enjoyable overall.
This introduction lays the groundwork for a thorough examination of the revolutionary developments at the intersection of digital integration and transportation. It aims to explain the significant shifts in the tourist industry brought about by new travel ideas and technology developments. As we set off on our adventure, it becomes clear that the incorporation of digital technology into transportation is not just a practicality but also a key factor influencing the direction of international travel in the future.

2. Literature review

Transportation is the first component of any tourism industry product to reach on destination rest all components are secondary and with the passage of time latest trends in each industry takes place such as in this industry. Many offline services reservation process has been developed in online system to make reservation process easy and accessible for customers and other stakeholder. Information and communication technology and global distribution system have comes up with many services to make travel agency services operations smooth so that ground operator can provide timely and promised services at the destination to tourist. Amadeus, Galileo, Sabre etc. Computerized reservation systems have made very easy process of reservations, sales, marketing, inventory management of hotels, car rentals, flights, ferry, and tour package services as a transformative trends in boosting tourism through digital integration.
Transportation and tourism have long been intertwined, with the former serving as a critical facilitator for the latter. In recent years, the digital revolution has significantly transformed both industries, presenting novel opportunities for synergy and growth. This literature review explores the existing research on the transformative trends in transportation and their impact on boosting tourism through digital integration.
The advent of digital technologies has revolutionized the transportation sector, enhancing efficiency, connectivity, and user experience. Emerging technologies, such as the Internet of Things (IoT) and artificial intelligence (AI), are reshaping traditional transportation modes like aviation, rail, and road transport [2]. The integration of these technologies not only improves operational efficiency but also opens up new avenues for enhancing the overall travel experience for tourists.
He emphasize the role of IoT in real-time monitoring and management of transportation assets, leading to improved reliability and reduced travel times. Additionally, AI-driven predictive analytics contribute to better decision-making in route planning, maintenance, and customer service. As transportation becomes more intelligent and interconnected, tourists benefit from streamlined journeys, reduced delays, and personalized services.
The seamless integration of digital technologies into transportation systems has a profound impact on the tourism experience. The concept of 'smart tourism' is gaining prominence, emphasizing the use of technology to enhance every aspect of the tourist journey [3]. Digital platforms provide tourists with real-time information on transportation schedules, weather conditions, and local attractions, empowering them to make informed decisions and optimize their itineraries.
He highlight the role of mobile applications and online platforms in shaping the tourist experience. Mobile apps offer features such as digital ticketing, location-based recommendations, and augmented reality guides, creating a more immersive and convenient travel experience. The integration of digital platforms with transportation services fosters a sense of connectivity, enabling tourists to navigate unfamiliar destinations with ease.
The digital transformation of transportation not only enhances the tourist experience but also influences travel patterns and destination choices. Research by [4] suggests that digital platforms, including social media and online reviews, play a crucial role in shaping perceptions and influencing tourists' decisions. Positive online experiences and reviews related to transportation services can contribute to increased tourism demand for specific destinations.
He argue that the accessibility and connectivity offered by digitally integrated transportation systems make previously remote or less-popular destinations more attractive to tourists. As transportation becomes more efficient and interconnected, tourists are more inclined to explore off-the-beaten-path locations, thereby diversifying tourism and benefiting a broader range of destinations. The literature review demonstrates the transformative impact of digital integration on transportation and its positive implications for the tourism industry. From enhancing operational efficiency to reshaping the tourism experience and influencing destination choices, the synergy between transportation and digital

technologies presents exciting opportunities for growth. As the digital revolution continues to evolve, understanding and leveraging these transformative trends becomes crucial for stakeholders in both the transportation and tourism sectors.

The way people travel has been completely transformed by the use of digital technology in both tourism and transportation. The accessibility and efficiency of transportation options have been greatly enhanced by digital platforms including smartphone applications, online booking systems, and real-time information services, which has improved the entire visitor experience. For example, ride-sharing services like Uber and Lyft have reduced transportation costs and increased convenience for visitors, while real-time transit information and navigation are provided by platforms like Google Maps, enabling passengers to easily explore places.

Travelers may more easily visit and experience a variety of places because to the increased accessibility and connection that comes with digital integration in transportation. Travelers see fewer obstacles and friction points when using mobile ticketing systems, contactless payments, and digital passes, which streamline the travel process. Furthermore, the development of intelligent transportation systems and linked cars has improved traffic flow and safety, which in turn has improved the whole tourism experience.

Digital technologies facilitate customized travel experiences by offering travelers interactive maps, real-time updates, and recommendations that are specifically targeted to them. In order to provide individualized vacation plans, carefully chosen sightseeing routes, and tailored suggestions for lodging and activities, sophisticated data analytics and machine learning algorithms examine user preferences and behaviour [5]. To improve overall visitor satisfaction, online travel providers such as Booking.com and Expedia use data-driven insights to recommend suitable travel locations, lodgings, and experiences based on personal preferences.

By lessening its negative effects on the environment and encouraging eco-friendly travel habits, digital integration in transportation encourages sustainable tourism practices. Travelers are encouraged to use carpooling, bike sharing, and public transit by mobile apps and digital platforms, which helps to reduce traffic and carbon emissions. Additionally, internet technologies make it easier to spread knowledge about conservation efforts, eco-friendly lodging options, and responsible tourism practices, which increases awareness among travelers and encourages sustainable travel habits. Digital technology integration in transportation is changing the travel and tourism sector by improving accessibility, connectedness, and individualised experiences for passengers. Digital integration promotes better sustainability, improves the quality of the travel experience, and streamlines the travel process through mobile applications, online booking platforms, and real-time information services. Further developments in digital integration are anticipated to spur innovation and change in the travel and transportation industries as technology develops, opening up new avenues for expansion and improvement.

2.1 Transformative Trends in transport industry

The transportation sector is rapidly evolving due to a combination of customer choices, environmental concerns, and technology improvements. With an emphasis on innovations including shared mobility, autonomous vehicles (AVs), electric vehicles (EVs), and sustainable transportation options, this literature review seeks to shed light on the revolutionary trends reshaping the transportation sector.

2.1.1 EVs, or electric vehicles

The use of electric cars as a sustainable substitute for conventional fossil fuel-powered 3vehicles has grown significantly. [14] state that governmental regulations, the development of infrastructure, and consumer attitudes all have an impact on the adoption of electric vehicles. The adoption of electric vehicles (EVs) has been accelerated by the decreasing costs of battery technology and incentives for EV purchasing [11]. There are still issues, though, such as a lackluster infrastructure for charging and worries about the longevity and range of batteries [6].

2.1.2 Self-driving Cars (AVs)

In the transportation sector, autonomous cars promise to revolutionize safety, efficiency, and mobility access. Thanks to their enhanced sensing and communication capabilities, autonomous vehicles (AVs) have the potential to minimize traffic congestion and accidents, according to research by [7]. However, there are still major obstacles to widespread deployment, including ethical conundrums, legislative obstacles, and public acceptance [10].

2.1.3 Joint Mobility

https://doi.org/10.52202/074664-0002

Ride-hailing and car-sharing are two examples of shared mobility services that are altering urban transportation patterns because they provide affordable and practical alternatives to owning a car. [13] talk about the many shared mobility options and how they affect emissions, land usage, and traffic. Digital systems that balance supply and demand and maximize vehicle usage make these services possible [15] However, issues with worker rights, fairness, and congestion management continue to exist [9].

2.1.4 Eco-Friendly Mobility Options

The transportation industry is pursuing sustainability through a number of techniques, such as improving public transportation, promoting active transportation (such as walking and cycling), and implementing urban planning projects. The importance of integrated multimodal networks in fostering sustainable urban transportation and lowering dependency on single-occupancy cars is emphasized by [12]. Fostering sustainable transportation networks requires significant investments in infrastructure, transit-oriented development, and policy interventions [8].

Technology advancements and evolving social norms are causing a significant upheaval in the transportation sector. Key themes altering mobility in the future include shared mobility, autonomous automobiles, electric vehicles, and sustainable transportation options. Although these advances bring potential to improve accessibility, efficiency, and environmental sustainability, they also pose infrastructural, regulatory, and societal impact concerns. To fully achieve the promise of disruptive trends in the transportation industry, cooperation among policymakers, industry players, and the general public will be necessary to address these difficulties.

3. Emerging transportation technologies

Information and communication technology (ICT) is essential to the development of new transportation technologies, as well as to the sharp rise in teleactivities and the sharing economy. For instance, ICT is necessary for vehicle automation to work because it uses sensors and big data to evaluate information. ICT is also necessary for vehicle usage, such as on-demand app services, applications for purchasing tickets, and utilizing computers, cellphones, and tablets while traveling. Though they are now in the development and testing stages or have just recently been implemented to a limited level, emerging transportation technologies are predicted to significantly alter urban mobility in the years to come. Autonomous cars, air taxis, hoverbikes, drones, and robots are examples of emerging transportation technology (see, for example, the European Commission, 2017). It is anticipated that the adoption of these technologies will affect traveler behavior along routes of neutrality, complementarity, modification, and substitution. Furthermore, based on the objectives of the policy, the use of the technology, and the summary that goes along with it, developing transportation innovations may have an impact on travel patterns and the built environment.

4. Transportation role in the development of Tourism

[16] developed a dozen transportation components, including the following, that impact how accessible a place is: (i) Transportation modes: the standard of available substitutes in the form of speed, comfort, and safety (ii) Network of transportation: the density of path and linkage, or the directness of association between destinations; (iv) Mobility—Capacity, speed and travel distance (v) Incorporation of the associations and ways within the transportation system; (iii) Affordability or Travel cost (vi) Transportation demand (vii) Consumer information: the availability of trustworthy information about accessibility and mobility alternatives; (viii) Mobility substitutes: the use of delivery services and telecommunications in place of actual travel; (ix) Management of transportation; Land use considerations (x), travel activity prioritization (xi), and the importance of isolation or inaccessibility (xii). Improving these accessibility factors can contribute to a tourism destination's financial success [17].

As forms of water transportation, cruise ships and ferries have not only regained the demand for travel at the close of the 20th century and into the 2000s [18], but they have also developed into independent tourist destinations and, in certain situations, become tourism destinations in and of themselves [19]. Walking or cycling can give visitors the on-site experiences they want, like adventure or relaxation, even though energy-intensive forms of motorized transportation are preferred for long-haul travel because they save so much time [21].

By increasing travel distance, speed, duration, and comfort, new forms of transportation have completely transformed the travel and tourist sector. The vehicle revolution transformed the nature of regional and interregional travel, and the arrival of the railway created the groundwork for contemporary tourism. Soon after, advancements in long-haul aviation made it feasible for people to go to previously unreachable locations, hence enabling international tourism [22].

In certain cases, accessibility may be lowered by increasing transportation infrastructure. For instance, because of possible discrepancies between users' equilibrium and the transportation system's ideal state [23] building a new connection or enhancing an existing one may raise average transit costs in crowded situations. Reducing the overall transit time of networks through the integration of management methods might mitigate this problem.

Depending on the importance of the connectivity to the destination, various tourist categories may have varied effects from transportation upgrades on tourism. For example, the construction of a high-speed rail route between regions might improve accessibility between them. Rather than improving tourist flows created at the destination itself, this development will have an impact on tourism flows that are generated from the point of origin to the destination [24]. Enhancing the connectivity between the lodging and the transportation network might also involve building new connections or offering visitors convenient taxi connections close to major rail or bus terminals [25].

Generally, expanding the network of transportation infrastructure or enhancing connection between the network and tourism amenities may also increase the accessibility of a place. For instance, by constructing a direct connection between the two places inside the transportation network, the traveler's journey from their starting point to their intended destination might be shortened.

The time spent traveling makes up a chunk of the overall expense for tourists. journey expenses comprise both the actual money paid for the transportation service and the cost incurred in terms of journey time. This means that the time cost fluctuates based on their value of time [26]. Enhancements in accessibility lead to reductions in travel expenses. One of the main factors influencing demand for a place is cost. The cost of living and other services at the destination, as well as the cost of travel, are typically included [27].

5. Methodology

This technique provides an organized way to examine papers about disruptive developments in transportation and how digital integration might promote tourism. A thorough list of all the references used in the review is included at the end of the paper to guarantee openness and reliability in the information's sourcing. It highlights how crucial it is to do a thorough literature research and add knowledge from expert interviews in order to offer a thorough overview of the topic. The references include academic papers, industry reports, case studies, and other pertinent resources that have advanced our knowledge of how digital integration in transportation may improve tourism.

6. Conclusion and Discussion

This research has demonstrated how important digital innovations are to changing travel experiences and tourist trends by doing a thorough assessment of the literature and analyzing new transportation technology. This paper's research has thoroughly examined the revolutionary developments that are taking place at the nexus of digital integration and transportation, clarifying their noteworthy effects on the travel and tourist sector, reservations more quickly, and navigating in real-time thanks to the rise of digital platforms and smart mobility solutions. Furthermore, customized travel experiences, personalized suggestions, and the promotion of sustainable travel practices have been made possible by technological breakthroughs like artificial intelligence, the Internet of Things, and augmented reality. The significant influence of digital integration on improving accessibility, convenience, and efficiency in transportation services is one of the main conclusions of this study. Travelers may now easily obtain travel information, make.

The study also emphasizes the complex connection between the growth of tourism and transportation infrastructure. Transportation improvements have a big influence on visitor flows and destination preferences because they improve accessibility, lower travel times, and improve connection between locations. The report also highlights how new transportation innovations, such shared mobility services, driverless vehicles, and electric vehicles, have the ability to completely change urban mobility and improve connectivity. These developments promise to enhance transportation systems' cost, efficiency, and safety while simultaneously providing ecologically friendly substitutes. The results imply that investments in digital integration and transportation infrastructure can boost inclusive growth throughout regions and support the financial viability of tourist destinations.

In summary, this study offers insightful information about the revolutionary developments reshaping the travel and tourist sector through the incorporation of digital technology into transportation. Adopting these revolutionary tendencies will be crucial for propelling future developments and guaranteeing the tourist industry's long-term sustainability as the digital revolution plays out. Through an awareness of the intricate relationships that exist between technology, tourism, and transportation,

https://doi.org/10.52202/074664-0002

interested parties may take proactive steps to seize possibilities for expansion, creativity, and sustainability in the rapidly changing travel industry.

7. Discussion

First of all, the conversation emphasizes how crucial it is for the public, industry participants, and legislators to work together to maximize the potential of digital integration in transportation to improve traveller experiences. Stakeholders may take advantage of chances for innovation and growth while addressing issues with infrastructure, regulation, and sustainability by establishing partnerships and using a comprehensive approach to planning and development. This research paper's discussion section considers the ramifications of the study's findings and offers perceptions into the more general implications for researchers, industry stakeholders, and policymakers.

To spot new possibilities and reduce risks, the tourist sector will need to continuously monitor and analyze how digital technologies are changing and upending established paradigms. The second point raised in the conversation is the necessity of ongoing research and modification to keep up with changing global trends, customer tastes, and technology breakthroughs.

Through the adoption of innovative practises, promotion of teamwork, and prioritisation of sustainability, stakeholders may effectively navigate the intricacies of the digital era and seize novel prospects for expansion and success within the tourist industry. The conversation as a whole focuses on how digital integration in transportation has the power to completely change the travel and tourism sector and improve the quality of travel.

References:

1. Gretzel, U., Sigala, M., Xiang, Z., & Koo, C. Smart tourism: Foundations and developments. Electronic Markets, 25(3), 179-188 (2015)
2. Sundarakani, B., Qasim, S. Z., Othman, M. F., & Kumar, V. Transportation 4.0: Transforming the future of smart transportation. Sustainable Cities and Society, 60, 102238 (2020)
3. Xiang, Z., Du, Q., Ma, Y., & Fan, W. A comparative analysis of major online review platforms: Implications for social media analytics in hospitality and tourism. Tourism Management, 58, 51-65 (2017)
4. Gretzel, U., Sigala, M., Xiang, Z., & Koo, C. Smart tourism: Foundations and developments. Electronic Markets, 25(3), 179-188 (2015)
5. Xiang, Z., & Gretzel, U. Role of social media in online travel information search. Tourism Management, 31(2), 179-188 (2010)
6. Albrecht, J., Schnell, J., & Larsen, E. R. Electric vehicle adoption: An analysis of driving factors in Europe. Transportation Research Part D: Transport and Environment, 79, 102235 (2020)
7. Anderson, J. M., & Sorensen, P. Autonomous vehicles: A learning approach to optimize policy and infrastructure. Transport Policy, 74, 215-222 (2019)
8. Banister, D. (2019). The sustainable mobility transition: Social, economic and technological dimensions. Transport Policy, 80, 1-9 (2019)
9. Clewlow, R. R., & Mishra, G. S. Disruptive transportation: The adoption, utilization, and impacts of ride-hailing in the United States. Institute of Transportation Studies, University of California, Davis, Research Report UCD-ITS-RR-17-07 (2017)
10. Fagnant, D. J., & Kockelman, K. Preparing a nation for autonomous vehicles: Opportunities, barriers and policy recommendations. Transportation Research Part A: Policy and Practice, 77, 167-181 (2015)
11. Gillingham, K., Knittel, C. R., Li, J., & Ovaere, M. The short-run and long-run effects of COVID-19 on energy and the environment. Joule, 4(7), 1337-1341 (2019)
12. Litman, T. Evaluating transportation equity. Transportation Research Record, 2673(9), 674-682 (2019)
13. Shaheen, S., Cohen, A., & Chung, A. North American carsharing: 10-year retrospective. Transportation Research Record, 2587(1), 15-21 (2016)
14. Sierzchula, W., Bakker, S., Maat, K., & van Wee, B. The influence of financial incentives and other socio-economic factors on electric vehicle adoption. Energy Policy, 68, 183-194 (2014)
15. Zervas, G., Proserpio, D., & Byers, J. W. The rise of the sharing economy: Estimating the impact of Airbnb on the hotel industry. Journal of Marketing Research, 54(5), 687-705 (2017)
16. Litman, Todd. "Evaluating Accessibility for Transportation Planning," March, (2008)
17. Currie, Christine, and Peter Falconer. "Maintaining Sustainable Island Destinations in Scotland: The Role of the Transport–tourism Relationship." Journal of Destination Marketing & Management, transport and destination management, 3 (3) 162–72 (2014). doi:10.1016/j.jdmm.2013.10.005
18. Davenport, John, and Julia L. Davenport. "The Impact of Tourism and Personal Leisure Transport on

Coastal Environments: A Review." Estuarine, Coastal and Shelf Science 67 (1–2) 280–92 (2006) doi:10.1016/j.ecss.2005.11.026

19. Tang, Liang, and Soocheong Jang. "The Evolution from Transportation to Tourism: The Case of the New York Canal System." Tourism Geographies 12 (3) 435–59 (2010) doi:10.1080/14616688.2010.494683
20. Millonig, A., and K. Schechtner. "City Tourism: Pedestrian Orientation Behaviour." International conference on walking and liveable communities, 7th, melbourne, victoria, australia, October (2006) http://trid.trb.org/view/2006/C/835625
21. Chew, Joseph. "Transport and Tourism in the Year 2000." Tourism Management 8 (2) 83–85 (1987) doi:10.1016/0261-5177(87)90003-3
22. Sheffi, Yosef. Urban Transportation Networks: Equilibrium Analysis with Mathematical Programming Methods. Englewood Cliffs, N.J.; London: Prentice-Hall (1985).
23. Pagliara, Francesca, Andrea La Pietra, Juan Gomez, and José Manuel Vassallo. "High Speed Rail and the Tourism Market: Evidence from the Madrid Case Study." Transport Policy 37 (January): 187–94 (2015) doi:10.1016/j.tranpol.2014.10.015
24. Rietveld, Professor Dr Piet, and Dr Frank Bruinsma. "Concepts and Data." In Is Transport Infrastructure Effective? Transport Infrastructure and Accessibility: Impacts on the Space Economy, 17–45. Advances in Spatial Science. Springer Berlin Heidelberg (1998) http://link.springer.com/chapter/10.1007/978-3-642-72232-5_2
25. Pagliara, Francesca, Andrea La Pietra, Juan Gomez, and José Manuel Vassallo. "High Speed Rail and the Tourism Market: Evidence from the Madrid Case Study." Transport Policy 37 (January): 187–94(2015) doi:10.1016/j.tranpol.2014.10.015
26. Martin, Christine A., and Stephen F. Witt. "Tourism Demand Forecasting Models: Choice of Appropriate Variable to Represent Tourists'Cost of Living." Tourism Management 8 (3) 233–46(1987) doi:10.1016/0261-5177(87)90055-0

 https://doi.org/10.52202/074664-0002

An Initial Contribution To Promoting India Gastronomic Tourism Is The Mapping Of Indian Food Culture

Dharmender Singh Rawat [1], and Rakesh Dani [2]

[1] Research Scholar, Departement of Hospitality Management, Graphic Era Deemed to be University, India
[2] Associate Professor, Departement of Hospitality Management, Graphic Era Deemed to be University, India

Abstract. India is famous for its culture, atmosphere, food, spices, various religions and its culture. India is the seventh-largest country in the world by land area. If we talk about population it will be the world's second most populous country in 2023, with a current population surpassing 1.425 billion and the largest democracy in the world. Indian cuisine is rich in variety and taste. Therefore, food can be employed as a strategic tool to enhance the tourism industry of the country and offer the opportunity for certain locations to become specialized in culinary tourism. In the last five years Indian government is also been working on developing the tourism sector in India various strategies are being followed up to develop Indian gastronomy not only in India but also worldwide. The study aimed to demonstrate the evolution, shaping, and preservation of Indian food culture as a social value passed down from one generation to the next. The purpose of this study is to investigate the development of Indian cuisine and the potential of Indian cuisine to attract tourists. To increase tourism, it is necessary to investigate millet's culinary contributions and provide suitable suggestions for enhancing cumulative development. To fulfill the purpose of this research, an extensive review of the literature was conducted, encompassing journals, books, articles, magazines, archives, and relevant online sources discussing Indian gastronomy. The key finding of the study regarding the importance of various food Commodities, Staple meals, local food, famous food preparation and promoting Millets from different geographical locations, disappeared for a long time in the taste palate of Culinary. So that Indian Gastronomic can promote tourism.

Keywords: Indian Gastronomy, Culinary Tourism, Food, Culture and Millets.

1. Introduction

Food is essential to everyone's existence since it meets fundamental physiological demands, promotes social connection, and allows people to express their emotions. As a symbol of identity and culture, it is currently one of the most treasured aspects of cultural tourism. There is no end on the discussion of food and the various ingredients used to make it so mouthwatering. It's a contribution of various commodities to make the food so delicious. Various recipes which were followed from grandmother to mother and then to us [1]. India is famous for its unity in diversity and is well known by the name 'Bird of Gold' in ancient times. Located in South Asia, India, formally referred to as the Republic of India, is ranked seventh in the world by land area. If we talk about population it will be the world's second most populous country in 2023, with a current population of over 1.425 billion with the largest democracy in the world. The commencement of Indian history traces back to the inception of the Indus Valley Civilization and the arrival of the Aryans. There are 28 states and 8 union territories in India with its enormous geographical, food, and cultural diversity[2]. India is a very big and vast country with various cultures that have their unique way of preparing food with unique cooking techniques. India is well known for its

https://doi.org/10.52202/074664-0003

university diversity but also for its famous food preparation, Spices, and tempering which enhance the aroma of the food and attract the guests. Culinary tourism is becoming its separate entity within the tourism industry [3]. The development of culinary tourism as a destination could serve as a crucial element of competitive advantage, contributing to the overall economic growth. Local for vocal is tag line which is becoming very popular. Since a nation's food reflects aspects of its national culture and identity, local cuisine becomes an important tool for setting one location apart from others in the global marketplace [4]. There is a variety of local and famous preparations of food in India from different regions. Different geographical conditions, temperatures, local products, Traditions, and eating habits make Indian food unique and different from other countries."Athithi Deva Bhava" is a famous tagline for India. India is not only famous for its food but also for its heart-warming Hospitality [5]. According to a review of the literature, most references to Indian cuisine concentrate on popular food preparations, recipes, regional cuisine, and cooking methods. The government of India is working on developing the Tourism sector. Ministry of Tourism is focusing more over on the Tourism sector.

2. Indian Food and Culture

Indian food is divided into two major aspects that is North Indian and South Indian food. Both the regions's food has its unique flavors, tempering, cooking ingredients, and methods. The cuisine of Kashmir differs significantly from that of Punjab or Himachal Pradesh, even though these two states share borders with Jammu and Kashmir. Similarly, South India exhibits a comparable diversity; for instance, the food of Andhra Pradesh shares similarities with that of Kerala or Tamil Nadu. The royal state of Bengal once comprised states such as West Bengal,Bihar,Orissa, and Assam. Even though these are different states now their food preparation influences Bengali food. Different states and union territories have their unique way of food preparation which has different techniques, seasonal availability, ingredients, and types of equipment. In different regions, people have their way of eating habits. Aryans first came to India and are known as Aryavrats they settled around in Indus Valley after that various rulers and invaders from ancient times came to India and just because of that also various states have different eating habits [6]. The 8,000-year history of the subcontinent's connections with other tribes and cultures is reflected in Indian cuisine, which has contributed to the variety of tastes and regional cuisines that are currently found in India. Later, the already varied Indian cuisine was further enhanced by trade with the Portuguese and British. The long period of Portuguese rule, besides that of Muslim and Hindu kingdoms had also influenced Indian cuisine.[7] The connections between colonial perceptions of India and its cuisine are evident in linking the colonial British "fabrication" of curry powder to the broader colonial construction of India. Vasco da Gama is believed to have brought chilies to India. Dietary customs relate to political and religious tensions in modern India. Indian cuisine has gained unprecedented popularity, especially in the Western world, owing to factors such as globalization, increased immigration, the availability of online recipes, and a surge in tourism. It has firmly established itself on the global culinary landscape, exemplified by "chicken tikka masala" being adopted as the national dish of Britain, and various Indian dishes featuring prominently on many international menus. Indian food has changed over time and has a close relationship to its location, history, and culture. Additionally, food habits have changed as a result of different religious practices.[8].

India doesn't have a specific "National food," however Khichdi is sometimes suggested. As a result, many eateries in Nashua produce this delectable dish. A large portion of the population is familiar with this food, which is also easily assimilated into the diverse cultures, faiths, and social classes of the nation [9]. In a given cultural group, food represents culture and has both utilitarian and symbolic meaning. Technically speaking, food satisfies the most fundamental physiological demands of humans. This type of eating behavior serves utilitarian or instrumental purposes, addressing both the need to satisfy hunger and meet the body's nutritional requirements, according

https://doi.org/10.52202/074664-0003

to the consumer behavior viewpoint linked to consumption values [10]. Food is examined in terms of hedonistic goals including pleasure, enjoyment, and sensory stimulation in addition to its nutritious value. Food can serve as a representation of social cohesion. When any group of persons and/or outsiders eat together, it can be utilized to foster familial ties, create friendships, and provide hospitality. National identities, regional, and ethnic are represented via food. Anthropologists frequently base their studies of communities on the important factor of food intake [11]

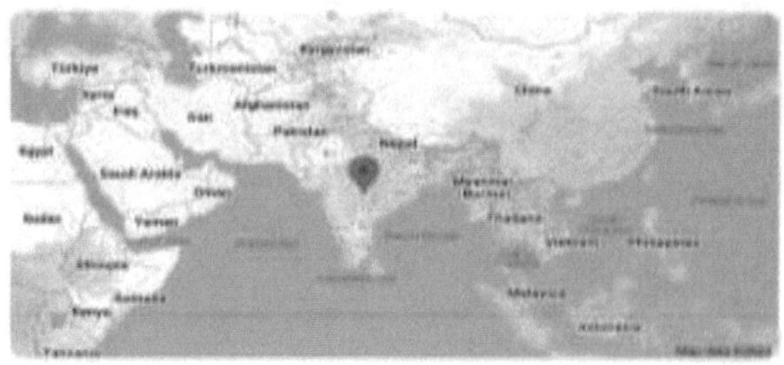

Fig.1. Source:https://rb.gy/lilnu0

3. Indian Culinary Phase & Development

Indian food is famous for its Tradition, customs, and religion. Which are changes from place to place in India? It is a place where climatic conditions also vary from place to place due to different geographical regions. In India people always respect food, they place the food in the form of god whenever some good things happens. It believes that Annapurna is a goddess of food. The name Annapurna comes from two wo3rds: "purna," which means "filled completely," and "anna," which means food. This goddess, an incarnation of Goddess Parvati, Lord Shiva's wife, is connected to food and the kitchen. As the goddess of sustenance, Annapurna is worshipped and makes sure her followers never go without food. Indian Food is born from the concept of Ayurveda. By about 7000 BCE, humped cattle, aubergines, and sesame had all been tamed in the Indus Valley. Numerous recipes originated during the early Vedic era. The ethical principle of Ahimsa, deeply rooted in ancient Hindu beliefs, prohibits the harming or killing of living creatures due to the negative karmic consequences associated with aggression. Consequently, this philosophy influenced certain segments of the population to adopt a vegetarian lifestyle. We are not talking about Kebab, Rich gravies, Pilaf, and Biryani because this food was influenced by the Mughals who came to India from Persia and Iran [12]. Fruits like apricots, melons, peaches, and plums were imported to India during the Islamic invasion in the Middle Ages. Ayurveda contains a wealth of knowledge in several medical specialties. Therefore, traditional foods and their dietary requirements are advised by Ayurveda. Many of India's traditional health foods can be referred to as Ayurvedic foods due to the tight relationship between Ayurveda dietetics and traditional cuisines. Ayurveda is derived from two words, Ayus, meaning life, and Vedas, meaning study or knowledge. Therefore, Ayurveda translates to the knowledge of life. This knowledge begins with the fundamental elements, including food. Ayurveda is Classified into six tastes Sweet, Salty, Pungent, Bitter, Astringent and Sour. Ayurveda also classified food based on its characteristics Rasa,Veerya, Prabhav. The human state of mind is also classified into three types according to Ayurveda Satyavik, Rajasik and Tamasik [13].

4. Classification and Characteristics of Indian Cuisine

India's food is as diverse and extensive as its multiethnic society. The deft use of herbs and spices distinguishes India's delectable and diverse cuisine. Regional Indian cuisines differ greatly from one another. Different cuisines are adopted by various regions. India takes pride in offering the most varied food available worldwide. Every location has its delicacies and distinctive cooking methods. With the changes in the landscape, so do the cooking techniques, flavors, and specialties. Despite the fact that the majority of Indians are vegetarian, Indians also like non-vegetarian dishes made with chicken, lamb,fish, goat, and other meats. In Indian culture, food has always had a significant role. [14]

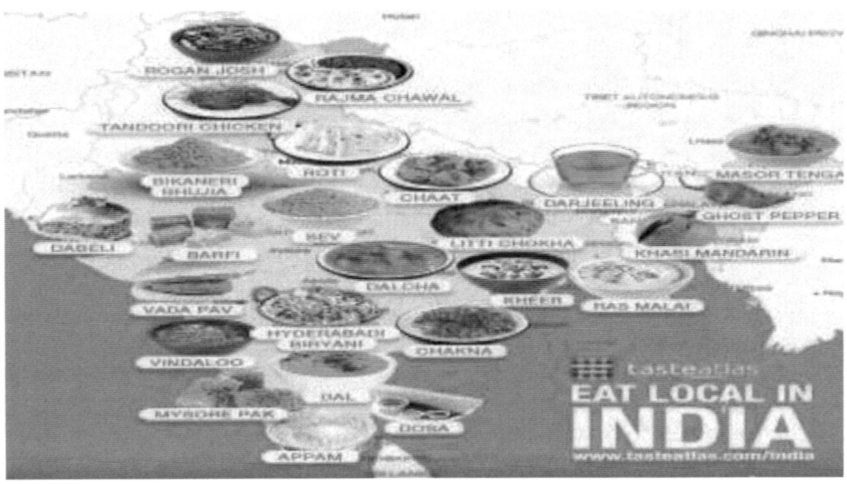

Fig:1.2

Source-https://images.app.goo.gl/tcafyeemmFbznwvY6

5. North Indian Cuisine

North Indian cuisine is notable for its frequent usage of curries with a butter base as well as dried fruits and nuts. The majority of North Indian meals are made from wheat, which is grown widely in northern India. In contrast to the northern and southern regions of India, Rice is the primary cereal, the Roti, Chappati, Paratha, and Tandoori are the staples in northern India. The Mughal style of cooking has had a significant influence on North Indian cuisine. They made several contributions to India throughout their nearly 500-year reign, particularly to its cuisine. The cuisine is very close to central Asian cuisines because the Mughals originated from that region. North Indian food has a very strong flavor because it is prepared using a lot of butter, ghee, and oil as well as rich spices. Additionally; meat is given a unique role in North Indian cooking. The Mughal era left behind a variety of Kebabs and Biriyanis, a mouthwatering rice and meat marinade.In north India, samosas may be the most popular snack. Another curd-based beverage with a strong flavor is called lassi. Popular candies in this region include gulabjamun and motichur laddoo. Rajasthani cuisine features a lot of spices. However, if you don't mind a little spice, you might be able to like Rajasthani food. North Indian specialties include Laal Maas, Mutton Rogan Josh, Safed Maas, Kundan Kaliya, Murgh Ke Sweta , Seekh Kabab, Shammi Kabab and Tandoori Chicken among others. Kashmiri Pulao is another noteworthy dish. [15]

6. South Indian Cuisine

 https://doi.org/10.52202/074664-0003

Other regions of India like South Indian food almost as much. For a change, ask people from various parts of the nation about their favorite substitute Indian dishes. Most people would say "South Indian."The southern region of India is made up of four distinct states:Tamil Nadu, Andhra Pradesh, Kerala and Karnataka. The southern part of India's food is therefore a vegetarian's dream. The usage of coconut oil in south Indian cuisine is another notable aspect. The south Indian dish has a very distinct flavor thanks to coconut oil. Coconut is another ingredient that is often used in most delicacies.Here, rice is another staple cuisine. Popular South Indian delicacies include Dosa, Idly, Uttapams and Vadas, which are made from rice and lentil paste and are primarily produced in Tamil Nadu. While visiting Tamil Nadu, make sure to sample the spiciness, oil, and scent of Chettinad cuisine. Biriyani made with rice and meat is a masterpiece in some regions of South India, such as Andhra Pradesh. The dish has a long history of Mughal ties.Similar to Kerala, the Malabar coasts produce amazing seafood dishes, including crab and prawn. Another appealing aspect of Malabar Coast cuisine is the powerful spice fragrance.[16]

7. East Indian Cuisine

States like West Bengal, Orissa, and Bihar are located in the eastern part of India. In India's eastern area, rice is the primary food source. Eastern India produces a lot of rice because of the region's favorable environment. In stark contrast to the north of India, where wheat predominates, rice is an essential component of every meal in eastern India. In addition, fish can be found in great numbers in the area's ponds and rivers. The inhabitants of eastern India also enjoy eating fish. Fish are used to produce a wide range of cuisines.[17] In eastern India, Hilsa is the most widely consumed fish. It is primarily available during the monsoon season and is renowned for its memorable flavor. Hilsa, as well as other well-known fish kinds like Rohu, Vetki, etc., are used to make several delicacies that are sure to please your palate. This region of India also produces a lot of veggies. Eastern Indian cuisine includes a lot of vegetables as well. Additionally, their use of spices in cooking differs significantly from that of other regions of India. In Eastern Indian cuisine, sweets are a prominent element. The people living in the eastern region have a preference for sweet flavors, it seems. The majority of the foods from eastern India have a propensity to be sweet. The famous delicacy from West Bengal that is Rasgulla is well-known throughout the world. While visiting the eastern part of India, you can also eat Mishti Doi (sweetened yogurt). [18]

8. West Indian Cuisine

The western region of India features a wide variety of meals due to its geographic diversity. The diverse landscape, which ranges from moist coastal areas to dry deserts, has contributed to the region's cuisine's adaptability. Western India is also a cultural mash-up of many different races and cultures. Therefore, a range of influences may be recognized in its cuisine. Additionally, rice is a major cuisine in western India. In places like Gujarat and Rajasthan, people also eat wheat, bajra, and jowar. Of course, Mumbai and the Gujarat region also have Parsi influences in their food. The vegetarian fare offered here, such as Panipuri,Bhel Puris, Pav Bhajis Dhoklas, and Khandvi is a gourmet's joy. Gujarat is well-known for its vegetarian food, which is influenced by the large Jain population in the state and is distinguished by a subtle spice combination and rich texture.[19]. Sea fish are widely available in Goa and Maharashtra's coastal regions. Sea fish dishes that have been marinated in mouthwatering spices are sure to tempt your palate. Geographical considerations influence the diversity of Maharashtrian cuisine. Coastal regions, like Goa, mostly consist of rice, coconut, and seafood. On the other hand, groundnuts—jowar (sorghum) and bajra (millet)—replace coconut in the mountainous areas of the Western Ghats and Deccan plateau. In the coastal Indian gastronomic scene of Konkani, Saraswat cuisine is significant. A popular form of dried fish is Bombay duck, also known as bombil. In this area, koshimbir (a coconut-pepper salad) and pickles, chutneys, and papad are common side dishes. There's potato or onion poha, sheera, and thalipith for breakfast. Common ingredients include curd, buttermilk, aamtis, and meals prepared with grains, veggies, and spices like goda masala, onions, ginger, and garlic. [20]. Goa, with its distinctive Portuguese flavour, has particularly captured the appetite and heart of food lovers with its diverse seafood selection, Along with pomfret, shark, tuna, and mackerel, kingfish (also known as visvan or venison) is the most popular dish. Favorite shellfish include prawns, crabs, squid, lobster, and mussels. Goan Hindu food is noted for having fewer spices and using a variety of vegetables, lentils, pumpkins, gourds, bamboo shoots, and roots. It also uses less onion and garlic. Beef dishes are a staple of Goan Christian cuisine, such as the well-known vindaloo, which the Portuguese first called "vinha d'alhos." [21]

9. North Eastern

The culinary traditions of North East India stand out distinctly from the rest of the country. North Eastern cuisine is notably influenced by the People's Republic of China and neighboring Burma, incorporating lesser-known Indian spices. In this part of India, yak meat is very common.Bamboo Shoots are usually are used in their food prepration which is rich source of Protein, Carbohydrates,Vitamins,Minerals, and Fibers. Few famous preparation food from Bamboo Shoot are Rep ,Ushoi,Soibum,Mesu,Ekhung ,Hirring and Eup etc.[22]

10. Indian Spices & Herbs

India is well known all over the world for its famous spices and herbs.It gives a nice taste, flavor, and aroma to the food. Spices also give nice food pairing to Indian Cuisine. Some spices are also widely known for their therapeutic properties, which have been utilized for many years [23]. Turmeric is used in Indian traditional medicine to treat rheumatic conditions, gastrointestinal issues, addressing rhinitis, Deworming and serve cosmetic purposes. Recently, several molecular targets for the therapeutic and preventative benefits of turmeric have been discovered. Fenugreek seeds, a dietary adjuvant for diabetes and a rich source of soluble fiber used in Indian cooking, lower blood sugar and cholesterol levels. Similarly, it has been discovered that ginger, garlic, and onions can positively influence the development of cancer [24]. Masala is a common name used in Indian cuisine. There are two types of masala that are used in Indian Cuisine i.e. Dry Masala and Wet Masala. Dry masala is in a powder form whereas wet masala is in a liquid form in which water is added in masala and then we have to a paste of it. Black mustard seeds, turmeric, chili powder, asafetida, garam masala, fennel, cumin, coriander, fenugreek, nigella, carom, black cardamom, green cardamom, saffron, and mace are a few of the spices used in Indian cooking. [25].

Table 1.India's Food Culture Mapping

SNO.	State/UT's	Cuisine
1	Jammu & Kashmir	"Gustaba, TabakMaz Dum Aloo, Haak or KaramkaSaag"
2	Ladakh	"Momos, Thukpa, Skew, Tigmo, Khambir"
3	Uttrakhand	"AlookeGutke, Kaapa, Jhangora (millets) ki , Kheer, Chainsoo"
4	Uttar	"Kabab, Biryanis, BedmiAlooKachori,

https://doi.org/10.52202/074664-0003

	Pradesh	BanarasiChaat"
5	Jharkhand	"Thekua, Pua, PitthaMarua-ka-roti"
6	Sikkim	"Momos, Thukpa, GundrukPhagshapa and Seal Roti"
7	Manipur	"Iromba, Kabok, Chakkouba"
8	Nagaland	"Momos, Rice Beer and , Cherry Wine"
9	Assam	"MasorTenga, Pitha"
10	Arunachal Paradesh	"Chinese Cuisine &, Apong (Local Beer)"
11	Meghalaya	"Jadoh, Kyat (Local Beer), Bitchi"
12	Bihar	"Litti-Chokha, SattuParathaKhaja, KhubiKa Lai, Anarasa, Tilkut"
13	West Bengal	"Rosogulla, MishtiDoi, BhapaIllish"
14	Tripura	"Chakhwi, MwkhwiMuitru"
15	Mizoram	"Zu (a special tea)"
16	Puducherry	"Kaduguyerra, Vendakkai, Patchaddy"
17	Andhra Pradesh	"Hyderabad Biryani, MirchiSalan, Ghongura Pickle Korikoora"
18	Telangana	"GonguraGhosht, Pappucharu, GonguraPappu, Hyderabadi Biryani"
19	Orissa	"Fish Orly, Khirmohan, Rasabali, Chhenapodapitha"
20	Tamil Nadu	"Appam, Dosai, Idli, Sambhar, Rasam, Chettinad Chicken Pongal"
21	Kerala	"Puttu-Kadala, Kappa-Meen Kari Sadya Meal, Avial, Malabar ParothaPayasam, Irachi Stew, Karimean Kari"
22	Karnataka	"BisiBeleBhaat, Kesari Bath, Mysore Pak, DharwadPedha, Chiroti"
23	Goa	"Vindaloo, Xacuti, Bibinca, Prawn Balchao"
24	Maharashtra	"Shrikhand, Thalipeeth, VadaPao, Modak, PaniPuri"
25	Madhya Pradesh	"Lapsi, Bafla, Bhutteki , Khees, BhopaliKabab"
26	Gujarat	"Thepla, DhoklaKhandvi, Handvo, Panki"
27	Chhattisgarh	"Bafauri, Kusli Red Ant Chutney"
28	Rajasthan	"Dal-Bati-Churma Ker Sangari, Lal Maas, Gatte"
29	Delhi	"Chaat, Tandoori Chicken Paranthe, CholeBhature"
30	Haryana	"Rabadi, BajrekiKhichdi, Cholia, Chaach-Lassi, KachrikiSabzi"
31	Chandigarh	"Butter Chicken, Tandoori , Chicken, Mutton Pulao"
32	Punjab	"Dal Makhni, Makke di Roti-, Sarson da Saag, ChanaBhature"

33	Himachal Pradesh	“Sidu, Aktori, DhamSeppuVadi, Badana, Babru”
34	Andaman & Nicobar Islands	“Fishes, Lobsters, Prawans Crabs”

Source: https://www.mapsofindia.com/maps/india/indian-cuisine-map.html

11. Contribution of Millet Year 2023 in Indian Gastronomy

In an effort to boost domestic and international demand while promoting wholesome food for people, the Indian government proposed to the UN that 2023 be designated as the International Year of Millets.

Government has taken a step ahead for the Millets and At the Global Millets (Shree Anna) Conference in New Delhi, India's Prime Minister, Shri Narendra Modi, gave the keynote address. The theme for the International Year of Millets 2023 (IYM) is "Harnessing the untapped potential of millets for food security, nutrition, and sustainable agriculture".India is the largest producer of Millet.Dr. Kadhar Ali is well known as the Millet Man of India.

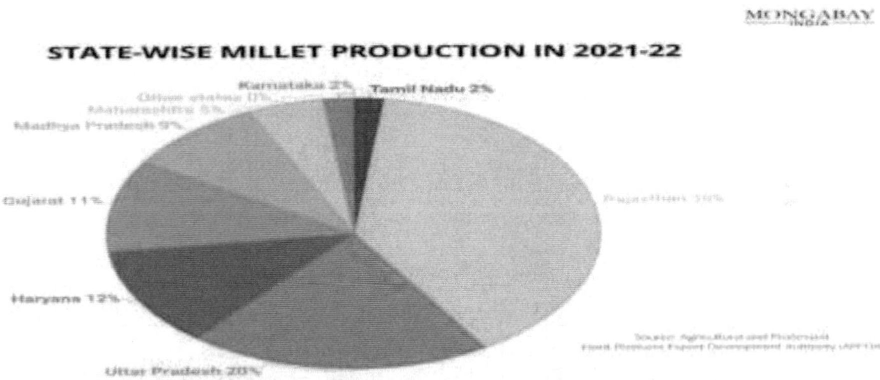

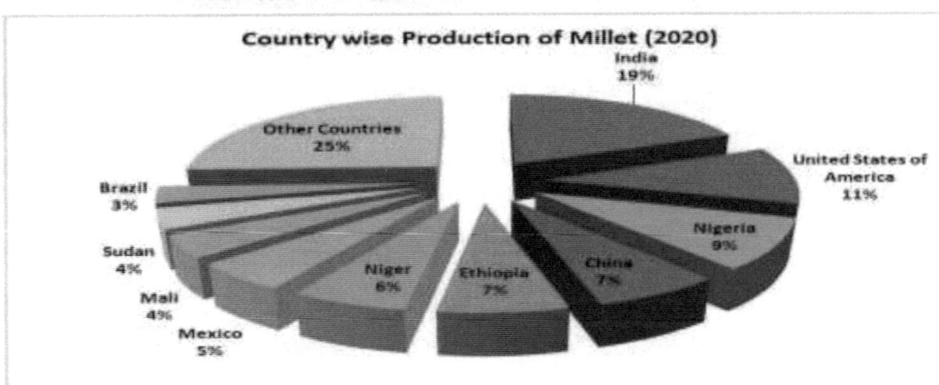

Fig: 1.3 & 1.4

Source:https://apeda.gov.in/milletportal/Production.html#:~:text=India%20is%20the%20largest%20producer%20of%20millets%20in%20the%20world.

https://doi.org/10.52202/074664-0003

Earlier Millet was seen as a poor man's meal and it was a neglected meal and underutilized crop spices. Millets have been consumed in India for more than 10,000 years. Famous millet varieties like Finger, Pearl, Foxtail, Barnyard, Proso, Kodo, and Little are grown around the world, and some of them are even grown in foreign countries. Millets are good sources of Protein, Dietry fiber,Minerals,Polyphenol and Antioxidant which makes to help fight against diseases.[26] In G-20 submit which was held in New-Delhi, India where various country representator came to India for the meet. In that meeting also a special meal was planned for them on Millet based. Jowar and Mushroom Khichda (in which Jowar and Pink oyster, chanterelles,shiitaemenoki,portbello and button mushroom were used), Gujrati 'Kea Na Cutlets' which was filled with Barnyard millet and coated with puffed amaranth, Millet medley with a millet thekua and a lemon Shrikhand mille-feuille dessert are the famous preparation made by Chef Kunal Kapoor, Chef Ananhita Dhondy and Chef Ajay Chopra on 9th September 2023.

Fig: 1.5 &1.6
Source: Instagram (Chef KunalKapoor)

12. Conclusion

The review and conversations that have taken place up until this point have shown that India has a long and illustrious past and that it has been shaped by a wide variety of civilizations. In addition to that, this research provides a historical overview of the many different types of Indian food. Since the Vedic period up till the present day, the nation's diverse population has exerted a significant amount of influence, which has contributed to the development of a wide variety of distinctive ethnic cuisines and spices. This study aims to improve consumer understanding of Indian food and its diverse options by researching socio-cultural elements. The purpose of this study is to lay the framework for enhancing consumer knowledge. Tourists are captivated by the wide variety of culinary gems that are available in each of the Indian states, which showcase their unique cuisine. To contribute to the strengthening of reputations for diverse places and tourist attractions across India, tourists are enticed to explore the distinctive flavors and culinary traditions that are associated with each region. His attention has been drawn to him by millets, which were completely lost, and they have come to his attention after the G20. Not only in India but all around the world has appreciated it in India. At the end of the investigation, the researchers concluded that Indian cuisine has a big impact on tourism, which is important for mapping Indian culinary culture.

The authors have no competing interests to declare that are relevant to the content of this article.

 https://doi.org/10.52202/074664-0003

References

[1]. J. C. Henderson, “Food tourism reviewed,” Br. Food J.,111(4), 317-326 (2009) doi:10.1108/00070700910951470.

[2]. M. R. Kukreti and R. Dani, An Analytical Study of Uttarakhand Cuisine.

[3]. P. S. Bali, “Quantity food production operations,” New-Delhi. Oxford University Press (2011)

[4]. G. W. Feenstra, “Local food systems and sustainable communities,” Am. J. Altern. Agric., 12(1), 28-36 (1997) doi:10.1017/S0889189300007165.

[5]. (G. P. Green and M. L. Dougherty, “Localizing linkages for food and tourism: Culinary tourism as a community development strategy” in Local Food and Community Development. Routledge, 53-63 (2014)

[6]. U. Narayan, “Eating cultures: Incorporation, identity and Indian food,” Soc. Identities, 1(1), 63-86 (1995) doi:10.1080/13504630.1995.9959426.

[7]. S. Zubaida, “The idea of ‘Indian food’, between the colonial and the global,” Food Hist., 7(1), 191-209 (2009) doi:10.1484/J.FOOD.1.100643.

[8]. S. Mangalassary, “Indian cuisine—The cultural connection,” Indigenous Cult. Educ. Global. Crit. Perspect. Asia, 119-134 (2016).

[9]. N. Negi et al., “Optimization of instant foxtail millet based khichdi by using response surface methodology and evaluation of its shelf stability,” J. Food Sci. Technol.,58(12), 4478-4485 (2021) doi:10.1007/s13197-020-04929-x.

[10]. M. W. Allen et al., “The interactive effect of cultural symbols and human values on taste evaluation,” J. Consum. Res., vol. 35(2) 294-308 (2008) doi:10.1086/590319.

[11]. M. Stankova and I. Vassenska, “Raising cultural awareness of local traditions through festival tourism,” Tourism Manag. Stud., 11,(1), 120-127 (2015).

[12]. V. Antani and S. Mahapatra, “Evolution of Indian cuisine: A socio-historical review,” J. Ethn. Foods, vol. 9(1) 15 (2022) doi:10.1186/s42779-022-00129-4.

[13]. P. Sarkar et al., “Traditional and ayurvedic foods of Indian origin” J. Ethn. Foods, 2(3), 97-109 2015 doi:10.1016/j.jef.2015.08.003.

[14]. C. T. Sen, Food Culture in India. USA: Bloomsbury Publishing. (BOOK) (2004).

[15]. K. G. Dubey, The Indian Cuisine. PHI Learning Pvt, Ltd. (BOOK) (2010).

[16]. G. E. Ferro-Luzzi, “Ritual as language: The case of south Indian food offerings,” Curr. Anthropol., 18(3), 507-514 (1977) doi:10.1086/201930.

[17]. S. Basumatary, “The Advent of culinary tourism, its scope & challenges,” Int. J. Creat. Res. Thoughts (IJCRT), 9(5) 120-128 (2021).

[18]. P. Dasgupta, “Defining Bengali cuisine: The culinary difference between West Bengal and Bangladesh” in A Comparative Development Narrative of Bangladesh and West Bengal of India, T. Bengals, Ed. Singapore: Springer Nature Singapore, 331-366 (2023).

[19]. O. J. Lakhani and A. Shaikh, “The sweet’truth’of Gujarat–Gujarati diet & lifestyle and diabetogenesis,” J. Soc. Health Diabetes, 6 (1), 022-026 (2018).

[20]. R. Billimoria, “Cuisine: A marketing tool for tourism development in Maharashtra,” Sansmaran Res. J., 6(1), 28-33 (2016).

[21]. M. Fernandes et al., Gastronomic Tourism Experiences of Goa: Tapping the Potential of Goan Cuisine as a Destination Attraction. Experiencias de TurismoGastronómico de Goa: Aprovechar el Potencial de la Cocina de Goa Como Atracción de Destino.

[22]. P. Nongdam and L. Tikendra, “The nutritional facts of bamboo shoots and their usage as important traditional foods of northeast India,” Int. Sch. Res. Not. (2014) doi:10.1155/2014/679073.

[23]. A. Jain and G. Bagler,, “Spices form the basis of food pairing in Indian cuisine,” arXiv Preprint ArXiv:1502.03815 (2015)

[24]. K. Krishnaswamy, “Traditional Indian spices and their health significance,” Asia Pac. J. Clin. Nutr., 17(1), 265-268 (2008).

[25]. A. Shukla and N. Yadav, “Role of Indian spices in Indian history,” Int. J. Manag. Res. Rev., 8(11), 1 (2018).

[26]. S. Singh et al., “Current status and promotional strategies of millets: A review,” Int. J. Environ. Clim. Change, 13(9), 088-3095 (2023), doi:10.9734/ijecc/2023/v13i92551.

A Case Study on Sustainable and Regenerative Community-Inspired Tourism in Garhwal Region of Uttarakhand, India

Tushar Uniyal[1] and Dr. Shikha Sharma[2]

[1,2] Chandigarh University, Mohali, Punjab, 140413, India

Abstract:

This research paper investigates sustainable and regenerative community-inspired tourism in the Garhwal region of Uttarakhand, India, focusing on the intersection of environmental sustainability, cultural preservation, and economic rejuvenation. Emphasizing the empowerment of local communities as key stakeholders, the study explores how participatory decision-making processes shape tourism development in alignment with cultural values, environmental sustainability, and socio-economic aspirations.

The methodology combines qualitative interviews, ecological and economic impact assessments, cultural mapping, and heritage documentation to capture diverse perspectives on community-inspired tourism's impact, challenges, and opportunities. Conversations with local organizations and government agencies evaluate environmental outcomes and economic benefits, while documenting cultural assets and traditions highlights their role in attracting responsible tourists.

The research demonstrates how community-led sustainable tourism catalyzes economic regeneration through job creation, entrepreneurship, and reinvestment of tourism revenues into community projects. It emphasizes the importance of preserving and showcasing local cultural heritage, fostering cross-cultural exchanges, and creating immersive experiences for visitors through storytelling and cultural events.

By providing a comprehensive understanding of community-led initiatives in the Garhwal region, this study contributes to the discourse on sustainable tourism, offering insights applicable to similar contexts worldwide. It showcases the transformative potential of empowering local communities to shape their tourism destinies and positions tourism as a force for positive change.

Keyword: Sustainable, Tourism, community, heritage, culture

1. Introduction:

Background of Garhwal Region and its Tourism Potential:

The Garhwal region, located in the state of Uttarakhand, India, is renowned for its breathtaking natural beauty, rich cultural heritage, and spiritual significance. Nestled in the lap of the majestic Himalayas, Garhwal is home to diverse flora and fauna, pristine rivers, snow-capped peaks, and ancient pilgrimage sites.

Historically, Garhwal has been a prominent destination for pilgrims and spiritual seekers due to its association with revered Hindu shrines such as the Char Dham (comprising Yamunotri, Gangotri, Kedarnath, and Badrinath), Hemkund Sahib, and the Panch Kedar temples. Additionally, the region boasts of vibrant folk traditions, traditional architecture, and indigenous knowledge systems, offering visitors a unique cultural experience.

In recent decades, Garhwal has witnessed a surge in tourism, attracting both domestic and international tourists seeking adventure, relaxation, and spiritual rejuvenation. However, this influx of visitors has also led to environmental degradation, cultural commodification, and socio-economic disparities, threatening the region's ecological integrity and social fabric.

Despite these challenges, Garhwal possesses immense potential for sustainable and regenerative tourism development. By leveraging its natural and cultural assets, empowering local communities, and adopting responsible tourism practices, Garhwal can foster tourism that benefits both the environment and the people.

 https://doi.org/10.52202/074664-0004

Sustainable and regenerative tourism approaches offer a holistic framework for addressing the adverse impacts of conventional tourism while maximizing the positive outcomes for host communities and ecosystems. Unlike mass tourism models focused solely on profit maximization, sustainable tourism prioritizes environmental conservation, socio-cultural authenticity, and equitable economic distribution.

In the context of Garhwal, embracing sustainable and regenerative tourism practices is essential for several reasons:

- Environmental Conservation: Garhwal's fragile Himalayan ecosystem is susceptible to degradation due to unsustainable tourism activities such as deforestation, waste generation, and pollution. Adopting sustainable tourism practices minimizes environmental footprint and promotes biodiversity conservation, ensuring the long-term health of the region's ecosystems.
- Socio-Cultural Preservation: Garhwal's diverse cultural heritage and traditional way of life are at risk of erosion as tourism commodifies local customs and displaces indigenous communities. Sustainable tourism initiatives empower local residents to preserve their cultural identity, share traditional knowledge, and actively participate in tourism decision-making processes.
- Economic Empowerment: Sustainable tourism has the potential to generate inclusive economic opportunities by creating livelihoods for marginalized communities, supporting local enterprises, and redistributing tourism revenues equitably. By diversifying income sources and promoting community-owned tourism enterprises, Garhwal can reduce dependency on seasonal employment and external investments.

2. Research Objectives and Methodology:

The primary objective of this case study is to investigate and analyze the implementation, impact, and potential of sustainable and regenerative community-inspired tourism initiatives within the Garhwal region of Uttarakhand, India. The study aims to provide a comprehensive understanding of the various facets involved in fostering sustainable tourism development that not only preserves the natural and cultural heritage of the region but also contributes to the socio-economic well-being of local communities.

Specifically, the research objectives are outlined as follows:

Assess Current Tourism Practices: Conducting an examination of the existing tourism practices in the Garhwal region, including their environmental, socio-cultural, and economic implications. This involves evaluating the extent to which conventional tourism activities have affected the local ecosystem, cultural integrity, and livelihoods of residents.

Identify Sustainable and Regenerative Initiatives: Identify and document existing sustainable and regenerative tourism initiatives implemented by local communities, non-governmental organizations (NGOs), government agencies, and other stakeholders within the Garhwal region. This includes initiatives focused on eco-tourism, community-based tourism, conservation efforts, and cultural preservation.

Analyze Stakeholder Involvement and Collaboration: Investigate the roles, perceptions, and levels of involvement of various stakeholders, including local communities, government bodies, tourism operators, conservation groups, and tourists themselves, in promoting sustainable tourism practices. Assess the effectiveness of collaborative efforts and partnerships in achieving sustainability goals.

Evaluate Environmental and Socio-Cultural Impacts: Evaluate the environmental conservation efforts undertaken by sustainable tourism initiatives in mitigating negative impacts such as habitat degradation, resource depletion, and pollution. Additionally, assess the socio-cultural effects of tourism on local communities, including changes in traditional practices, cultural identity, and social structures.

Measure Economic Benefits and Livelihood Impacts: Quantify the economic benefits generated by sustainable tourism initiatives, including revenue generation, employment opportunities, and income distribution among local communities. Assess the resilience of community livelihoods to external pressures and the extent to which tourism contributes to poverty alleviation and socio-economic development.

Explore Visitor Perceptions and Experiences: Investigate the perceptions, motivations, and experiences of tourists visiting the Garhwal region, particularly those participating in sustainable tourism activities.

Understand their preferences, expectations, and satisfaction levels regarding eco-friendly accommodations, cultural exchanges, and immersive experiences.

Identify Opportunities and Challenges: Identify key opportunities for scaling up and replicating successful sustainable tourism models within the Garhwal region and beyond. Furthermore, analyze the challenges and barriers hindering the adoption and expansion of sustainable practices, including policy gaps, infrastructure limitations, market demands, and socio-economic disparities.

Provide Recommendations for Policy and Practice: Based on the research findings, formulate actionable recommendations for policymakers, local authorities, community leaders, and other stakeholders to enhance the sustainability and resilience of tourism development in the Garhwal region. Propose strategies for mainstreaming sustainable tourism principles into policy frameworks, capacity-building initiatives, and destination management plans.

By achieving these research objectives, this case study aims to contribute valuable insights and practical solutions for promoting sustainable and regenerative tourism practices that benefit both the environment and local communities in the Garhwal region of Uttarakhand, India, while also offering lessons and inspiration for similar contexts globally.

2.1 Methodology:

This research will employ a mixed-methods approach, combining qualitative and quantitative techniques to gather and analyze data. The primary methods include:

- Literature Review: A comprehensive review of academic articles, government reports, and industry publications on tourism in Garhwal and sustainable tourism practices globally.
- Qualitative Interviews: Semi-structured interviews will be conducted with key stakeholders involved in tourism management, including government officials, community leaders, tour operators, and local residents. These interviews will provide insights into perceptions, experiences, and challenges related to sustainable tourism development in Garhwal.
- Participant Observation: Field visits and participant observation will be conducted in selected tourism destinations and community-based tourism projects in Garhwal. This will involve observing tourism activities, interacting with visitors and locals, and documenting observations and insights.
- Data Analysis: Qualitative data from interviews and observations will be analyzed thematically to identify recurring patterns, emerging themes, and critical issues related to sustainable tourism in Garhwal. Quantitative data, such as tourism statistics and economic indicators, will be analyzed using descriptive and inferential statistical techniques.
- By employing a rigorous research methodology, this study aims to generate evidence-based insights into the challenges and opportunities of sustainable tourism development in the Garhwal region, thereby informing policy and practice for promoting responsible tourism in this ecologically and culturally sensitive area.

3. Literature Review:

Concepts of Sustainable Tourism and Regenerative Tourism:

Sustainable tourism is a concept that emphasizes responsible travel practices that minimize negative impacts on the environment, preserve cultural heritage, and provide socio-economic benefits to local communities. According to the World Tourism Organization (UNWTO), sustainable tourism "takes full account of its current and future economic, social, and environmental impacts, addressing the needs of visitors, the industry, the environment, and host communities."

Regenerative tourism goes a step further by focusing on restoring and revitalizing ecosystems, cultures, and economies. It seeks to create positive impacts by actively contributing to the regeneration of natural resources, cultural traditions, and community resilience. Regenerative tourism aims not only to sustain but to enhance the well-being of host communities and ecosystems.

Both sustainable and regenerative tourism paradigms advocate for principles such as minimizing carbon footprint, conserving biodiversity, promoting local culture and heritage, supporting community participation and empowerment, and fostering economic equity. These concepts provide a framework for guiding tourism development in a manner that ensures long-term viability and benefits for all stakeholders.

https://doi.org/10.52202/074664-0004

The literature on sustainable and regenerative tourism in the Garhwal region of Uttarakhand, India, underscores the importance of community involvement, environmental conservation, and cultural preservation in shaping tourism initiatives. Scholars have documented the challenges posed by conventional tourism models, which often lead to environmental degradation and socio-economic disparities. They highlight the need for a paradigm shift towards community-led approaches that prioritize the well-being of both local residents and the environment.

Several studies emphasize the role of local communities as stewards of their natural and cultural heritage. By engaging residents in decision-making processes and empowering them to participate in tourism development, these initiatives aim to ensure that tourism activities are aligned with community values and aspirations. Participatory approaches, such as community-based tourism planning and management, have been recognized as effective strategies for fostering sustainable tourism development in the region. [1]

Researchers have also explored the environmental impacts of tourism in the Garhwal region, including deforestation, habitat loss, and pollution. They emphasize the importance of adopting eco-friendly practices, such as responsible waste management, energy conservation, and biodiversity conservation, to minimize negative environmental consequences. Moreover, studies highlight the potential of sustainable tourism to contribute to ecological restoration and conservation efforts, such as reforestation projects and wildlife conservation initiatives.[2]

Cultural heritage and identity are central themes in the literature on sustainable tourism in the Garhwal region. Scholars highlight the significance of preserving and showcasing local traditions, rituals, and craftsmanship as a means of attracting culturally sensitive tourists and promoting cross-cultural exchanges. Community-based initiatives, such as cultural festivals, heritage walks, and artisan workshops, are seen as vital components of sustainable tourism development, fostering cultural pride and socio-economic empowerment among local communities.

Pilgrimage travel bridges cultural divides, foster spiritual growth, and stirs strong emotions in its participants. A wealth of information is available to help those interested in pilgrimage travel better comprehend the holy locations found all over the globe. Despite its spiritual and cultural value, pilgrimage tourism also plays an important economic function. This phenomenon has greatly aided industrial development, particularly at the world's most important pilgrimage sites, where demand and supply have multiplied. Thus, there is great social and economic importance to pilgrimage travel.[3]

In terms of economic benefits, scholars have documented the positive impacts of community-led tourism initiatives on local livelihoods and economies. By creating employment opportunities, supporting local businesses, and generating revenue for community development projects, sustainable tourism has the potential to alleviate poverty and improve standards of living in the region. Additionally, studies highlight the importance of equitable distribution of tourism benefits and the need to address issues of social inclusion and empowerment within local communities.

According to Singh, pilgrimage is "an act that brings strong sentiments, faith, belief, respect, and reverence to the almighty, and above all, the sincerity of devotee" (s). Equally, a pilgrim travels to a sacred site for spiritual reasons. [4]

Overall, the literature underscores the transformative potential of community-led sustainable tourism in the Garhwal region, offering valuable insights for policymakers, practitioners, and researchers interested in promoting responsible tourism practices and fostering inclusive and sustainable development in mountainous regions.

Community-Based Tourism Models and Practices:

Community-based tourism (CBT) refers to tourism initiatives that are owned, managed, and operated by local communities, with the aim of promoting socio-economic development while preserving cultural and environmental integrity. CBT models prioritize community participation, cultural authenticity, and equitable distribution of benefits.

Examples of CBT practices include homestays, community-run guesthouses, guided tours led by local residents, and artisanal workshops showcasing traditional crafts. These initiatives not only provide authentic and immersive experiences for visitors but also create opportunities for local residents to generate income, preserve cultural heritage, and take pride in their identity.[5]

CBT projects often involve capacity-building activities, such as training programs for community members in hospitality, guiding, and sustainable resource management. By empowering local communities to actively participate in tourism decision-making and benefit-sharing, CBT models contribute to poverty alleviation, social inclusion, and cultural revitalization.

However, CBT initiatives also face challenges such as limited access to market networks, inadequate infrastructure, lack of financial resources, and potential conflicts over land and resources. Addressing these challenges requires collaboration between community stakeholders, government agencies, NGOs, and private sector partners to provide support and resources for sustainable tourism development.

Challenges and Opportunities in Sustainable Tourism Development:

Sustainable tourism development faces various challenges, including:

- Environmental Degradation: Unsustainable tourism practices, such as overdevelopment, pollution, habitat destruction, and resource depletion, can degrade natural ecosystems, diminish biodiversity, and exacerbate climate change impacts.
- Socio-Cultural Disruption: Tourism can disrupt local communities and cultures by commodifying traditions, displacing indigenous peoples, and fostering dependency on tourism revenues. Cultural appropriation, loss of cultural authenticity, and social inequalities are common challenges associated with tourism development.
- Economic Vulnerability: Reliance on tourism as the primary economic driver can make communities vulnerable to external shocks such as economic downturns, natural disasters, or pandemics. Seasonality, leakage of tourism revenues, and unequal distribution of benefits can exacerbate poverty and socio-economic disparities.[6]

However, sustainable tourism also presents opportunities for positive change, including:

- Economic Diversification: Sustainable tourism diversifies income sources, creates employment opportunities, and stimulates entrepreneurship in diverse sectors such as agriculture, handicrafts, eco-tourism, and community-based enterprises.
- Cultural Revitalization: Sustainable tourism celebrates and preserves cultural diversity, traditions, and heritage, fostering pride, identity, and intercultural understanding among host communities and visitors.
- Environmental Stewardship: Sustainable tourism practices promote environmental conservation, resource efficiency, and carbon neutrality, contributing to the protection and restoration of natural ecosystems and biodiversity.

By addressing these challenges and harnessing these opportunities, sustainable tourism can play a transformative role in promoting inclusive growth, environmental resilience, and cultural vitality in destination communities like the Garhwal region of Uttarakhand, India.

Community-Inspired Tourism Initiatives in Garhwal:

Overview of Community-Based Tourism Projects:

In the Garhwal region of Uttarakhand, several community-based tourism (CBT) projects have emerged to harness the region's natural and cultural assets while empowering local communities. These initiatives are often led by grassroots organizations, non-governmental organizations (NGOs), or community cooperatives, with the support of government agencies, private sector partners, and international development organizations.

Common features of CBT projects in Garhwal include homestays with local families, guided treks and cultural tours led by community members, participatory workshops on traditional crafts and cuisine, and sustainable agriculture and eco-tourism ventures. These initiatives aim to provide authentic and immersive experiences for visitors while generating income and employment opportunities for residents.

Key principles of CBT projects in Garhwal include community participation in decision-making, equitable benefit-sharing, respect for local culture and environment, and capacity-building for sustainable tourism management. By engaging with visitors directly and showcasing their cultural heritage and natural beauty, communities in Garhwal can take ownership of tourism development and preserve their way of life.[7]

https://doi.org/10.52202/074664-0004

4. **Case Studies of Successful Initiatives:**

- The Sarmoli Village Tourism Cooperative: Located in the Munsiyari region of Garhwal, the Sarmoli Village Tourism Cooperative is a pioneering example of community-led tourism development. Founded in 2004 by local women, the cooperative offers homestays, guided treks, and cultural experiences that highlight the region's biodiversity, traditional knowledge, and sustainable lifestyle practices. The cooperative has empowered women as decision-makers and entrepreneurs, providing them with a source of income and a platform to showcase their skills and heritage.
- The Binsar Wildlife Sanctuary Community Eco-Tourism Project: Situated in the Binsar Wildlife Sanctuary, this community eco-tourism project is managed by local villagers in partnership with the forest department and NGOs. Visitors can stay in eco-friendly accommodations, participate in nature walks and birdwatching tours, and engage with community members in cultural activities. The project has helped reduce human-wildlife conflict, promote forest conservation, and create alternative livelihoods for villagers dependent on forest resources.
- The Malari Tourism Initiative: Malari, a remote village near the Indo-Tibetan border, has embraced tourism as a means of economic revitalization and cultural preservation. Community members offer homestays, trekking expeditions, and cultural exchanges to visitors interested in exploring the region's pristine landscapes and indigenous traditions. The initiative has boosted local incomes, revived traditional practices, and fostered cross-cultural understanding among visitors and residents.

5. **Impact Assessment on Local Communities and Environment:**

The impact of community-inspired tourism initiatives on local communities and the environment in Garhwal can be assessed in several ways:

- Economic Empowerment: CBT projects provide supplemental income and livelihood opportunities for local residents, particularly women, youth, and marginalized groups. Revenue generated from tourism activities supports community development projects, education, healthcare, and infrastructure improvements.
- Socio-Cultural Preservation: CBT initiatives promote cultural pride, intergenerational knowledge transfer, and cultural exchange between visitors and locals. By showcasing traditional arts, crafts, music, and cuisine, communities preserve their cultural identity and heritage while fostering cross-cultural understanding and appreciation.
- Environmental Conservation: CBT projects prioritize environmental stewardship and sustainable resource management. Through eco-friendly practices, waste reduction, and conservation initiatives, communities in Garhwal contribute to the protection of biodiversity, watersheds, and natural habitats.
- Overall, community-inspired tourism initiatives in Garhwal have the potential to catalyze inclusive growth, environmental conservation, and cultural revitalization, making them valuable models for sustainable tourism development in mountainous regions worldwide. Continued collaboration between stakeholders and investments in community capacity-building are essential to ensure the long-term success and resilience of these initiatives.[8]

6. **Challenges and Solutions:**

Environmental Degradation and Conservation Measures:

Challenges:

The Garhwal region is vulnerable to environmental degradation due to unsustainable tourism practices, including deforestation, waste generation, pollution, and habitat destruction. Rapid urbanization, infrastructure development, and increasing visitor numbers exacerbate these challenges, threatening the region's biodiversity, water resources, and fragile ecosystems.

Solutions:

- Implementing Sustainable Tourism Practices: Encouraging responsible tourism behaviors such as waste reduction, energy conservation, and water management can mitigate environmental impacts. Adopting eco-friendly accommodations, promoting low-impact recreational activities, and enforcing strict regulations on waste disposal and pollution control are essential measures.
- Biodiversity Conservation: Protecting sensitive habitats, wildlife corridors, and biodiversity hotspots through the establishment of conservation areas, wildlife sanctuaries, and protected reserves is crucial. Engaging local communities in participatory conservation initiatives, such as

community-managed forests and ecotourism enterprises, can foster stewardship and ensure the sustainable management of natural resources.

- Sustainable Infrastructure Development: Prioritizing green infrastructure solutions such as renewable energy, sustainable transportation, and eco-friendly construction techniques minimizes the ecological footprint of tourism infrastructure. Investing in eco-certification programs, green building standards, and sustainable tourism planning processes promotes environmentally responsible development.

7. Socio-Cultural Impact and Preservation Efforts:

Challenges:

Tourism can exert socio-cultural pressures on local communities, leading to the commodification of culture, erosion of traditional practices, and social disruptions. Cultural appropriation, loss of authenticity, and conflicts over resource use and land tenure pose significant challenges to socio-cultural preservation in Garhwal.

Solutions:

- Community Engagement and Empowerment: Involving local communities in tourism planning, decision-making, and benefit-sharing enhances their sense of ownership and agency. Establishing community-based tourism enterprises, cooperatives, and cultural heritage committees empowers residents to actively participate in tourism development and preserve their cultural identity.[9]
- Cultural Revitalization Programs: Promoting cultural festivals, heritage trails, and cultural exchange programs enriches visitor experiences while safeguarding intangible cultural heritage. Supporting traditional artisans, performers, and craftsmen through training, marketing, and entrepreneurship initiatives sustains traditional livelihoods and fosters cultural resilience.
- Responsible Visitor Behavior: Educating tourists about local customs, etiquette, and cultural sensitivities fosters mutual respect and understanding between visitors and hosts. Encouraging immersive cultural experiences, homestays, and guided tours led by local residents promotes authentic interactions and cross-cultural dialogue, fostering cultural exchange and appreciation.

Economic Sustainability and Livelihood Diversification:

Challenges:

Garhwal's economy is heavily reliant on tourism, making communities vulnerable to fluctuations in visitor numbers, seasonal demand, and external shocks such as natural disasters or pandemics. Limited economic opportunities, income disparities, and over-reliance on tourism revenues pose challenges to long-term economic sustainability.

Solutions:

- Economic Diversification: Supporting diversified livelihood options such as agriculture, agroforestry, handicrafts, and alternative tourism ventures reduces dependency on tourism as the sole income source. Encouraging entrepreneurship, skill development, and value-added enterprises empowers local communities to diversify their income streams and build resilience against economic shocks.
- Community-Based Tourism Enterprises: Establishing community-owned and managed tourism enterprises, including homestays, eco-lodges, and adventure tourism companies, ensures equitable distribution of tourism benefits. Providing training, access to finance, and marketing support strengthens the capacity of local entrepreneurs to run successful tourism businesses while retaining control over resources and decision-making.
- Destination Management Planning: Developing sustainable tourism policies, strategies, and action plans in collaboration with local stakeholders, government agencies, and tourism industry representatives ensures a coordinated approach to destination management. Balancing tourism development with environmental conservation, socio-cultural preservation, and economic prosperity maximizes the long-term benefits of tourism for host communities and ecosystems.[10]
- By addressing these challenges through collaborative efforts and innovative solutions, Garhwal can harness the transformative potential of sustainable and regenerative tourism to promote inclusive growth, environmental resilience, and cultural vitality.

Outcomes and Lessons Learned:

Economic Empowerment of Local Communities:

Outcomes:

- Increased Incomes: Community-inspired tourism initiatives in Garhwal have led to diversified income sources for local communities, including homestays, guided tours, artisanal crafts, and

https://doi.org/10.52202/074664-0004

agro-tourism ventures. By engaging in tourism-related activities, residents have been able to supplement their traditional livelihoods and improve their economic well-being.

- Job Creation: Tourism development has generated employment opportunities for local residents, particularly women, youth, and marginalized groups. From hospitality and guiding services to handicraft production and culinary experiences, tourism-related jobs have provided avenues for skill development, entrepreneurship, and economic self-reliance.
- Poverty Alleviation: Sustainable tourism initiatives have contributed to poverty reduction by redistributing tourism revenues more equitably among community members. Through community-based tourism enterprises and revenue-sharing mechanisms, marginalized households have gained access to income and social services, enhancing their standard of living.[11]

Lessons Learned:

- Empowering Local Ownership: The success of community-based tourism projects hinges on empowering local communities to take ownership of tourism development processes. By involving residents in decision-making, resource management, and benefit-sharing, initiatives can foster a sense of ownership, accountability, and pride among community members.
- Building Capacity: Investing in capacity-building programs, training workshops, and skill development initiatives enhances the capabilities of local entrepreneurs to participate effectively in tourism enterprises. Equipping community members with business management skills, marketing strategies, and quality standards improves the sustainability and competitiveness of tourism businesses.
- Fostering Partnerships: Collaboration between government agencies, NGOs, private sector actors, and community organizations is essential for supporting sustainable tourism development. By fostering partnerships and leveraging resources, stakeholders can address common challenges, share expertise, and scale up successful initiatives for greater impact.

Strengthening of Cultural Identity and Heritage:

Outcomes:

- Cultural Revitalization: Tourism initiatives in Garhwal have provided platforms for preserving and showcasing traditional arts, crafts, music, and cuisine. Through cultural festivals, heritage trails, and interactive experiences, residents have been able to celebrate their cultural heritage, transmit traditional knowledge, and instill pride in their cultural identity.
- Intercultural Exchange: Tourism facilitates cross-cultural dialogue and exchange between visitors and locals, fostering mutual understanding, empathy, and respect. By engaging with diverse cultures and perspectives, residents gain insights into global trends, alternative lifestyles, and sustainable practices, enriching their cultural experiences and worldview.
- Community Cohesion: Tourism has strengthened social bonds and solidarity among community members, fostering cooperation, reciprocity, and collective action. By participating in tourism-related activities, residents forge connections with one another, share experiences, and build social networks that contribute to community resilience and cohesion.

Lessons Learned:

- Safeguarding Intangible Heritage: Preserving intangible cultural heritage, such as oral traditions, rituals, and performing arts, requires concerted efforts to document, safeguard, and transmit traditional knowledge to future generations. Integrating cultural preservation into tourism programming, education, and policy-making ensures the continuity and vitality of cultural practices.
- Promoting Authentic Experiences: Authenticity is a key driver of visitor satisfaction and cultural exchange in tourism. By offering genuine, immersive experiences rooted in local traditions, values, and lifestyles, communities can differentiate themselves from mainstream tourism destinations and attract discerning travelers seeking meaningful interactions and authentic encounters.
- Balancing Tradition and Innovation: Embracing innovation while respecting tradition is essential for cultural sustainability in tourism. By adapting traditional practices to contemporary contexts, communities can revitalize cultural heritage, stimulate creativity, and respond to changing market demands without compromising authenticity or integrity.[12]

Contribution to Biodiversity Conservation and Natural Resource Management:

Outcomes:

- Habitat Protection: Community-based tourism initiatives in Garhwal promote responsible stewardship of natural resources, including forests, water bodies, and wildlife habitats. By raising awareness about the importance of biodiversity conservation, initiatives encourage residents to protect and restore ecosystems, reduce human-wildlife conflicts, and mitigate anthropogenic threats to biodiversity.
- Sustainable Resource Use: Tourism enterprises prioritize sustainable resource management practices, such as energy efficiency, waste reduction, and water conservation, to minimize environmental impacts. By adopting eco-friendly technologies, renewable energy sources, and sustainable agriculture practices, communities reduce their ecological footprint and promote resource efficiency.
- Ecotourism Development: Ecotourism initiatives in Garhwal provide alternative livelihood options for communities while promoting biodiversity conservation and ecosystem restoration. By offering nature-based experiences, such as wildlife safaris, birdwatching tours, and trekking expeditions, communities raise awareness about the value of natural ecosystems and the need for their protection.

Lessons Learned:

- Engaging Communities in Conservation: Integrating local communities into conservation efforts fosters a sense of ownership, responsibility, and stewardship over natural resources. By involving residents in biodiversity monitoring, habitat restoration, and ecotourism management, initiatives build community support for conservation goals and enhance the effectiveness of conservation interventions.
- Promoting Responsible Visitor Behavior: Educating tourists about environmental conservation, wildlife protection, and sustainable tourism practices is essential for minimizing negative impacts on biodiversity. By providing interpretive signage, guided tours, and visitor guidelines, communities raise awareness about conservation issues and empower visitors to make responsible choices.
- Strengthening Protected Areas: Collaborating with government agencies, NGOs, and conservation organizations to establish and manage protected areas strengthens biodiversity conservation efforts. By advocating for the expansion of protected areas, habitat connectivity, and wildlife corridors, communities contribute to the long-term viability and resilience of ecosystems in Garhwal.
- Overall, community-inspired tourism initiatives in Garhwal have demonstrated the potential to generate positive outcomes for local communities, cultural heritage, and biodiversity conservation. By embracing sustainable practices, fostering community engagement, and promoting responsible tourism behaviors, initiatives contribute to the well-being and resilience of both people and ecosystems in the region.

8. Conclusion:

Summary of Findings and Contributions:

In this research paper, we have explored the landscape of sustainable and regenerative tourism in the Garhwal region of Uttarakhand, India. Through an examination of community-inspired tourism initiatives, we have identified key outcomes and lessons learned in three main areas: economic empowerment of local communities, strengthening of cultural identity and heritage, and contribution to biodiversity conservation and natural resource management.

Our findings highlight the significant positive impacts of community-based tourism projects in Garhwal, including increased incomes, job creation, cultural revitalization, habitat protection, and sustainable resource use. These initiatives have empowered local communities to take ownership of tourism development processes, celebrate their cultural heritage, and conserve natural ecosystems.

9. Implications for Sustainable Tourism Development in Garhwal and Beyond:

The success of community-inspired tourism initiatives in Garhwal provides valuable insights and lessons for promoting sustainable tourism development not only in Garhwal but also in other mountainous regions and rural areas worldwide. Key implications include:

https://doi.org/10.52202/074664-0004

- Importance of Community Engagement: Engaging local communities in tourism planning, decision-making, and benefit-sharing is critical for fostering ownership, accountability, and resilience in tourism development processes.
- Emphasis on Cultural Preservation: Preserving cultural identity, traditions, and intangible heritage is essential for enhancing visitor experiences, promoting intercultural exchange, and fostering social cohesion in destination communities.
- Integration of Conservation and Development: Integrating biodiversity conservation, natural resource management, and sustainable development goals into tourism initiatives ensures the long-term viability and resilience of ecosystems and communities.

Recommendations for Policy, Practice, and Future Research:

Based on our findings, we propose the following recommendations for policymakers, practitioners, and researchers:

- Policy Support: Governments should enact supportive policies, regulations, and incentives to promote community-based tourism, cultural preservation, and environmental conservation in Garhwal and beyond.
- Capacity Building: Investing in capacity-building programs, training workshops, and skill development initiatives strengthens the capabilities of local communities to participate effectively in tourism enterprises and natural resource management.
- Collaboration and Partnerships: Facilitating collaboration between government agencies, NGOs, private sector actors, and community organizations fosters inclusive and sustainable tourism development in Garhwal and other rural areas.
- Research Agenda: Future research should focus on monitoring and evaluating the long-term impacts of community-inspired tourism initiatives on local communities, ecosystems, and economies. Additionally, research on innovative financing mechanisms, market linkages, and technology solutions for sustainable tourism development is needed.

By implementing these recommendations and building on the successes of community-based tourism initiatives in Garhwal, stakeholders can work together to create a more sustainable, equitable, and resilient tourism industry that benefits both people and the planet. Garhwal can serve as a model for sustainable tourism development that prioritizes community well-being, cultural preservation, and environmental conservation in mountainous regions worldwide.

References:

1 Chaudhary, S., Kumar, A., Pramanik, M., & Negi, M. S:Land evaluation and sustainable development of ecotourism in the Garhwal Himalayan region using geospatial technology and analytical hierarchy process. Environment, Development and Sustainability .23,115-120 (2022).

2 Joshi, R., & Singh, S. P. Community based tourism development: A case study of Sarmoli Village in Munsiyari, Uttarakhand. International Journal of Humanities & Social Sciences (2015)

3 Paudyal, K., Bastola, S., & Nepal, S. K. Prospects and challenges of ecotourism for biodiversity conservation in the Kedarnath Wildlife Sanctuary, Garhwal Himalaya, India. Journal of Mountain Science (2020)

4 Singh, S. Secular pilgrimages and religious tourism in the Indian Himalayas, GeoJournal.64,215–223 (2005a,b &c).

5 Gurung, D. B. Community-based tourism: A case study from Garhwal Himalaya, India. Journal of Mountain Science (2019).

6 Acharya, L., & Bennett, N. J. The role of community in sustainable tourism development: highlighting the significance of social capital. Journal of Sustainable Tourism (2016)

7 Himachal Pradesh State Council for Science, Technology & Environment (HPSCSTE). Sustainable Tourism in Garhwal Region, Uttrakhand, India: A Report (2018)

8 Reddy, B. S. Community-based tourism in the Indian Himalayas: Case study of Auli, Garhwal Himalayas. International Journal of Innovative Knowledge Concepts (2018)

9 Saikia, S. P. Community participation in sustainable tourism: A study of Dhanaulti tourism in Uttrakhand. Advances in Economics and Business Management (2017)

10 Tourism Development Council of Garhwal Region (TDCGR).Sustainable Tourism Development Plan for Garhwal Region: A Strategy Document (2021)

11 UNWTO. Tourism and Rural Development, (2020): https://www.unwto.org/tourism-rural-development

12 Uttrakhand Tourism Development Board (UTDB). Uttrakhand Tourism Statistics Report (2019).

https://doi.org/10.52202/074664-0004

Sustainability of Homestay Operators in West Jaintia hills, Meghalaya: Opportunities and Challenges

Nagy Asiad Son Langstang[1] and Dr. Trilochan Kumar[2]

[1]Department of Tourism and Hospitality Management, Chandigarh University

[2]Department of Tourism and Hospitality Management, Chandigarh University

Abstract. This research examines the obstacles and prospects for homestays that are sustainable in West Jaintia Hills, Meghalaya, with an emphasis on the viewpoints of home stay hosts. In-depth interviews with stakeholders and field observation were key components of the study technique, which was also supported by secondary data sources including government policy documents, academic books, and journals. The data were categorized using thematic analysis into "external challenges" and "internal challenges."The results suggest that the goal of showcasing local culture, promoting community development, and diversifying revenue sources motivated the creation of home stay programmes. The main services and attractions provided to visitors are comfortable lodging with organic meals, guided tours, and cultural encounters. The management of daily operations involves meticulous planning, collaboration, and guest communication. However, issues including poor resource management, a lack of ecotourism practices, and inadequate education and training stand in the way of home stays' sustainable expansion. Moreover, inadequate facilities, such as dependable power and internet access, present difficulties for visitors. All things considered, this study gives insightful information about the variables affecting the sustainability of home stays in West Jaintia Hills, Meghalaya, as well as suggestions for improving their viability and tourist appeal.
Keywords: Homestay, Community-Based Tourism, Sustainability, Development, Challenges.

1. Introduction

Accommodation is the fundamental part of tourism product, it could benefit indigenous economies more than the establishment of sizable hotels controlled by outsiders, as they could supplement the meagre earnings of the hosts, who are mostly from rural agrarian economies [5]. In majority of India's rural tourist programmes, homestays are a key component of the tourism product offering, according to MoT India/UNDP (Ministry of tourist (MoT) India/United Nations Development Programme, 2008). To solve the growing shortage of accommodations for tourists (ibid) and to boost the rural economy, policymakers are actively pushing homestays. As this study illustrates with a case from the West Jaintia Hills, Meghalaya, developing appropriate lodging options to support sustainable tourism does present unique difficulties. Most rural households in West Jaintia Hills, Meghalaya, rely mostly on agriculture for their income, albeit not all of them do so. While rural households employ a multitude of livelihood methods, one prevalent approach is for household members to engage in multiple jobs that collectively address one or more household need. "Sustainable development is development that satisfies current needs without jeopardising the ability of future generations to satisfy their own needs." As stated on page 43 of the 1987 World Commission on Environment and Development. The phrase "sustainable development" refers to addressing the needs of the underprivileged and marginalised while preserving the environment and using little in the way of resources and technology. At first, the growth of the tourism industry was linked to economic expansion; later, environmental, social, and local empowerment factors were included [16]. Local residents' aspirations for growth, as well as their expertise and experience, can best be taken into consideration when developing such a sustainable development policy [9]. Many countries are using the leisure-travel hybrid idea of homestay to promote community-based tourism that stops the economic boom of rural inhabitants and supports sustainable tourism that preserves the environment. The homestay programme was established, tourists have had the opportunity to live in a community and learn about the traditions and manner of life of the local people [14].

Due to its operations being designed to allow visitors to become acquainted with the customs and biodiversity of a typical rural village under the supervision and guidance of hosts/owners of the house, homestays [1] article, "homestay tourism adds authentic sociocultural richness to the tourists' experience, viewing tourism as an alternative product for a nation that possesses an abundance of tourism opportunities in remote communities but cannot make extensive infrastructure investment a priority." The research by [17] confirms the importance of ecotourism in terms of how it affects business strategies and how conflicts

https://doi.org/10.52202/074664-0005

and obstacles might prevent ecotourism from being sustainable in terms of the economy, society, and environment. The Theory of Constraints (TOC), developed by [20] is significant in this regard according to this reasoning. According to this notion, an organisation is a controllable system with few obstacles preventing it from accomplishing its objective. The Homestay Programme, as it is applied to this study, is made up of organisations and service providers or operators. It can be thought of as a manageable system with goals to accomplish and constraints in the form of operational difficulties that could prevent the system from reaching its objective of sustainable development.

As a topic of cross-disciplinary study, homestays are represented in a literature that exhibits a range of viewpoints and levels of research quality. Although postgraduate dissertations and NGO project development reports are occasionally used because of their value, the paper primarily draws from academic studies on the following topics: gender, entrepreneurship, poverty reduction, geography, ecology, sustainability, and mountain research and development. The development of the travel and tourism industry in Meghalaya and across India has not been given much thought, with regards to home-stay initiatives. People's ideas about homestays and how they operate are imprecise. When developing the concept, the authorities were unable to see the real value of homestay from the first. Higher officials ignore a lot of glaring issues with homestay performance and operation. In an effort to make homestays in West Jaintia Hills, Meghalaya, more sustainable, this essay looks at the many challenges, problems, and facets of development that they face.

1.1 A Look at Tourism in Meghalaya

The state of Meghalaya, also called the "abode of clouds," is a beautiful region in northeastern India. Famed for its verdant surroundings, undulating hills, and tumbling waterfalls, Meghalaya presents a distinctive fusion of scenic splendor, opulent culture, and exhilarating adventures. An outline of Meghalaya's tourism attractions is provided below:

Shillong: Owing to its remarkable resemblance to the Scottish Highlands, Shillong, the capital city of Meghalaya, is frequently referred to as the "Scotland of the East". It is well-known for its colonial architecture, lively culture, and nice temperature. The Don Bosco Museum, Ward's Lake, Elephant Falls, and Shillong Peak are a some of the city's well-known sights.

Known for having some of the world's highest rainfall, Cherrapunji (Sohra) provides breathtaking views of lush landscapes, misty valleys, and magnificent waterfalls. The Seven Sisters Falls, Mawsmai Cave, Nohkalikai Falls, and the live root bridges of Mawlynnong Village are among the attractions. The town of Mawlynnong, which is frequently referred to as Asia's cleanest, embodies eco-tourism that is driven by the local community and sustainable living. Its pathways are kept up beautifully, and visitors can marvel at the inventiveness of its live root bridges and its distinctive bamboo constructions. Meghalaya is home to many sacred groves, which are conserved forests that are important to the indigenous Khasi and Jaintia populations on a religious and cultural level. These groves provide a window into the customs and conservation methods of the area and are hotspots for biodiversity.

The largest river island in Meghalaya is called Nongkhnum Island, and it is situated in the West Khasi Hills district. It provides peaceful surroundings for those who enjoy the outdoors, along with chances for camping, birdwatching, and wilderness exploration. The Jaintia Hills are renowned for their breathtaking scenery, sacrosanct forests, and vibrant tribal cultures. Trekking enthusiasts can experience the indigenous Khasi culture while exploring the stunning scenery and secret waterfalls. Dawki is known for its glistening Umngot River, where guests can go boating and take in the breathtaking sight of the riverbed. Dawki is located close to the border with Bangladesh. Views of the surrounding countryside can be seen from the suspension bridge that spans the river.

The UNESCO Biosphere Reserve Nokrek National Park is well-known for its abundant biodiversity, which includes uncommon plants and animals. Hikers may walk to the peak of Nokrek Peak for breathtaking panoramic views, explore lush woods, and see species including the endangered red panda and the unusual stump-tailed macaque. Known as the "abode of perpetual winds," Balpakram National Park is well-known for its awe-inspiring scenery, soaring canyons, and eerie folklore. Numerous varieties of wildlife can be found there, such as clouded leopards, elephants, and unusual orchid species. Popular activities in this area include trekking and animal safaris. Local Food: Meghalayan food is a delicious fusion of Garo, Jaintia, and Khasi traditional flavours. Pork salad (Doh Khlieh), sweet rice cake (Pumaloi), and jadoh (rice and meat dish) are among the must-try foods.

1.2 Tourist Footfall in Meghalaya

 https://doi.org/10.52202/074664-0005

About 12.7 lakh tourists visited the State in 2019 (including 25,000 foreign visitors). This is greatest among the Northeastern states following the regions of Sikkim and Assam. Over the past ten years, there has been a consistent increase in foot traffic, which is currently experiencing a comeback after the COVID-19 pandemic. By 2024, it's anticipated that the number of visitors will surpass 15 lakhs annually.

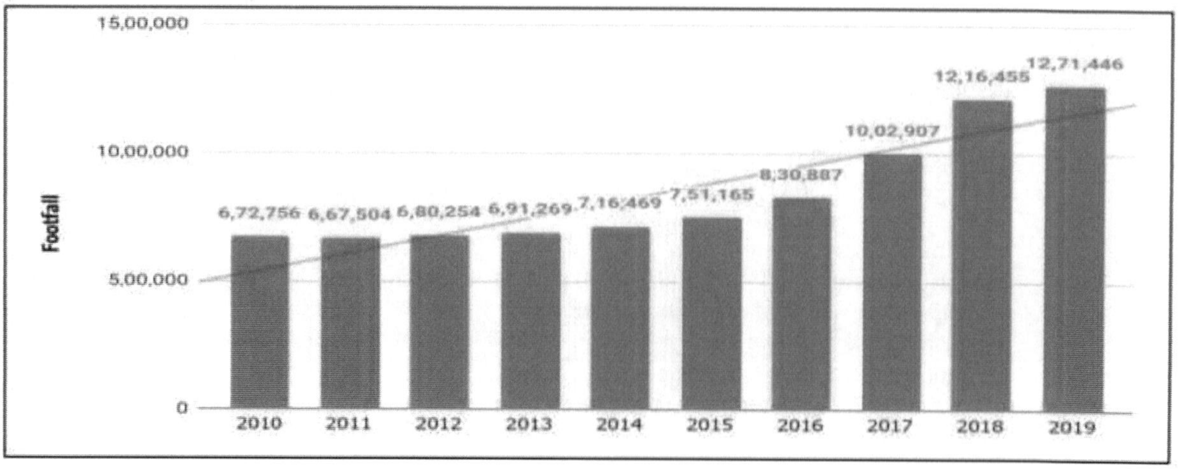

Figure 1: Tourist footfall in Meghalaya

Source: Meghalaya Tourism Policy 2023

1.3. Research Objectives

- By conducting in-depth interviews with stakeholders and field observations, the aim of this study is to determine the potential and obstacles that contribute to sustainable homestays in West Jaintia Hills, Meghalaya.
- To investigate the motivations for the creation of homestay programmes, day-to-day activities, attractions and goods provided, and challenges encountered by homestay operators in managing their businesses.

2. Review of Literature

The current study has identified the internal barriers that prevent community members from participating in the West Jaintia hills homestay programme, including a lack of trained human resources, issues with leadership, a lack of capital, and a lack of basic infrastructure development. The external barriers include threats to safety and security, a lack of monitoring from the state and ministry government, an informal organisational structure within the homestay programme, and a lack of marketing and promotional efforts. Sustainable tourism is acknowledged as a critical industry for accomplishing the Sustainable Development Goals (SDGs) of the United Nations since it has been boosting the economies and jobs of poor nations (UNWTO 2017). It's common knowledge that greater community involvement can make CBT a driving force behind both economic expansion and the growth of the tourism industry [12; 15]. In many nations, community homestays have been an effective means of promoting rural community development by allowing local residents to earn a living from their own houses. However, in other places, these homestays have been forced to close due to various difficulties [7; 8].

According to [5], there aren't many studies looking into ICT use during homestays. With the use of ICT tools, CBT may enhance their worldwide visibility [4], produce new goods, market tourism resources, attract more tourists, and generate more foreign income [6]. Furthermore, it is expected to improve community engagement; raise involvement to maintain community norms and structures; foster leadership qualities in community members; and enable community organisers to maintain authority over the choices and elements influencing their communities and lives (WHO 2009). An analysis of ICT tools used to support sustainable tourism reveals that GIS, web apps, augmented reality, the Internet of Things (IoT), and social media can all be leveraged to leverage ICT technologies [10]. major challenges that current home-stay operators are facing are as follows: inadequate infrastructure facilities; lack of skilled human resources; absence of legal regulations for home-stay operations; lack of marketing and promotional tactics; poor coordination among various tourism stake holders with home-stay operations; lack of awareness regarding conservation of natural and cultural resources; improper resource management system; and poor maintenance of security and existing facilities, according to [2].

Poor Infrastructure facilities	- Poor road conditions - Transportation problems - Electricity - Lack of healthcare - Lack of communication facility
Lack of skilled human resources	- Guides - Entrepreneurs - Hospitality professionals - Lack of educational institutes
Lack of legal regulations	- Registration of home-stay - Booking of home-stay
Lack of Marketing & Promotions	- Poor marketing network - Insufficient funds
Poor ordinations	- Government bodies - Tourism intermediaries - Non-government bodies
Less awareness	- Conservation - Upgrading
Improper Management	- Physical resources - Human resources - Facilities

Source: [2]

Acc

SoSource [2]

According to [13], there are three (three) components to a sustainable perspective: economic, social, and cultural elements. As to [3] the economic aspect of sustainable homestay development pertains to the proper policies and tactics that ensure the homestay owner's gain and lessen any adverse effects. The sociocultural aspect pertains to the manner in which homestay proprietors retain their own social identity [1]. Sociocultural sustainability in this study refers to maintaining the homestay owner's lifestyle, long-term social relationships (like effective communication), and quality of life (like happiness) [3]. Homestay owners' involvement in tourism can be established by giving them the freedom to make their own decisions and by offering possibilities for mutual gain [18]. The planning phase of creating a tourism village is when their position as a subject in their area starts [5]. The village's community must support tourism growth during the planning phase. Providing information about both advantages and disadvantages encourages support from the neighbourhood [4]. Acceptance by the community during the planning stage boosts community involvement throughout the execution stage of tourism activities [5]. Numerous studies have demonstrated that the local community's favourable impression of the tourist development in their village

 https://doi.org/10.52202/074664-0005

motivates them to be aware [3]. The local community should be considered a subject in tourism development projects.

3. Research Methodology

This study aims to identify the challenges and opportunities contributing for sustainable homestay in West Jaintia Hills, Meghalaya. Field observation and in-depth interviews with significant stakeholders were used as primary data gathering methods. In addition to using primary data collections, this study also made use of secondary data sources to learn about opportunities and challenges for sustainable homestay operators in the context of homestay tourism development. These sources included journals, published academic books, unpublished reports from the local management committee and local tourism organisation, government policy documents, and the internet. The background information gathered during the interview session mostly included the evolution of the homestay programme and the difficulties faced by homestay operators in running their daily operations, which is gradually nearing the end of its life. Purposive sample technique, the informants chosen for the interviews were individuals with experience running homestays, particularly those with such experience. These interviews were conducted primarily in order to gather more information about the following topics: the reasons behind the program's establishment; daily operations; attractions and products offered; homestay operators' involvement; and, last but not least, the difficulties they encountered while running the programme.

4. Findings

In the northeastern Indian state of Meghalaya lies a district known as West Jaintia Hills. Rich cultural heritage, varied communities (chiefly the Jaintia tribe), and scenic beauty are its main draws. West Jaintia Hills offers the following main features and draws:

Geography: The West Jaintia Hills are distinguished by their hilly topography, numerous rivers and waterfalls, and strong forests. The Garo-Khasi range includes the Jaintia Hills, which provide hiking and exploration opportunities along with stunning scenery.

Culture: Prominent for its distinct festivals, customs, and traditions, the Jaintia tribe resides in the region. Behdienkhlam is a traditional festival that is celebrated with great fervour, and visitors can participate in it to fully immerse themselves in the native way of life.

Many natural attractions may be found in West Jaintia Hills: Beautiful waterfall, perfect for leisurely picnics and relaxation, Krang Suri Falls is surrounded by thick foliage. Another stunning waterfall, Thlumuwi Falls, provides a serene setting amidst the area's natural splendour. Ancient monoliths from the Jaintia rulers called Nartiang Monoliths shed light on the rich history and cultural legacy of the area. Adventure tourism: With its rough landscape, the West Jaintia Hills are a great place to go trekking, caving, and rock climbing. Spelunkers from over the world are drawn to the area's caverns, which include the well-known Krem Liat Prah. Sites of Cultural Heritage: West Jaintia Hills is home to a number of historical and cultural landmarks besides the Nartiang Monoliths, including: The Nartiang Durga Temple is a noteworthy Hindu temple that honours the goddess, Durga. It has elaborate architecture and holds religious significance. Ialong area is a famous recreational area that attracts both locals and tourists. It offers sweeping views of the surrounding hills and valleys.

The average rural community in the West Jaintia hills, Meghalaya has little interest in engaging in homestay tourism since there are no incentives in the marketing environment. And there are no incentive programmes designed by the authorities in order to motivate and attract the rural community for homestay operations. Inadequate education and training are now impediments to both the successful operation of already-existing enterprises and the entry of new ones. The viability of the homestay programme might also be seriously hampered by improper resource management at the location. A significant obstacle to the sustainable growth of homestay in Meghalaya is the lack of ecotourism practices and a lower level of local awareness regarding the preservation of natural and cultural resources. In the state, homestay tourist development and marketing are still in their infancy. Even though the Meghalaya government has done a great job promoting homestays over the past ten years, much more work has to be done. Basic conveniences like reliable electricity, hot water, and adequate sanitary facilities can not be available in certain homestays. This may provide a problem for visitors used to greater metropolitan luxuries. Because of Meghalaya's sometimes

rough and isolated terrain, it might be challenging to get to some homestays, particularly during the monsoon season when roads may get muddy or washed away.
It's possible that homestays don't provide the same amenities or services as hotels, so visitors should be ready for more basic lodging and potentially fewer food options. Meghalaya has a lot of rain, especially in the monsoon season, which can cause travel schedules and outdoor activities to be thrown off. This could have an impact on visitors' entire experience when staying at homestays. Internet connectivity may be spotty or nonexistent in isolated places. It is advisable for visitors to anticipate restricted internet access throughout their visit. Although the safety of its guests is the top priority for most homestays, guests staying in rural areas may find it difficult to get to emergency services or medical facilities in the event of an unexpected situation. Cultural exchange can provide benefits, but there are drawbacks as well because hosts and guests may have different expectations, habits, and lifestyles. Effectively addressing these obstacles requires mutual respect and clear communication.

5. Result of Interview and Thematic Analysis

This study's use of purposive sampling provided insightful information about the difficulties and prospects facing sustainable homestays in West Jaintia Hills, Meghalaya. The in-depth information obtained from the homestay operators' interviews included the motivations for the program's creation, the day-to-day activities involved, the attractions and goods provided to visitors, and the degree of homestay operators' involvement in it. Furthermore, the interviews highlighted both internal and external issues that homestay operators confront in managing their businesses.

1. Can you describe the reasons behind the establishment of your homestay program?

The motivation for starting our homestay programme was to introduce tourists to the local way of life and cultural heritage. In addition to giving visitors a genuine taste of West Jaintia Hills, Meghalaya, we aimed to establish a stable revenue stream for our locality. We also believed that the homestay programme offered a chance to foster understanding and cross-cultural exchange between visitors and residents. **(Interviewee 1)**

We needed to diversify our sources of revenue, which is why we decided to start a homestay programme. Our village's main source of income, agriculture, was becoming more and more erratic as a result of shifting weather patterns and market dynamics. We saw homestay tourism as a means of maintaining our cultural heritage while also raising our level of life and augmenting our income. **(Interviewee 2)**

Our homestay programme was started as a part of a bigger project for community development. We were aware that tourism may benefit our village economically, but we also wanted to make sure that the advantages were distributed fairly among the locals. We were able to give local youth job opportunities and enable women to engage in income-generating activities by implementing a homestay programme. **(Interviewee 3).**

Result: The primary finding of this inquiry indicates that the inception of homestay initiatives in West Jaintia Hills, Meghalaya, was propelled by multiple considerations, including the aspiration to exhibit indigenous customs and traditions, broaden the range of revenue streams, and foster community advancement. Homestay operators saw their businesses as a way to give visitors genuine experiences, augment revenue from traditional ways, and give locals work chances. The programmes were also viewed as a means of promoting cross-cultural communication and understanding between visitors and residents, which would aid in the community's general growth and sustainability.

2. What are the key attractions and products that you offer to guests?

Our homestay's main draw is the chance for visitors to take in the breathtaking scenery of the West Jaintia Hills. We provide guests with guided visits to neighbouring caverns, waterfalls, and hiking routes so they may experience the breathtaking scenery of the area. We also provide cultural activities to give visitors a better understanding of the local way of life, like cooking workshops and traditional dance performances. (Interviewee 1)

Our guesthouse places a strong emphasis on giving visitors a calm and relaxing space to unwind in. Our goal is to provide guests with a comfortable stay by providing them with modern conveniences in traditional

 https://doi.org/10.52202/074664-0005

cottages. To bring visitors a taste of real regional food, we also serve organic meals prepared using products that are sourced locally. (Interviewee 2)

Our homestay's main selling point is the chance it gives visitors to participate in local events. We arrange village walks so visitors can talk to locals and discover their way of life. Additionally, we provide courses on traditional crafts like weaving and pottery so that visitors can learn about the rich cultural history of our neighbourhood. (Interviewee 3).

Result: This question's primary result indicates that giving visitors immersive experiences with the area's natural beauty and cultural legacy is the main draw and product supplied by homestay operators in West Jaintia Hills, Meghalaya. Along with cultural activities like traditional dance performances and cookery lessons, these attractions include guided visits to waterfalls, caverns, and hiking routes. In addition, homestays provide cosy lodging in historic cottages outfitted with contemporary conveniences, organic meals prepared with foods that are acquired locally, and chances for visitors to participate in community events and classes on traditional crafts. All in all, these services are meant to give visitors a distinctive and genuine taste of West Jaintia Hills living, enriching their visit and supporting the program's long-term viability.

3. How do you manage the daily operations of your homestay?

The day-to-day management of our homestay necessitates meticulous organisation and preparation. Using fresh products from nearby shops, we prepare meals for our guests at the beginning of each day. Additionally, we make sure the rooms are kept up to date and clean to give our visitors a comfortable stay. To ensure that our visitors have a wonderful time while visiting, we also plan events and trips for them. (Interviewee 1)

It takes a team to run our homestay on a regular basis. Our hardworking staff assists with cleaning, cooking, and serving guests. Every employee is assigned distinct duties and responsibilities, which guarantee the effective completion of all tasks. Additionally, we keep lines of communication open with our visitors, immediately resolving any problems or complaints they may have. (Interviewee 2)

We adhere to a rigid timetable and rank chores according to priority in order to oversee the day-to-day operations of our homestay. Every day, we begin by making sure the rooms are available and ready for incoming visitors. Additionally, we make sure to restock on necessities like food and toiletries. We keep in constant contact with our visitors throughout the day to make sure their needs are satisfied and to handle any criticism or grievances they may have. (Interviewee 3).

Result: The primary finding of this inquiry indicates that homestay providers in West Jaintia Hills, Meghalaya, oversee their day-to-day operations by means of meticulous preparation, synchronisation, and collaboration. Creating meals for visitors using local, fresh products, keeping rooms tidy and comfortable, and planning enjoyable activities and trips are all part of the daily operations. Homestay hosts stress how crucial it is to communicate with visitors in order to quickly resolve any problems or concerns. They also emphasise how their hardworking employees help with cooking, cleaning, and guest services, making sure that everything is done quickly and effectively. In general, it is believed that successful management of day-to-day operations is essential to guaranteeing both the success and longevity of the homestay programme and to give guests an unforgettable experience.

6. Recommendations

Numerous difficulties arise in setting up and running a sustainable homestay in West Jaintia Hills, Meghalaya. The long-term viability of the homestay and its sustainability initiatives depend on resolving these issues. Following a thorough analysis and investigation of the subject, the following suggestions can be made to significantly advance Homestay Operator Sustainability:

- Preparing a sufficient plan and strategy for the homestay business and the local community is the first and most important piece of advice. Homestay development cannot be accomplished without planning and technical support.

- To encourage waste reduction, water and energy conservation, and appropriate behaviour, thoroughly educate visitors through welcome packets, signs, and in-person orientations. In order to justify initial investments, conduct a thorough cost-benefit analysis that demonstrates the long-term financial benefits of sustainable practices, such as energy savings and lower operational expenses.

- To encourage voluntary participation, create interesting and dynamic instructional materials, workshops, or guided tours that emphasise the social and environmental advantages of sustainable practices. Set a good example for your guests by acting sustainably yourself. Include messages about sustainability in all of your communications and interactions with them to normalise and encourage sustainable behaviour.

- In order to develop trust and support, it is important to engage stakeholders in decision-making processes by addressing concerns, getting feedback, and collaborating openly and collaboratively with locals, community leaders, and organisations. To win over the acceptance and cooperation of the locals, highlight the good effects of the homestay on the environment, culture, and economy. Examples of these include the creation of jobs, the preservation of local heritage, and support for community initiatives.

- Offer off-season packages, focus on niche markets, and advertise alternative travel experiences or activities that appeal to eco-aware tourists or those looking for genuine cultural immersion to diversify revenue streams. Establish alliances with nearby companies, tour guides, and travel bureaus to collectively advance eco-friendly travel programmes, lengthen the travel season, and lessen the effects of seasonality.

- It might be difficult to strike a balance between the year-round sustainability measures required and the varying demand for lodging during tourist seasons. Using creative marketing techniques may be necessary to stand out from the competition in a congested market and persuade prospective guests of the value of sustainability. Tourism entrepreneurs should spearhead the effort to create a marketing campaign for homestay tourism, as well as tourism stakeholders should assume responsibility for promoting homestays. It is advised to set up a community fund from which locals can make investments for infrastructure development and capacity building.

- To guarantee long-term sustainability of technology and infrastructure, regular upgrades and maintenance are necessary. Continuing education and assistance to ensure that staff members preserve sustainable practices and promote the sustainability objectives of the homestay.

- To address these issues and remove obstacles to the sustainability of homestay operations, a comprehensive strategy comprising cooperation with visitors, nearby communities, governmental organisations, and industry partners is needed. Sustainable homestays can overcome obstacles and flourish as socially conscious, environmentally conscious, and financially successful tourism businesses by implementing these measures and utilising the combined efforts of stakeholders.

7. Conclusion

The study on sustainable homestays in West Jaintia Hills, Meghalaya, emphasises how important homestay programmes are for advancing community development, encouraging local culture, and diversifying sources of income. The interview data from homestay operators was interpreted using thematic analysis in the framework of the study on sustainable homestays in West Jaintia Hills, Meghalaya. The process of study involves identifying topics pertaining to the motivations for the creation of homestay programmes, the day-

https://doi.org/10.52202/074664-0005

to-day activities involved, the products and attractions provided to visitors, and the challenges encountered by homestay operators. Through the examination of these themes, researchers were able to get a more profound comprehension of the obstacles and prospects for homestays that are sustainable in the area. Homestay hosts show tenacity and ingenuity in running their businesses in spite of obstacles like poor education and training, poor resource management, and inadequate infrastructure.Targeted actions are necessary to address these issues and improve the sustainability of homestays. This entails offering homestay hosts opportunities for education and training, putting ecotourism principles into effect, and enhancing infrastructure like electricity and internet access. Furthermore, fostering mutual understanding and cultural interchange between hosts and visitors can enhance the homestay experience even more and support its long-term viability.In general, eco-friendly homestays in West Jaintia Hills have the ability to conserve regional history, generate new business prospects, and give visitors unforgettable experiences. Homestay programmes have the potential to become a vital and sustainable component of Meghalaya's tourism industry by tackling the recognised obstacles and capitalising on the distinctive cultural and natural resources of the area.

References

1. Acharya, B. P., & Halpenny, E. A.: Homestays as an alternative tourism product for sustainable community development: A case study of women-managed tourism product in rural Nepal. *Tourism Planning & Development*, *10*(4), 367-387, (2013).
2. Bhan, S., & Singh, L.: Homestay tourism in India: Opportunities and challenges. *African Journal of Hospitality, Tourism and Leisure*, *3*(2), 1-5, (2013).
3. Fong, L. H. N., Gursoy, D., & Sigala, M.: Experimental research in tourism. *Asia Pacific Journal of Tourism Research*, *25*(7), 707-709, (2020).
4. Gan, C., Wang, K., Voda, M., Ye, J., & Chen, L.: Analysis on the impact of technological innovation on tourism development: The case of Yangtze River Delta Urban Agglomeration in China. *Tourism Economics*, *29*(5), 1251-1271, (2023).
5. Janjua, Z. U. A., Krishnapillai, G., & Rahman, M.: A systematic literature review of rural homestays and sustainability in tourism. *Sage Open*, *11*(2), (2021).
6. Jiang, L., Wu, H., & Song, Y.: Diversified demand for health tourism matters: From a perspective of the intra-industry trade. *Social Science & Medicine*, *293*, (2022).
7. Kunjuraman, V.: Community-based ecotourism managing to fuel community empowerment? An evidence from Malaysian Borneo. *Tourism Recreation Research*, *47*(4), 384-399, (2022)..
8. Kunjuraman, V., & Hussin, R.: Challenges of community-based homestay programme in Sabah, Malaysia: Hopeful or hopeless?. *Tourism Management Perspectives*, *21*, 1-9, (2017).
9. Labadi, S., Giliberto, F., Rosetti, I., Shetabi, L., & Yildirim, E.: Heritage and the sustainable development goals: Policy guidance for heritage and development actors. *International Journal of Heritage Studies,* (2021).
10. Lama, S., & Pradhan, S.: Community Homestay Platform to enhance Sustainable Tourism in Developing Countries, (2022)..
11. Liu, A., & Wall, G.: Planning tourism employment: a developing country perspective. *Tourism management*, *27*(1), 159-170, (2006).
12. Mishra, S., & Mishra, J. M.: Self-help groups and community-based tourism: a study of interrelationship and convolutions. *Journal of Tourism*, *24*(1), 71-78, (2023).
13. Park, D. B., Nunkoo, R., & Yoon, Y. S.:Rural residents' attitudes to tourism and the moderating effects of social capital. *Tourism Geographies*, *17*(1), 112-133, (2015).
14. Sen, V., & Walter, P.: Community-based ecotourism and the transformative learning of homestay hosts in Cambodia. *Tourism Recreation Research*, *45*(3), 323-336, (2020).
15. Stone, L. S., & Stone, T. M.: Community-based tourism enterprises: challenges and prospects for community participation; Khama Rhino Sanctuary Trust, Botswana. *Journal of Sustainable Tourism*, *19*(1), 97-114, (2011).
16. Telfer, D. J.: The evolution of tourism and development theory. *Tourism and development: Concepts and issues*, 35-80, (2002).
17. Thompson, B. S., Primavera, J. H., & Friess, D. A.: Governance and implementation challenges for mangrove forest Payments for Ecosystem Services (PES): Empirical evidence from the Philippines. *Ecosystem services*, *23*, 146-155, (2017).

18. Vivek, K., Mishra, S., & Pradhan, R. C.: Characterization of spray dried probiotic Sohiong fruit powder with Lactobacillus plantarum. *Lwt*, *117*, 108699, (2020)..
19. Wallace, J. B., & Webster, J. R.: The role of macroinvertebrates in stream ecosystem function. *Annual review of entomology*, *41*(1), 115-139, (1996).
20. Tulasi, C. L., & Rao, A. R.: Review on theory of constraints. *International Journal of Advances in Engineering & Technology*, *3*(1), 334, (2012).

 https://doi.org/10.52202/074664-0005

Analyzing growth prospects for the Indian wine Industry: A Comprehensive study of Domestic and International markets considering evolving consumer preferences and global demand.

Balraj Benedict[1] [0009-0003-6907-1501] and Dr. Kiran shashwat[2] [0009-0009-9176-4113]

[1,2] Chandigarh University, Punjab, India

Abstract

The potential for growth in the Indian wine industry. The focus will be on understanding market preferences within India and exploring the demand for Indian wines on a global scale. The study aims to provide a thorough analysis of factors contributing to the industry's growth, shedding light on both domestic consumer preferences and international market trends. The principal aim is to examine and comprehend the wine consumption habits of the Indian market. Through a detailed examination of consumer behavior, regional trends, and changing tastes, to learn more about the factors that influence the nation's wine demand. Events and conferences in the industry provide forums for professionals to exchange ideas and talk about new developments, which makes them important sources of information on how the market is changing. Government publications and papers pertaining to trade, agriculture, and commerce can offer further perspective. In addition, study investigates the demand for Indian wines abroad, broadening its scope to include the entire world. By combining a rigorous investigation of domestic market dynamics with a global vision, this study attempts to provide a comprehensive picture of the variables driving the growth trajectory of the Indian wine business.

Keywords: Perspective, Dynamics, Pertaining, comprehend, consumption.

1.Introduction

The Indian wine market size is estimated to grow at a CAGR of 30.92% between 2022 and 2027. The size of the market is forecast to increase by USD 688.16 million. In 2017 the size of the market was valued at USD 68.80 million. This wine market in India report extensively covers market segmentation by Type (domestic and imported) and Product (red, fortified, white, and sparkling). It also includes an in-depth analysis of drivers, trends, and challenges. The wine market in India is growing due to the availability of online stores and speciality retailers, providing consumers with a wide range of choices and convenient online shopping experiences. Strong online distribution channels and platform providers like Wine Well and Liquor Kart are boosting online sales. This expansion creates opportunities for vendors to target customers without geographical limitations, enhance operational efficiencies, and offer customized products. The evolving retail landscape allows vendors to develop online strategies and eliminate middlemen, resulting in competitive pricing and increased profit margins. Factors such as rising disposable income, changing attitudes towards alcohol consumption, wine's perception as a sophisticated drink, and the growth of wine bars and tourism contribute to this market growth. The Indian wine market is expected to continue its upward trajectory and become a significant player in the global industry. n recent years, the wine industry in India has witnessed a remarkable transformation. Traditionally, India has been renowned for its vibrant array of spices and culinary diversity, but now it is increasingly gaining recognition as a producer of high-quality wines. This shift reflects changing consumer preferences and growing acceptance of wine as a beverage of choice, both domestically and internationally. Understanding market preferences within India and exploring the demand for Indian wines on a global scale presents a compelling narrative of evolution and opportunity within the wine industry. This exploration involves delving into various aspects, including consumer behavior, cultural influences, economic factors, and technological advancements shaping the production and distribution of Indian wines. (As per survey done by Technavio in feb 2023 on India Wine Market by type and Product - Forecast and Analysis 2023-2027) [23].

India's domestic wine market is undergoing a significant transformation, driven by evolving consumer tastes and preferences. Traditionally, wine consumption in India has been relatively low compared to other alcoholic beverages. However, changing lifestyles, urbanization, and a burgeoning middle class have contributed to a growing interest in wine consumption. Understanding the preferences of Indian consumers involves analyzing factors such as taste preferences, price sensitivity, brand perception, and cultural influences. India's diverse cultural landscape plays a crucial role in shaping preferences for wine consumption. While wine has historical associations with certain regions and communities in India, its acceptance and consumption have expanded beyond traditional boundaries.[24] Exploring the cultural nuances surrounding wine consumption can provide valuable insights into consumer behavior and preferences in different parts of the country. Economic factors such as income levels, purchasing power, and disposable income significantly influence the demand for wine in India. As disposable incomes rise and the purchasing power of consumers increases, there is a growing propensity to spend on premium and luxury goods, including wine. Analyzing economic indicators and consumer spending patterns can help understand the dynamics of demand within the Indian wine market. The global demand for Indian wines is also on the rise, presenting opportunities for Indian wineries to expand their presence in international markets. Factors such as globalization, changing perceptions of Indian products, and increased exposure to global cuisines contribute to the growing interest in Indian wines abroad. Exploring avenues for export, understanding international consumer preferences, and navigating regulatory frameworks are essential considerations for Indian wineries seeking to tap into global markets. Understanding market preferences within India and exploring the demand for Indian wines on a global scale offer valuable insights into the evolving dynamics of the wine industry. By analyzing consumer behavior, cultural influences, economic factors, and internation l trends, stakeholders can identify opportunities for growth and innovation, driving the continued success of the Indian wine market both domestically and internationally. Conducting a comprehensive analysis of consumer preferences, purchasing behavior, consumption occasions, and brand perceptions is essential for stakeholders in the wine industry. By gaining a deeper understanding of the Indian market dynamics, wine producers, distributors, retailers, and policymakers can devise informed strategies to capitalize on emerging opportunities and address challenges effectively.

2. Literature Review

2.1The Indian wine

The study's primary goal is to determine and investigate variables related to the wine related lifestyle technique in the Indian setting. Six factors were extracted based on the results of the EFA. It was determined that drinking routine was the first factor. The quality of drinking rituals is what makes them distinctive from other wine-related rituals; they serve to reiterate the cultural norms around drinking. It is crucial to comprehend not only people's understanding of the wine but also how they perceive it to be prepared and presented. Since wine is seen as a social beverage, rituals surrounding its consumption are significant. Consumption motivation (social) was shown to be the second component. This component accounts for wine preference at social gatherings and special occasions. [20]. The wine may be the oldest known fermented product to humans, as the Rigveda clearly attests. The term "wine" describes the fermented byproduct of grapes. The wine industry is very young in India. In India, there are just a few institutions and universities that conduct more than twelve. [9]. Wine product distribution and sales with limited scope to identify the variables influencing the wine price. Hence, further investigation can be carried to understand the supply chain management practices, the innovative business models as the need for social gathering and wine tourism expected to rise.[22].

2.2 The Indian wine Industry

The wine business has been present in the Nashik District of India for approximately 15 years. The qualities are appealing as they support the local community's way of life. The creation of jobs and the diversification of rural economies are perhaps the two main advantages of the wine sector development. By combining cutting-edge ideas with creative business models and wine production, rural economies may be stimulated, and jobs can be created. Appropriate infrastructure is probably going to have a big impact on drawing tourists to wine-producing regions, which is good for the local economy. An analysis of the emerging Indian wine business in the Nashik district points to the possibility of synergistic effects. [22]. Indian wine has made an amazing journey across the ages, passing through Persia, Europe, the Mughals, Portugal, Britain, and India. The emerging Indian wine industry is prepared to take its rightful place in the vibrant international wine market.[6] The current Indian wine market is tiny but expanding; the country's yearly per capita wine consumption is around 9 millilitres, or around 1/8000th of France's.

 https://doi.org/10.52202/074664-0006

The primary cause of this is the preference of Indians for hard liquor and beer, which together account for about 98% of the market share, whereas low-ABV wine only makes up 2%.[Grapevines are thought to have been brought to India from Persia during the Indus Valley civilization, when viticulture in the country began. Although winemaking has been practiced for most of the India's history, it was especially promoted during the Portuguese and British colonization of the subcontinent. Recovering from the devastating events at the close of the 1800s proved to be an arduous task for the Indian wine sector. Alcohol was seen negatively by the public and religion, and this culminated in the prohibition of alcohol in many Indian states throughout the 1950s. Vineyards were either forced to make way for table grape and raisin production or encouraged to do so. While wine production persisted in some regions, such as Goa, the resulting beverage was typically exceedingly sweet and very alcoholic. The establishment of The Tonia Group in the state of Goa in the early 1980s marked a turning point for the contemporary Indian wine industry. The Tonia Group started importing Vitis vinifera grape varietals such Cabernet Sauvignon, Chardonnay, Pinot blanc, Pinot noir, and more with help from French winemakers.[5]

2.3 Consumer Preferences

As India's burgeoning middle class enjoys the fruit of the vine, wine consumption is rising quickly. However, for the time being at least, wine firms who find themselves far ahead of demand are afraid. The capacity of these festivals to produce significant long-term income and maintain consistent growth in wine and wine tourism also contributes to an understanding of the shifting consumer dynamics. If not, it will put an excessive amount of pressure on the players in the wine market, which will degrade the product's quality and erode its already small clientele. [21] Aside from the variables found in this study, a few more risk/benefit perception factors might possibly exist in the wine service in Indian restaurants and cafés. Therefore, more research may be needed to examine the risk/benefit aspects in greater detail customer's non-purchase behavior or decision regarding the wines, but that knowledge may not be thorough enough, and further studies may shed more light on those specifics and contributing variables. a nation where the tradition of wine and dining is very new and limited to a few contemporary cities. Furthermore, the data were only gathered in Delhi, thus they might not accurately represent the consumption habits of the entire nation. Furthermore, there are regional variations in the methods that restaurants and cafes serve wine, so one cannot generalize about them all. Upcoming [11]. The study also explores the role of government support in promoting the wine industry and facilitating its expansion. The findings provide insights into the importance of favourable government policies and the need for policy reforms to foster the growth of the Indian wine industry. [29]. Paper investigates consumer preferences and market demand for Indian wines. It presents a comprehensive study that examines the factors influencing consumer choices and the market dynamics in the Indian wine industry. The study analyzes data from surveys and interviews conducted with wine consumers and industry experts to gain insights into the preferences and behaviours of Indian wine consumers. [30}. Paper examines the global demand for Indian wines and explores the export potential of the Indian wine industry. It provides a comprehensive analysis of the factors influencing the export market for Indian wines, including market trends, consumer preferences, trade barriers, and competitive landscape. The study investigates the challenges faced by Indian wineries in expanding their presence in international markets and identifies strategies to enhance export potential.[31].

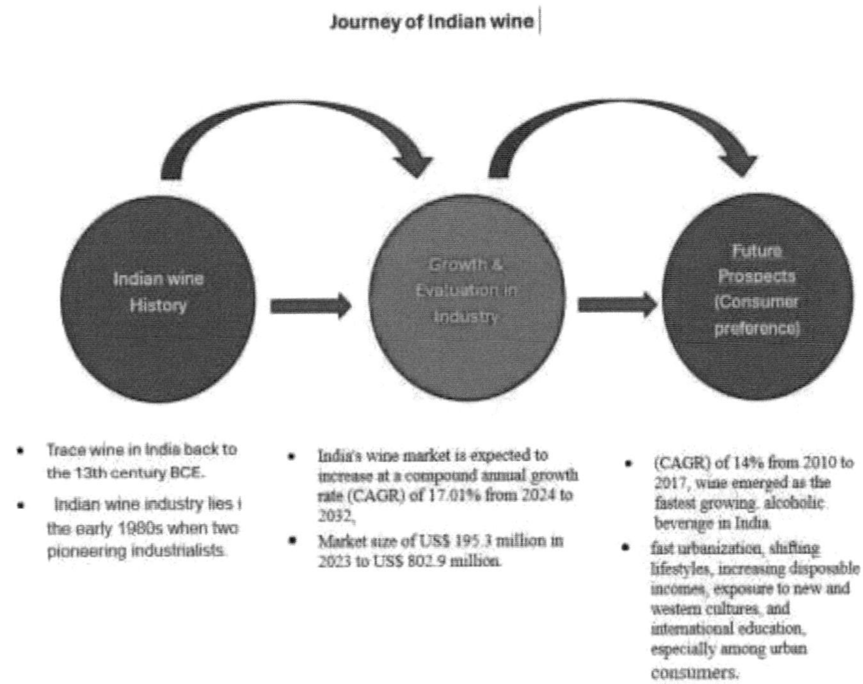

Figure 1 Journey of Indian Wine

3. Research Objectives & Methodology

This is research is based on qualitative analysis adopted a comprehensive review of various literature available on different data base, reports literature articles. For the analysis for this study thematic analysis has been implemented. The objective of the research is to Evaluate the current market landscape of the Indian wine industry, including production, consumption trends, distribution channels, and regulatory frameworks. The study aims to provide a complete knowledge of the variables affecting the growth potential of the Indian wine business, both domestically and globally, through a rigorous qualitative analysis that incorporates insights from diverse sources andviewpoints.

4. Discussion & Analysis

The wine sector in India is growing at a promising rate because to changing customer tastes and rising demand abroad. Due to a combination of favourable domestic market conditions and rising worldwide demand, the Indian wine business is expected to experience significant expansion. Through the utilization of consumer behavior, regional preferences, and global trends, market participants can effectively traverse hurdles and leverage emerging possibilities. With a thorough grasp of the growth-promoting variables, the Indian wine industry is well-positioned to become a significant force in the world wine market. This study tries to identify the drivers driving the industry's explosive expansion through a qualitative investigation of market dynamics, customer behavior, and global trends.

4.1 Evolution of wine culture in India.

India's history with grape cultivation and winemaking dates to the Bronze Age, when Persian traders introduced these traditions to the area. Soon, fermented grain beverages and wine produced from grapes were commonplace throughout the region. In the 19th century, winemaking was very common in areas ruled by the British. But government opposition and phylloxera at the start of the 20th century almost destroyed the sector. Although wine has been produced in India since the 13th century BCE, according to historical and literary sources, the modern Indian wine industry began in the early 1980s when two entrepreneurial and industrialist pioneers, Shymarao Chowgule and Kanwal Grover, set out to produce

 https://doi.org/10.52202/074664-0006

their own wines on Indian soil. Ancient India's varied and pre-existing drinking culture included grape wine. Since the early Indian wine trade was centered around the trade in wine, it is unlikely that people engaged in the kind of micro-calibrated connoisseurship found in modern wine culture. Although wine was not as ancient, canonical, holy, or symbolic as locally made grain sura, it was nevertheless highly valued in South Asia. [1]. The growth and evolution of global wine tourism. In Routledge Handbook of Wine Tourism. Grape wine was incorporated into the diverse and pre-existing drinking culture of ancient India. Wine was not created in the same place as other drinks, and its variety got their names from foreign sources. Certain aromatics' high price and unusual provenance were a a well-known aspect of scents' attraction in ancient India. Undoubtedly, an exquisite caṣaka cup filled to the overflowing with Kāpiœāyana wine was also shared with an air of exotic status. Our research aligns with the prevalent Indic discourse of wine as an import, even if some previous scholars have focused on wine produced in peripheral parts of South Asia—at least marginal in terms of liquor production patterns. [3]

4.2 Key Milestones in The Indian Wine Industry

This study looked at the reasons why people go to local wine festivals in the larger scheme of wine tourism. This study is significant in two ways. It is the first known empirical study to look at push and pull factors in addition to traveler motives to visit a wine festival. Second, the study created a thorough survey instrument that considered the reasons for wine tourism as well as festivals. Investigations were also conducted into significant trip and demographic factors related to participants of wine festivals [15]. findings lend credence to the claim that festivalscape plays a significant role in predicting emotional experience and contentment. It should be noted that this study only looks at a particular wine and culinary event in a specific region. So, it would be beneficial to extend this research to additional events to generalize the findings to a larger population. There are additional forms of tourism (such as wine tourism, cultural tourism, heritage tourism, etc.) that can use the paradigm that this study produced.[18]. Businesses understand the importance of small analytics for strategy and operations related to customer service, as well as innovation and product development. Businesses have recently begun using social media analytics. Although wine did not have the same historic, canonical, theological, or symbolic history as locally brewed grain surā, it was nonetheless highly valued throughout South Asia. In this way, South Asia resembled ancient Mesopotamia more than wine-producing locations like ancient Persia or the Mediterranean, from which Botero could speak of "wine in a civilisation of beer. However, because to its relative novelty, wine was given considerable cultural and legal leeway, allowing the elite to treat it with respect. It may also be no coincidence that wine was less strictly forbidden for Hindus than surā.113 Another comparison to early modern England would be the tavern, where the "upper" classes drank wine and the alehouse, where the poor people drank ale.[16].

4.3 Impact of Historical Factor on Present Market Landscape

Given the importance and range of fruit wines, comprehensive and methodical research on "Oenology" is needed, particularly on the various aspects of wine research. Reviewing studies conducted in India has made it evident that a solid foundation has already been set, but given the breadth of the topic, this would be insignificant in and of itself. Significance, role, nutritional value, current situation, and upcoming tactics for fruit wines in India. It would be beneficial to focus and coordinate the workforce in one location with specific goals and a set timeframe to gather ongoing research input for the industry's optimal growth.[19]. Public health laws can also be a vehicle for the interaction of politics and regulation. An alcoholic beverage is wine. It has several detrimental social externalities as a result. Like with tobacco, the legislation might become more restrictive when it comes to wine. Increased prohibition would have a big effect on the wine trade and demand. The main comparative benefits of wine-producing countries are covered in this article, along with some future perspectives on their development. There are two parts to our process. Using Porter's diamond, we first examine comparative advantages across 16 nations. Next, we present the findings of a study that questioned experts in wine economics to predict these nations' future trade performance. Regarding the prospective "heavy weights," such as China, but also New Zealand and Chile, the results are comparatively comparable across techniques in terms of which nations have the most potential to flourish in the global wine trade. To a lesser degree, it is also anticipated that Australia, the United Kingdom, and Georgia would play significant roles. Our research shows that relative advantages in the wine trade are neither constant nor consistent, and that terroir is no longer adequate.[7]

4.4. Major Player and Wine Regions

Breaking the Monopoly of Wine Imports in India. The expansion of the wine industry is strongly linked to the history of wine grape farming. Rebuilding the Indian wine business after the devastating events of the late 1800s was a protracted process. Negative public and religious attitudes on alcohol evolved to a climax in the 1950s when many Indian states outlawed alcohol. Either table grape and raisin production was encouraged, or vineyards were uprooted. Some regions, like as Goa, still produced wine, but it was typically highly sweet and heavily alcoholic. Following a lengthy hiatus of a few years, the establishment of the 1985-founded Château Indage winery marked a turning point in the history of the contemporary Indian wine industry. The Indian wine market has enormous potential, and Indian wine is becoming more and more popular abroad. India's wine exports are increasing annually. As news travels quickly, more people are becoming aware of Indian wines on the global scene. The market for wine is expanding by 25–30% annually. The amount consumed by each person in India uses just 0.07 litres per person annually; however, major cities like Mumbai (39%), Delhi (23%), Bangalore (9%), and the state of Goa, which is heavily visited by foreigners (9%), are verified to consume up to 80% of the country's total. The other 20% of India uses just 20% of this. In addition to the exponential rise in imported wine, Indian producers have also launched a variety of wine genres and bottle labels. Maharashtra is a leading state in production of Grapes in the whole country. Regarding agriculture land under grapes cultivation and grapes production, Nashik and Sangli districts are at forefront in the State. Maharashtra ranks first in wine grape cultivation with an area of 8,000 acres, which formed a proportion of 91.95 per cent in the countries. wine grape area during the year 2007. It is one of the important states growing different varieties of wine grape. Cultivation of wine grape is extensively carried out in the belts of Nashik, Pune, Sangli, Solapur, Latur, Buldhana, Osmanabad and Ahmednagar districts [13].

Table No.1 India's wine consumption is mostly centered in its cities.

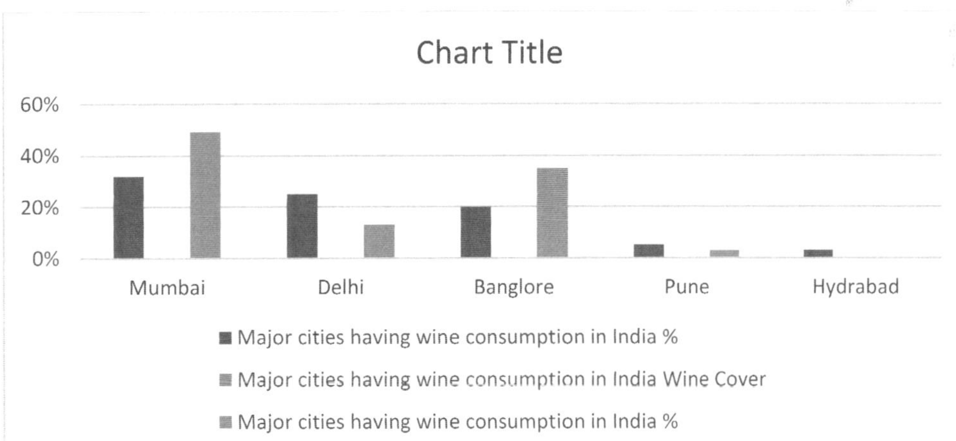

Figure 2International Journal of Management (IJM)

https://doi.org/10.52202/074664-0006

4.5 Popular wine verities and styles

India's wine business has grown from a few brands to over a dozen competitors. Known for its high quality, Fratelli is an Indian wine produced from a huge vineyard that has been producing various varieties since 2009. Chardonnay pairs nicely with grilled fish, and Sangiovese red pairs well with biryani. Sula, arguably one of the most well-known wine labels in India. the Dindori Reserve Shiraz because it goes so well with creative mutton dishes. (Sharma, A. et al.,2021). Despite its widespread adoption, grape farming is not always suitable in each region due to factors such as climate and location. Over the last few decades, there has been a shift in the global wine production landscape, with emerging nations gaining prominence in the wine market. Fermented beverage consumption dates back thousands of years. According to Ramishvili (1983), the area including northwest Turkey, northern Iraq, Azerbaijan, and Georgia is recognized as the grape's center of origin. But grapevine domestication also happened on its own in Spain. The primary shift, on the other hand, happened in the domestication of grapes when it was discovered that their blossoms were hermaphrodite and that domesticated grape varieties self-pollinated. Larger berries, bigger clusters, and fuller blossoms are the products of domesticating grapes. The vintage effect refers to how the variations in weather from year to year affect the production of high-quality grapes. A region's climate variability—such as its extremely harsh winters, severe spring and fall frosts, and hot summers—determines grape yield and wine quality. Severe rainfall incidents and widespread drought conditions in space and time. Grapevines belong to Vitaceae family and genus Vitis, that exhibit a climbing habit. In addition to emphasizing the health benefits of moderate wine consumption, wine manufacturers can also benefit the wine industry by highlighting the importance of intermediaries in the supply chain management process and by promoting wine as a drink that is acceptable in a family setting.[17]

Table-2, Wine verities in India

	Popular wine verities and styles in India	
S. No	Wine Name	Grapes Varieties
1	Fratelli	Sangiovese red
2	Sula	Dindori Reserve Shiraz
3	YORK	Bangalore Blue
4	KRSMA	Cabernet Sauvignon
5	Vallonne	Pinot noir
6	Big Banyan	Gulabi (Muscat): Red Globe Grapes
7	Grover Zampa	Anab-e-Shahi or Dilkhush
8	Rhythm	Shiraz or Syrah
9	Dia	

Figure 3, Wine verities and styles in India.

4.6 International perception of Indian Wine

The data collection was carried out in a nation where the culture of wine and dining is still relatively new and only found in a few contemporary cities. Furthermore, the data were only gathered in Delhi, thus they might not accurately represent the consumption habits of the entire nation.[13]. there are regional variations in the methods that restaurants and cafes serve wine, so one cannot generalize about them all. [12]. Six factors—two benefits and four risks—that affect Indian wine consumers' attitudes and behavior intentions have been covered in the study. In addition to the characteristics covered in this study, there may be other aspects that have been overlooked but still have a big impact on how wine consumers feel. Subsequent investigations could go deeper into these variables to produce a thorough result for the study. [10]. India's wine market is expanding, and there is room for newer labels. Cost may be a significant determinant. Larger production facilities allow wine makers to cut average costs, which can then allow for reduced prices, more marketing spend, or capital investment. Argentina, Chile, and Australia are becoming more popular wine destinations. This can be because of the product's accessibility and cost.

4.7 **Indian consumer preference**

The preferences of consumers in the Indian wine market. It examines factors such as taste preferences, price sensitivity, brand recognition, and the influence of marketing strategies. The literature highlights the importance of understanding and catering to the specific taste preferences of Indian consumers, as well as the challenges faced by domestic and international wine producers in meeting these preferences. This paper provides a comprehensive analysis of market preferences for Indian wines. factors influencing consumer preferences such as price, quality, brand reputation, and grape variety. Through a survey of wine consumers in India, the research identifies the key drivers of wine purchasing decisions and highlights the most preferred wine attributes. The findings suggest that Indian consumers prioritize quality and value for money when choosing wines, with a preference for red wines and those produced by established domestic wineries.[29]. This research paper investigates the factors influencing the global demand for Indian wines. The study examines the role of factors such as wine quality, price, brand image, country of origin, and marketing strategies in shaping the international demand for Indian wines. The findings highlight the importance of building a strong brand image, improving wine quality, and implementing effective marketing strategies to enhance the global demand for Indian wines . This paper explores consumer preferences and market segmentation in the Indian wine industry. The study analyzes the factors influencing consumer choices, including demographic characteristics, wine knowledge, price sensitivity, and consumption occasions. Through a survey of wine consumers in different regions of India, the research identifies distinct market segments based on consumer preferences and behavior. The findings reveal variations in wine preferences among different consumer segments, highlighting the need for targeted marketing strategies and product offerings to cater to diverse consumer preferences [30] This research paper examines the impact of government policies on the growth of the Indian wine industry. The study analyzes the regulatory framework, taxation policies, and trade barriers affecting the production, distribution, and consumption of wines in India. Through a combination of qualitative analysis and case studies, the authors assess the influence of government policies on industry growth and identify the key challenges faced by Indian wineries. The findings highlight the need for supportive policies that promote domestic wine production, reduce trade barriers, and create a favourable business environment for the Indian wine industry. This paper examines the growth prospects and challenges faced by the Indian wine industry. It analyzes the current market preferences and global demand for Indian wines. The study identifies key factors influencing the growth of the industry, such as government policies, consumer preferences, and international competition. The paper provides valuable insights into the potential growth opportunities and the challenges that need to be addressed for the Indian wine industry to thrive. [25]. This paper focuses on analyzing the global demand for Indian wines by studying consumer preferences and market trends. It examines the factors influencing consumers' choices, such as price, quality, brand reputation, and cultural perceptions. The study also explores the impact of marketing strategies and promotional activities on the demand for Indian wines in international markets. The findings provide insights into the potential growth opportunities and strategies to enhance the global demand for Indian wines.[28]. This paper presents a comprehensive study of consumer behavior and preferences regarding Indian wines. It investigates the factors influencing consumers' choices, such as taste preferences, price sensitivity, brand loyalty, and perceived quality. The study also examines the impact of demographic factors and cultural influences on market preferences for Indian wines. The findings provide valuable insights into understanding consumer behavior and developing effective marketing strategies for the Indian wine industry.[33]

https://doi.org/10.52202/074664-0006

5.1 Growth prospects for the Indian wine industry

The increasing demand for Indian wines worldwide can be attributed to various factors, including the expanding appeal of Indian food, the growing recognition and admiration for Indian wines, and the continuous efforts made by Indian wine producers to enhance their quality. It highlights how crucial brand development, effective marketing, and partnerships with foreign distributors are to increasing the demand for Indian wines around the world. There may be a knowledge gap about regional variations in consumer preferences and market dynamics, even though certain studies on regional variations highlight the significance of taking into account various regions within India. Researching regional variations in consumption patterns, taste preferences, and market trends in greater detail could yield insightful information for focused marketing campaigns. As digital marketing channels and e-commerce platforms expand, there might. Based on current projections, the wine market in India is expected to be worth US$182.6 million in 2023, with imported wines accounting for 30% of the market share and domestic wine supply accounting for the remaining 70%. From 2023 to 2027, the industry is expected to increase at a compound yearly growth rate of 14.99%. The establishment of the Champagne Indage factory in 1984 marked the beginning of India's association with wine. But wine drinking did not become popular until much later. Wine consumption has increased dramatically in recent years; in 2001, just one million liters were consumed annually, while in 2019, over 30 million liters were used annually. The expansion of the wine business in India. To succeed in the Indian wine market, it is essential to enhance consumers' knowledge to create a demand for higher quality and premium wines costing over US$20. The report conducted a thorough analysis of the Global Wine Market and identified several key drivers that are fuelling its growth potential. These drivers are shaping the industry and creating new opportunities for producers and consumers alike.[4].

The growing demand for wine in emerging economies is one of the main factors driving the industry. More individuals can now buy wine thanks to rising disposable incomes in nations like Brazil, India, and China, which boosts wine sales and consumption. It is anticipated that this trend will continue as these economies expand and their inhabitants rise in income. The move toward premiumization is another important motivator. Sales of premium and super-premium wines are rising as consumers show a greater willingness to pay for fine wines. Growing consumer interest in wine quality and provenance, along with shifting tastes for organic and biodynamic wines, are the main drivers of this movement. Another significant factor driving the wine market is the growth of e-commerce. Online purchases are rising quickly. Opportunities for the Wine Market The expansion of e-commerce, the increase of premiumization, the trend toward organic and biodynamic wines, growing markets, and technological advancements are a few of the major prospects facing the wine industry. Wine producers now could reach a larger audience and boost sales thanks to the expansion of online wine sales. High-quality wine producers have potential to stand out in the market thanks to the trend toward premiumization. Winemakers get a chance to show their dedication to sustainability by showcasing their organic and biodynamic winery. As disposable incomes rise, emerging economies like China, India, and Brazil are offering substantial development potential. Finally, technological developments present creative chances for manufacturers to enhance their goods and procedures. [5]

The Wine Market Restraints and Challenges report offers a comprehensive examination of the obstacles confronting the wine industry. Consumer tastes are shifting toward lighter, lower-alcohol wines as well as organic and biodynamic wines, which presents a big problem. Producers now face the task of modifying their products and marketing approaches to appeal to younger consumers considering this shift in consumer preferences. Government rules, such as taxation and limits on the sale of alcoholic beverages, present another difficulty for the wine business. These laws may hinder the expansion of the wine industry in some areas and present growers with new expenses and difficulties. Regulations pertaining to labeling standards and other matters present difficulties for producers, especially those who export wine to several countries. The phenomenon of climate change. For the wine sector to continue growing, several issues must be resolved in the market. To adjust their methods, producers need to be aware of shifting consumer tastes, regulatory requirements, climate change, and heightened competition. To do this, you'll need to be creative, adaptable, and ready to change as the market does. An extensive summary of the most recent developments in the sector is given in the Wine Market report. The rising popularity of e-commerce for wine sales is one of the biggest trends. To meet the needs of the growing number of customers who prefer to purchase online, shops are providing a greater selection of products and delivery alternatives. The growing desire for premium and ultra-premium wines is another trend, particularly in developing regions like Asia and South Asia.[27].

5.2 Globalisation marketing

The Study covers nine old and new world wine producing nations and highlights the dynamics driving the global wine industry. Globalization may present new obstacles for some wineries, but it is less likely to force those that take a market-oriented approach out of business. The existence of small and medium-sized wineries that can adapt and compete successfully is ensured by the presence of diversified consumer preferences and plenty of opportunities for experimentation. These wineries can even develop over time. Smaller wineries must adjust to the changing global landscape of mergers and acquisitions, increased international trade and investment, and greater potential for cross-border spillover effects if they are to survive. Advantage does not ensure a nation's long-term prosperity.[2] To align a winery's unique qualities with what customers, want, it must cultivate a competitive advantage. The driving forces suggest that controlling trade barriers, increasing export aid, and placing a greater emphasis on market orientation can all be advantageous for wineries, regardless of size and location. To survive in a demanding climate and adjust to the rapid speed of globalization, wineries of various sizes and nationalities can benefit from the recommendations. [8].

5.3 Recommendations & Suggestion

To give a thorough grasp of its dynamics, the wine market can be divided into different segments depending on several different criteria. These segmentation criteria include the kind of wine (red, white, or rosé), as well as the type of product (still, sparkling, or fortified). enabling companies to target customer preferences. The distribution channel, which comprises off-trade (retail outlets and internet platforms) and on-trade (restaurants, pubs, and hotels), is another criterion for segmentation. It offers information on where customers generally buy wine. Due to regional variations in wine production and consumption, the market can be divided geographically into regions and nations. Furthermore, the division of wines into price points—premium, mid-range, and discount wines, for example—allows for the examination of customer purchasing patterns according to their financial constraints.

The previous five years have seen encouraging growth in the nascent Indian wine business. However, there are still a ton of fascinating hurdles to overcome. The storage, transportation, wine exports, marketing through media campaigns and wine festivals, genetic resources and improvement, production technology, plant health management, pre- and post-harvest technology, cascading effect due to climatic changes on grape cultivation, grape juice composition, yeast ecology of wine production, fermentation kinetics, and wine character are the areas that the Indian market and research institutes should focus on. The Indian government has recognized the wine industry as one that is thriving and expanding, and it has helped by lowering taxes and financing processing facilities. The government ought to fund research and development and promote wine consumption at educational institutions. An environmental shock is the first. Accurately predicting the degree of climate change and how it will affect vines is extremely challenging. Climate change includes both global warming and extreme weather events that have an impact on vine health.

The long-term health of soils and vines is put at risk by the increased usage of synthetic phytosanitary products that could result from a second shock to preferences and demand. On the one hand, sustained demand growth in emerging markets may quicken the already noted trends and pave the way for the United States and China to emerge as the two dominant nations in the world's wine trade in the Linder-style. Even though the Indian wine industry is still in its infancy, there is a great deal of promise for its future development. Indian wine producers can establish themselves as strong competitors in the international wine industry by utilizing changing customer preferences, taking advantage of global demand patterns, and resolving innate obstacles through creativity and cooperation. India's wine industry has the potential to generate employment opportunities, value creation, and socio-economic development by means of strategic investments in product development, branding, distribution, and sustainability initiatives. These efforts will also help India become a prominent player in the global wine arena. To achieve wealth, recognition, and long-term success on the global stage, the Indian wine business needs to demonstrate persistence, tenacity, and a dedication to quality. The Indian wine business is in a unique position to grow and change significantly in the face of shifting consumer tastes and patterns of demand around the world. A careful analysis of consumer behaviour, market dynamics, and worldwide trends reveals a world full of opportunities and difficulties, The wine industry in India has grown remarkably in the last several years due to rising disposable incomes, urbanization, and changing consumer preferences. Industry sources predict that between 2022 and 2027, the Indian wine market would develop at a compound annual growth rate (CAGR) of more than 20%, with a market size of USD X billion at the conclusion of the forecast period.

 https://doi.org/10.52202/074664-0006

References

1. Yadav, M.K., Dixit, S.K.: The growth and evolution of global wine tourism. In *Routledge Handbook of Wine Tourism,* Routledge. 15-28 (2022)
2. Sheth, J.N.: Impact of emerging markets on marketing: Rethinking existing perspectives and practices. *Journal of marketing*, *75*(4), 166-182 (2011).
3. McHugh, J.: Grape wine in ancient and early Medieval India: The view from the centre. *The Indian Economic & Social History Review*, *58*(1), .113-144 (2021).
4. Jha, J.: Wine in India: Changing Consumer Perceptions. *ICFAI Journal of Consumer Behavior*, *2*(2) (2007).
5. Adsule, P.G.: October. Grape wine industry in India: development, status and prospect. In *II International Symposium on Beverage Crops 1274*, 57-64 (2018)
6. Bheemathati, S.: An Overview: Recent Research and Market Trends of Indian Wine Industry. J. Food Process. Beverages, 3(1), 1-5 (2015).
7. Bargain, O., Cardebat, J.M., Chiappini, R. and Laffitte, C.: Long-term prospects of the international wine trade. Journal of Wine Economics, 13(4), 442-450 (2018)
8. Hussain, M., Cholette, S., Castaldi, R.M.: An analysis of globalization forces in the wine industry: implications and recommendations for wineries. *Journal of Global Marketing*, *21*(1), 33-47 (2008).
9. Joshi, V.K., Attri, D.: Panorma of research and development of wines in India. (2005).
10. Srivastava, R.K.: Perception Study of Wine Market Culture of an Emerging Market-India-Will it Emerge from Infant to Growth Stage. Journal of Hotel & Business Management, 3(01), 1-6. (2013).
11. Gupta, V., Duggal, S.: Impact of perceived risks and benefits: a case study of wine consumption patterns in the Indian context. Journal of wine research, 31(1), 67-83 (2020).
12. Gupta, V., Sajnani, M.: Risk and benefit perceptions related to wine consumption and how it influences consumers' attitude and behavioural intentions in India. *British Food Journal*, *122*(8), 2569-2585 (2020).
13. Khot, V.: Breaking the Monopoly of Wine Imports in India. (2007).
14. Spawton, L.L.T.: International Journal of Wine Marketing. (2008).
15. Morrison, A., Rabellotti, R.: Gradual catch up and enduring leadership in the global wine industry. *Research Policy*, *46*(2), 417-430 (2017).
16. Costopoulou, C., Ntaliani, M., Ntalianis, F.: An analysis of social media usage in winery businesses. Adv. Sci. Technol. Eng. Syst, 4, 380-387 (2019).
17. Shinde, G.S., Singh, D.K., Jaju, D.R.: Wine Consumption Patterns of Indians with Special Reference to The Nashik District: A Review. International Journal of Management (IJM), 14(1) (2023).
18. Mason, M.C., Paggiaro, A.: Investigating the role of festivalscape in culinary tourism: The case of food and wine events. Tourism management, 33(6), 1329-1336 (2012).
19. Joshi, V.K., Kumar, V.: Importance, nutritive value, role, present status and future strategies in fruit wines in India. Bioprocessing of Foods. (Eds. PS Panesar et al.), 39-62 (2011).
20. Sharma, P., Singh, S., Misra, R.: Wine-related lifestyle segmentation in the context of urban Indian consumers. *International Journal of Wine Business Research*, *32*(4), 503-522 (2020)
21. Sheth, H., Pereira, K.A., Thomas, T.S., Kumar, V.K., Kelapanda, M.C., Dant, M.G.: Wine festival and wine consumerism-A perspective study on influence and awareness in the growth and acceptance of wine in Indian market. *Int J Eng Technol Manag Appl Sci*, *2*, 20-36 (2014).
22. Kansara, S.: Exploring the wine sector in the Nashik district of India. *International Journal of Wine Business Research*, *32*(2), 203-217 (2019).
23. India Wine Market by Type and Product - Forecast and Analysis (2023-2027).
24. Shinde, G.S., Singh, D.K., Jaju, D.R.: Wine Consumption Patterns of Indians with Special Reference to The Nashik District: A Review. *International Journal of Management (IJM)*, *14*, (2023).
25. Roy, D.: GRAPES TO GLASSES: UNVEILING INDIA'S WINE ODYSSEY-CHALLENGES AND OPPORTUNITIES IN GRAPE CULTIVATION AND WINE PRODUCTION. International journal of social sciences, 3(08), 01-05 (2023).
26. Insel, B.: The evolving global wine market. *Business Economics*, *49*(1), 46-58 (2014).

27. Campbell, Guibert., N.: Introduction: Old World strategies against New World competition in a globalising wine industry. *British Food Journal*, *108*(4), 233-242 (2006).
28. Gade, A.D.: WINE INDUSTRY IN INDIA A REGIONAL SURVEY. Ashok Yakkaldevi (2022).
29. Payini, V., Mallya, J., Piramanayagam, S.: Indian women consumers' wine choice: A study based on conjoint analysis. International Journal of Wine Business Research, 34(4), 469-494 (2022).
30. Sharma, P., Singh, S., Misra, R.: Wine-related lifestyle segmentation in the context of urban Indian consumers. *International Journal of Wine Business Research*, *32*(4), 503-522 (2020).
31. Ray, P., Chowdhury, S.: Popularizing grape cultivation and wine production in India–challenges and opportunities. International Journal of Social Sciences, 4(1), 9-28 (2015).
32. Castaldi, R., Cholette, S., Hussain, M.: A country-level analysis of competitive advantage in the wine industry (2006).
33. Sharma, P., Singh, S., Misra, R.: Wine-related lifestyle segmentation in the context of urban Indian consumers. International Journal of Wine Business Research, 32(4), 503-522 (2020).

 https://doi.org/10.52202/074664-0006

Digital Transformation and Emerging Trends in Tourism and Hospitality Industry of India: A Revolution

Neha Dubey[1(0000-0001-9037-5866)] Dr. Sandeep Guleria[2(0000-0001-5836-1691)]
[1]Research Scholar, UITHM, Chandigarh University, Punjab.
[2]Professor, UITHM, Chandigarh University, Punjab.

Abstract

Tourism significantly impacts the Indian economy through the generation of foreign currency and the creation of employment opportunities. The tourism industry in India has been significantly enhanced by the integration of travel technologies, which has enabled the achievement of the 2030 Agenda for Sustainable Tourism Development. In 2022, the tourism sector in India contributed almost Rs.15.7 trillion to India's Gross Domestic Product (GDP), as reported by the World Travel and Tourism Council. The objective of this study is to determine the constituent elements of intelligent travel technology that contribute to the progress of India's tourism and hospitality sector. This study utilized a qualitative approach to examine the latest developments in the travel industry and the need to integrate modern travel technologies with sustainable tourism practices. The research findings indicate that digitalization significantly affects the tourist business, particularly by improving its overall efficiency and stimulating its expansion. The survey analysis indicates that implementing digital transformation has significantly enhanced the performance of the tourist and hospitality business. However, the process of digitizing the tourism business has led to a combination of advantageous and disadvantageous consequences that impact the progress of tourism. The tourism industry is leveraging digital advancements such as online travel portals, Artificial Intelligence, Voice search, chip-enabled passports, online check-in& check-out facilities blockchain technology, and more to enhance consumer service. The progress of technology is vital in the growth of India's tourism and hospitality industry.

Keywords: Digital Transformation, Online Travel portals, Trends in Tourism & Hospitality.

1. Introduction

Tourism has a substantial role in India's gross domestic product. The government of India has taken various substantial majors to strengthen the country's tourism and hospitality industry, including enabling 100 percent foreign direct investment, simplifying visa application procedures, and enhancing infrastructure. The tourist sector is globally renowned and rapidly expanding. It is a crucial factor in driving economic growth due to its capacity to enhance a country's economic status and create numerous job possibilities. Advancements in information and communication technology have led to a growing significance of artificial intelligence in the tourism business. The tourist sector's marketing efforts heavily rely on machine learning technology, using AI to personalize the process of planning and booking tours and vacations. Moreover, it is increasingly beneficial in situations such as smart hotel rooms, as it identifies the anticipated needs of guests and tailors the environment and

https://doi.org/10.52202/074664-0007

facilities to their wishes and preferences. Artificial intelligence is widely used in different fields, such as customer service and security. In the future, artificial intelligence will integrate tourist principles, including virtual tour guides. Technology has accelerated operations and improved the efficiency and experience of travel in the hotel and tourism sector. Technology is crucial for the progress of tourism. It has enabled the replacement of expensive human labor with technology tasks. It helps decrease labor expenses while also addressing customer service problems.Tablets and smartphones have mostly supplanted desktop computers. Mobile bookings are a dominant trend in today's society. Hospitality organizations should efficiently inform consumers about changes in reservations and delays and provide them with discounts.Digital technology is significantly changing the social and economic culture of the world, especially in the hospitality and tourism industries. Digital technology has fundamentally transformed the functioning of modern companies. Previously, clients had to visit a travel agency to purchase their tickets. Currently, they can purchase their tickets through online booking portals. Ride-sharing applications are now widely used. Passengers are provided with transparent pricing before boarding, and vehicles can pick up multiple passengers along the same route, benefiting riders, drivers, and platform owners. Travel companies have shifted from using geographical boundaries to utilizing online methods to interact with customers due to the advancement of new technology and the swift pace of digitalization. Mobile reservations rose as a result of digitization.

2. Literature Review

Digital transformation has become a worldwide phenomenon, altering several industries and sectors, Digital transformation is the complete integration of new technology into traditional organizational procedures and frameworks to enhance efficiency, innovation, and customer experiences. The subject encompasses a broad range of technologies such as data analytics, artificial intelligence (AI), smartphone apps, and the Internet of Things (IoT). Digital transformation has far-reaching effects on societies by influencing the way they function in a more interconnected global environment.[1]Social networking sites have transformed people'sinteractions in the hotel and tourist industries due to the increasing use of communication technologies [2]Organizational drivers such as IT infrastructural availability and investment in digital innovation have a direct impact on enterprises' attitudestoward integrating digital innovation.[3]When examining digitaltransformation, two main viewpoints are considered: the technical perspective, which emphasizes the role of information technology in business operations, and the value perspective, which argues that digital transformation alters the value proposition and operational framework through technology utilization.[4]Recent technology progress has greatly contributed to enhancing the efficiency of businesses. Emerging technology has provided advantageous business prospects to supply chain companies, enhancing the firm's competitive edge. The blockchain enables companies to leverage business competitiveness [5] The advancement of blockchain technology is considered the second-generation internet, revolutionizing the commercial landscape with its decentralized and transparent nature.[6]These days, digitization has become an integral aspect of every company due to its lightning-fast platform for client interaction, it has revolutionized the way organizations operate. The advent of digital technology has revolutionized the travel business, making it more modern, inventive, and technically sound than ever before.[7] Service organizations like travel agencies and the hospitality industry have utilized social media as a crucial communication tool for their marketing and strategic decisions due to the

 https://doi.org/10.52202/074664-0007

advancement and widespread use of Web 2.0 and mobile technology by transforming consumer buying experience and by offering customized services.[8]. Advanced software and communication technologies enhance operational efficiency, such as enabling the processing of orders by improving in terms of efficiency, speed, and cost-effectiveness. Furthermore, utilizing cutting-edge technology and decision-support tools has expedited and simplified the decision-making process for the service industry[9]The Metaverse provides opportunities for tourism businesses to improve their consumer interaction and operational effectiveness. The platform provides unique and customized travel experiences, which could expand the market reach and enhance consumer satisfaction. it could result in cost savings and higher profits for tourism businesses.[10]Through digital transformation, hospitality and tourism organizations have expanded their online offerings and enabled virtual tours of various places such as museums, galleries, exhibitions, castles, zoos, and aquaria. In the future, the hospitality business will increasingly rely on digitization and innovative technologies.[11]

3. Objectives

- To ascertain the emerging trends in tourism
- To investigate digitization in the tourism industry.
- To understand various emerging smart technologies used in tourism

4. Research Methodology

The phenomenological approach is employed which enabled a comprehensive understanding of the primary experiences and perspectives of stakeholders and tourists on digital transformation and developing patterns. An exploratory approach is employed to acquire a deeper comprehension of the situation. The selection of participants is contingent upon their experience and active engagement within the tourism and hospitality sector. This includes those holding professional positions in hotels, travel agencies, tour operators, government entities, and technology providers. Semi-structured interviews are carried out with important individuals involved to investigate their viewpoints, encounters, and understandings regarding digital transformation and developing patterns. In the realm of digital transformation and emerging trends, thematic analysis is employed as a methodological approach to find repeating themes, patterns, and categories. The analysis aligns with the research objectives to extract significant insights, 6 consequences, and suggestions regarding the influence of digital transformation and emerging trends on the tourism and hospitality sector in India

5. Results & Discussions

The study's analysis of primary and secondary data reveals both emerging trends inthe tourism and hospitalitysector in India due to technological advancements as below

5.1 Metaverse Travel

Metaverse travel refers to a virtual realm whereby individuals have the opportunity to explore and interact with one another in a manner that holds significance. It resembles virtual reality technology. In addition to conventional virtual reality, the metaverse integrates elements derived from blockchain technology, social media, and augmented reality. Virtual reality has the capability to transfer users to a metaverse through the utilization of digital devices such as smartphones, tablets, laptops, or game consoles. This facilitates opportunities for consumers to immerse themselves in unique travel experiences. An instance of this type of interaction could manifest as a virtual experience that assists someone in organizing their next vacation. However, the metaverse can also be employed as a means of communication with hotel front desk personnel or travel agents. For example, a user can virtually explore a hotel, browse through its features, and make a room reservation while physically present.

5.2 Block chain Technology

The blockchain is a sophisticated technological innovation that serves to authenticate and record the transfer of assets within a network including interconnected individuals. A blockchain refers to a distributed database that securely records and maintains a comprehensive record of digital transactions. The utilization of blockchain technology enables the establishment of a decentralized peer-to-peer network, thereby enabling clients with varying levels of trust to engage in secure communication without dependence on a central authority. The implementation of blockchain technology in the tourism sector has the potential to enhance operational efficiency, reduce costs, and enhance consumer happiness. The responsibilities encompass overseeing reservations, tickets, hotel reservations, insurance, auto rentals, passenger identification, customization, luggage monitoring, loyalty programs, price reductions, incentives, digital currency transactions, business-to-business exchanges, and authentication tokens.

5.3 Virtual Reality

Virtual reality, at its fundamental level, refers to a technological innovation that generates a synthetic world within a computer system, hence providing users with the perception of being physically present in that environment. This suggests that it has the potential to significantly modify an individual's environment to the extent that they detect novel auditory and visual stimuli. Virtual reality enables travel agents to provide potential clients with a tactile experience of the hotel and its facilities before making a reservation. Certain tour operators are currently providing virtual reality (VR) tours, allowing consumers to virtually explore well-known locations and tourist destinations through the use of specialized VR goggles and 360-degree film. Hotel reservation businesses are employing virtual reality (VR) technology to provide virtual tours of hotel accommodations.

5.4 Chatbots

A chatbot is a computer program designed to mimic and understand human interaction, enabling people to communicate with electronic gadgets as if they were talking to a real person.chatbots can serve as intelligent virtual agents for various businesses such as travel agencies, airlines, hotels, and other organizations. The chatbot can function as a travel agent and respond to significant consumer inquiries. When human operators are not present, numerous consumers use online chatbots, specifically trained AI, to book their flights and accommodations, as these chatbots can answer inquiries and offer clients useful information.

5.6 Gamification with AR

Augmented reality offers significant gamification benefits, particularly in the tourism sector. For instance, augmented reality (AR) could enable hotel visitors to capture images of themselves alongside virtual representations of well-known individuals or fictional characters. It might also enable guests to virtually redesign a hotel room. Additionally, augmented reality can improve environments beyond the hotel premises. Visitors may use their phones to see evaluations of a restaurant by pointing their devices at it. They could even visualize an antique building as it looked many years ago, which is particularly useful for exploring ruins or historical sights.

https://doi.org/10.52202/074664-0007

5.7 Internet of Things (IoT)

The Internet of Things is a network of everyday physical objects and equipment that are capable of connecting and sharing data via the Internet. IoT gadgets encompass smart speakers, home control devices, wearable devices, smart meters, and smart health monitoring. The tourism industry has enthusiastically embraced IoT technology. Many hotels are already incorporating smart devices in their rooms to provide visitors with more control. Airports utilize mobile applications to enable clients to obtain current flight details through their smartphones. The process of digitization is significantly contributing to the expansion and advancement of the tourism sector.

6. Conclusion

The tourism and hospitality business in India is currently experiencing a favorable environment for innovation and expansion due to digital transformation and emerging trends. Business enterprises within this particular sector are prepared to enhance customer experiences by optimizing operational processes and enhancing overall efficiencies through the adoption of technology-driven solutions such as artificial intelligence, virtual reality, and big data analytics. The advent of digital platforms and mobile applications has revolutionized the manner in the way tourists engage in the exploration, reservation, and enjoyment of vacation destinations and choices for accommodation. Moreover, the emergence of sustainable tourism, experiential travel, and personalized services highlights the changing preferences of modern tourists, providing industry stakeholders with opportunities to adapt and cater to these evolving demands. However, it is imperative to acknowledge and tackle serious problems such as the digital gap, cybersecurity risks, and a shortage of a skilled employee base. To ensure ongoing growth and competitiveness in the digital era, it is imperative for stakeholders to continue allocating resources towards infrastructure development, talent acquisition, and cybersecurity measures. The utilization of digital technology and the adoption of new trends in the tourism and hospitality business in India have the potential to foster growth, innovation, and economic development. This, in turn, can enhance the travel experiences for worldwide tourists.

7. Acknowledgment

I would like to extend my appreciation to the participants involved in this study, as their invaluable contributions and perspectives have greatly enriched my understanding of the subject matter.I would like to extend my deep appreciation to our supervisors and mentors for their tremendous guidance, unwavering assistance, and motivation during the duration of our project.

8. Disclosure of Interests

The researcher in this work does not possess any conflicts of interest that need to be disclosed. During our investigation, we have consistently adhered to principles of transparency and integrity. We confirm that there are no monetary or non-monetary vested interests, associations, or connections that could potentially impact the objectivity, neutrality, or honesty of the conclusions offered in this research.we affirm that this research paper is free from any undue influence or bias arising from conflicting interests.

9. References

[1] Almasoodi, M. F., Rahman, S., Basendwah, M., & Alfarra, A. N.: Leveraging Digital Transformation to Enhance Quality Tourism Services in Babylon City, Iraq. International Journal of Sustainable Development and Planning. 18(10), 3195-3211 (2023). doi: 10.18280/ijsdp.181020.

[2] Kamalaveni, M. S., Suriya, A., Sakthivel, P., & Vignesh, S.: Issue 4 (I). [Online]. Available: www.googleimages.com
[3] Egala, S. B., Amoah, J., Jibril, A. B., Opoku, R., & Bruce, E.: Digital transformation in an emerging economy: exploring organizational drivers. Cogent Social Sciences. 10(1), (2024). doi: 10.1080/23311886.2024.2302217.
[4] Nanda, W. D., Widianingsih, I., & Miftah, A. Z.: The Linkage of Digital Transformation and Tourism Development Policies in Indonesia from 1879–2022: Trends and Implications for the Future. Sustainability (Switzerland). 15(13), (2023). doi: 10.3390/su151310201.
[5] Sarfraz, M., Khawaja, K. F., Han, H., Ariza-Montes, A., & Arjona-Fuentes, J. M.: Sustainable supply chain, digital transformation, and blockchain technology adoption in the tourism sector. Humanities and Social Sciences Communications. 10(1), (2023). doi: 10.1057/s41599-023-02051-9.
[6] Dash, S. P.: HR Digital Transformation: Blockchain for Business. Intelligent Systems Reference Library. 237, 59-87 (2023). doi: 10.1007/978-3-031-22835-3_3.
[7] Anuj, Upadhyay, R. K., Kargeti, H., & Sharma, A.: Adoption of digital marketing among tourism industry of Uttarakhand in India. International Journal of Business Excellence. 29(1), 80-97 (2023). doi: 10.1504/IJBEX.2020.10033064.
[8] Lee, M., Ahn, J., Shin, M., Kwon, W., & Back, K. J.: Integrating technology to service innovation: Key issues and future research directions in hospitality and tourism. Journal of Hospitality and Tourism Technology. 12(1), 19-38 (April 08, 2021). doi: 10.1108/JHTT-01-2019-0013.
[9] Brako, P. & Joseph, S. T.: Emerging Trends in Hospitality and Tourism in India Pro Poor Tourism View project Project Management View project. [Online]. Available: https://www.researchgate.net/publication/317643394
[10] Prados-Castillo, J. F., Torrecilla-García, J. A., & Liébana-Cabanillas, F.: Metaverse as a booster of tourism transformation towards virtual management strategies. Tourism Review. (2024). doi: 10.1108/TR-10-2023-0750.
[11] Zeqiri, A., Dahmani, M., & Ben Youssef, A.: Digitalization of the Tourism Industry: What Are the Impacts of the New Wave of Technologies?

 https://doi.org/10.52202/074664-0007

The Impact of Technology on Bakery and Confectionery Operations in Hotel Industry: A Systematic Review

NirajKumar[1[0009-0005-7465-8854]]and Dr. Praveen Kumar[2[0000-0003-0294-3830]]

[1]Research Scholar, UITHM, Chandigarh University.NH05, Mohali, Punjab.

[2]Assistant Professor, Guide, UITHM, Chandigarh University.NH05, Mohali, Punjab.

Abstract:
The three crucial steps in the manufacturing chain—milling, kneading, and baking—have the biggest effects on the quality of bread and bakery goods. Previous research has thoroughly examined the effects of milling and kneading; however, the state of knowledge about baking, its impact on product quality, and strategies for improvement has not been addressed. This gap is addressed in the current work. Range and methodology: The first goal is to provide an overview of what is currently known about baking technology and how it affects the properties of bread and other bakery goods. The second goal is to make recommendations for ways to enhance baking and ovens, boosting effectiveness, efficiency, and product quality, all of which will benefit output, earnings, and the environment. Important discoveries and conclusions: This review emphasizes how crucial it is to choose the best baking method based on the intended end results. By combining steam and suction methods, conventional ovens could be made better. Hybridization with other methods, such infrared and infrared-visible heating, which combine the benefits of infrared (IR) browning with microwave time-saving, could greatly enhance microwave ovens. There would be a decrease in baking time, energy use, and environmental effects with each of these hybridizations. In conclusion, even with the tremendous progress that has been accomplished, if we are to meet the enormous issue of environmental sustainability, we will require completely new baking techniques and ovens that can effectively employ renewable energy sources.

Key Terms: technology, bakery, confectionery, hotels, automation, digitalization, culinary operations, systematic review

1. Introduction

Around the world, baked goods are a basic diet. They are a vital source of vitamins, minerals, carbs, protein, dietary fibre, and antioxidants [6], making them necessary for human nutrition [6]. Among the earliest baked goods is bread, in particular [3].According to a survey of the literature, bread was first manufactured about 10,000 years ago BCE [7]. The beginnings of bread-making can be traced back to prehistoric times, when prehistoric humans learned that some plant seeds could be ground up, combined with water, and cooked over a fire to produce sustenance [2]. The "white art" of bread making was first practiced by the Egyptians, who popularized it across the globe [7].

While the fundamental formula only calls for a few basic ingredients—wheat flour, water, yeast, and salt—current production takes into account regional customs and raw material availability[16]. Various types of bread can be found in Europe. Many nations, areas, or municipalities—for instance, in Italy. Additionally, recipes might change depending on the altitude; for instance, rye bread varieties like Schüttelbrot and Roggenbrot are prevalent in alpine settings[3]... Another illustration originates from China and other Asian nations, where steamed bread, also known as Chinese mantou, is incredibly well-liked. In the latter case, baking is done with steam [20]. The three main steps in the production chain—milling, kneading, and baking—have the greatest impact on the quality of bread and bakery products, despite their differences in characteristics [16][20]. Regarding milling, previous research has compiled the state of knowledge about the two most popular methods, roller and stone milling[12]. The aforementioned review emphasised that the milling technique possessed the greatest influence over flour quality, dough rheology, and ultimate bread characteristics. It also proposed particular tactics to enhance these techniques. The two most intriguing improvement initiatives for stone milling are the rediscovery and upgrading of historic stone watermills[13] and the proper regulation of wheat conditioning and stone rotating speed [10]. Other strategies for the roller mill have been identified as follows: developing and improving automatic and adaptive mill plants[12]; using break, sizing, and reduction systems to improve milling technology, flour differentiation, and minimize environmental impacts[10]; and debranning wheat before milling in conjunction with the stabilization of bran, middlings, and germ [12].

An earlier study also provided a summary of the kneading procedure [6]. Although milling has the greatest influence on the rheology of the dough and the properties of the bread[12], kneading is also crucial[6]. In order to prevent over-softening and achieve the ideal dough rheology and consistency, Cappelli and Cini[7]. emphasize the significance of controlling the following critical parameters: proper kneading time, which prevents over- and under-mixing; proper dough temperature and mixing speed, which prevents warming and excessive weakening; and, last but not least, proper dough aeration, which ensures proper oven spring during baking and optimal bread crumb.

The last and most important stage in producing bread is baking. It needs extremely high temperatures, usually between 160 and 250 °C, and there are a few methods that make use of various oven types[3]. Dough experiences a variety of physical and chemical changes during baking, such as water evaporation, a transition from dough to crumb, the production of crust and crumb structures, volume expansion, protein denaturation, and starch gelatinization[5]. These changes demonstrate how delicate the procedure is. All of the advancements achieved by using milling and kneading technologies may be compromised by choosing an oven temperature and baking method that is not ideal [17]

It is evident that, in contrast to milling and kneading, there is currently no comprehensive review of the baking process, its impact on the quality of the finished product, or any recommended techniques for improvement. The current investigation was inspired by this discovery. Thus, this review's primary goal is to provide an overview of what is currently known about baking methods and how they affect bread and other bakery goods. The second goal is to make recommendations for ways to make ovens and baking better. The main objective is toimprove the end product's efficacy, efficiency, and quality, which will benefit the environment, profitability, and manufacturing.

2. Search strategy:

Utilizing the following search terms in PubMed and the Web of Science: - (bread OR bakery items) AND steam AND baking. Here were no limitations on publication status, time, or language. Every duplicate was eliminated. Reading the title and abstract of the first dataset was used to filter it; articles with just an abstract and/or index were discarded at this stage. Next, the

https://doi.org/10.52202/074664-0008

complete text was read. Any article that discussed baking in any way was included; articles that didn't explain equipment, how they work, or how they affect the features of the finished product were excluded. A flow chart was created for each database to provide an overview of the findings.

3. Modes of heat transfer during baking:

Transport phenomena and structural alterations must be taken into account for a thorough understanding of the baking process and, consequently, for strategies to maximize the quality of the finished product. The three primary heat transmission mechanisms involved in the dough-bread transition are convection, conduction, and radiation [1]. According to Lucas [21] radiation originates from the oven's heating sources, specifically the burner flames, as well as all of its hot metal components. It is in charge of altering the dough's colour and absorptive capacity. Over 50% of all heat transport occurs through conduction. It is the most common method for baking goods on an oven's hearth (like Indian chapatti), or its wall (for example, tandoor-baked Indian naan bread). According to the baking method, oven layout, and operational parameters, each mode's percentage and impact on quality differ [13]The ensuing subsections (3.1, 3.2, and 3.3) provide a more thorough explanation of these three modalities.

3.1. Conduction

Thermal conductivity is the capacity of a substance to carry out energy transmission. When a product comes into direct touch with a heated solid surface, conduction takes place [5] There are several surfaces that can be utilized as baking supports, such as grates, conveyor bands, trays, and stones [13]. The giving and receiving objects' temperatures must range significantly for conduction to take place. The ability of stones to quickly transfer heat to dough placed on them when they are directly over coals is why ovens still utilize this method today [21]

Observes that crackers, cookies, and biscuits are typically baked in tunnel ovens on continuous bands. The band is heated by a gas flame, and the tiny, thin biscuits bake quickly—typically in less than three minutes. In this instance, speedy leavening and the release of water as steam are caused by rapid heat transfer into tiny portions of dough [5]. Because conduction within the dough is slower than the rate at which heat can reach the surface, the end product is dry and crisp.It is highly challenging to calculate and predict heat transmission inside certain meals because there are many variables involved. First, there is a large shift in the heat transfer coefficient based on the product's composition. Second, distinct changes occur within the same product due to non-homogeneity in the chemical/physical composition (e.g. between crumb and crust). Lastly, the product's composition may alter noticeably from the start to the finish of baking.

3.2. Convection:

Convection is covered in two important studies [5]. Convection is the term for forced or natural heat transfer that occurs when air and steam circulate as a result of temperature and density variations. Cooler gases fall and are reheated, whereas hotter gases rise and become less dense as they expand. This causes air to circulate vertically inside the oven. Most ovens use both conduction and natural convection, but the latter is less effective and often distributes heat unevenly. Fan-forced convection ovens are more efficient since they don't require hot gases to circulate vertically and instead compel them to do so quickly. In this instance, controlling temperature, pressure, moisture content, and numerous other crucial factors is necessary in addition to air velocity and flow pattern, which have an impact on baking. The convective heat transfer coefficient is also influenced by the surface shape of the product, particularly in the case of natural convection [5]

Since the mid-1970s, forced convection has been employed most frequently in tiny machines designed for pizza shops. Rapidly baked food is targeted by hot gas jets directed perpendicularly by a unique form known as an impediment. Little nozzles are used to force circulating air through, which affects the product's surface. Shorter baking periods and lower temperatures are the outcome of the process, which sweeps away the comparatively cool, moist, viscous, dense boundary layer near to the product surface. It works especially well with thin dough that resembles pizza. [5]. Small nozzles, short tubes, perforated plates, or thin slits are used in current impingement ovens [5]

The idea of forced convection has been used in industrial-scale ovens because it often produces relatively uniform heat [4]. Computational fluid dynamic techniques can be used to forecast temperature profiles and air velocity. In an industrial continuous indirect-fired oven, the wall temperatures vary linearly while the air temperature profile follows a bell-shaped curve along its length (from the entrance to the end). The reduction of pre-heating time may be significantly impacted by the geometry of the baffle plate [16]

3.3. Radiation:

Another significant heat transfer method in ovens is radiation. Since method uses energy from the electromagnetic spectrum, no medium is needed. The baking product transfers energy when it comes into contact with the heat source, which is typically incandescent. However, the most efficient wavelengths are typically in the infrared (IR) range, which is invisible to the human eye. Although the method works wonders for browning, the energy only goes so deep below the surface of the food. Internal heating is therefore negligible[20]. A "browning" area is sometimes seen at the end of big ovens, where IR components are utilized to control the final surface color[5].

Only a tiny layer of the product's exterior is directly impacted by IR heating. Nevertheless, processes like conduction cause the energy that is absorbed at the surface to go inward. The red lights above restaurant counters that keep food warm before it is served are one application for infrared heaters. This idea is also frequently applied to small toasters and pizza ovens because they are inexpensive to build and function effectively with relatively thin goods[13]... Thick things like bread loaves cannot be baked using this method because the surface browns before the heat reaches the centre. This is due to the quick heating of the surface. Additionally, irregular surfaces are challenging to bake in infrared because any exposed areas will cook (heat) before the remaining areas. Even while heat radiation plays a significant role in almost all ovens, as is currently the case, it must be combined with conduction and convection[5].

4. Dough transformation:

Despite the fact that baking has been done for millennia, we still don't fully grasp the procedure (Mondal&Datta, 2008). Bread is especially intricate. Heat and mass transfer happen at the same time, and major structural alterations are needed to turn fermented dough into bread [16].The high oven temperature causes a number of physicochemical events (such as the coagulation of gluten and proteins, the creation of CO_2, the gelatinization of starch, etc.) that lead to the construction of the crust layer, the formation of the crumb's alveolar structure, and the product's expansion. A progressive loss of mass is concurrent with these events [1]

Numerous issues arise during baking, which can be broken down into four categories: (i) technical (cleanliness, energy management, packaging, etc.); (ii) chemical (molecular bond and structure changes, etc.); (iii) nutritional (denaturation of protein and starch gelatinization); and (iv) organoleptic and sensorial (quality-determining characteristics of the finished product). [5].

The rise in loaf volume, which results from four basic processes, is a significant physical change. Heat first causes the gas inside the cells to expand by raising its pressure. The solubility of dissolved gases is decreased by heat, causing them to be liberated and migrate into already-

 https://doi.org/10.52202/074664-0008

existing cells. This increases the internal pressure of the cells and, as a result, increases their size. Third, the dough's volatile liquids, such as ethanol, change into gaseous form and aid in the expansion of cells and an increase in gas pressure. Fourth, yeasts produce CO2 (until the rising heat inactivates them), which causes the gasses and vapor inside the cellular structure to expand[5].. These changes are summed up in the ensuing subsections.

4.1. Gas vaporization and increase in cell volume:

The dough's water content changes in a few significant ways throughout baking. Initially, the dough surface begins to evaporate as soon as it is placed in the oven and continues to do so throughout the baking process. Then, starch gelatinization is made possible by the transfer of water from free and protein-held solutions to starch, which begins at about 60 °C often referred to as oven spring, is caused by gas production. Air, ethanol, CO2, and vapour are the four gases that are involved [5]. Loaf volume is mostly determined by oven-rise [16].For bread[7]. and sponge cake [21].this can grow by 16–100% and 50–80%, respectively.

As the temperature rises during baking, vapour expands thermally and the water saturation pressure in the dough rises as well. This causes compression and higher densities elsewhere in addition to local expansion. CO2 is also a significant factor. It is released when pressure causes bubble walls to burst, increasing the continuity of the porous structure [7].It is well recognized that resistance to extension, also referred to as inadequate extensibility or high dough toughness, can contribute to poor expansion [15]. Cell membranes in dough can generally withstand expansion to varying degrees [3]

When the temperature is below 73 °C, evaporation is at least as significant as CO2, and when the temperature is over 95 °C, it is at least twice as significant. As long as the bubbles stay closed, the creation of gas inside them adds to the pressure rise. Conversely, the forces that the dough applies to the gas limit the size of the bubbles [15].According to Lucas[13],[10], these forces are only effective in cases where the dough can be viewed as a suspension of bubbles inside a continuous and incompressible viscous bulk liquid phase, or when they result from the elastic and viscous characteristics of a viscoelastic material[16].Lastly, it's critical to emphasize that, in some bakery goods, like sponge cake or puff pastries, water vapour causes cell inflation [21].

4.2. Structural changes in starch and protein

Protein denaturizes, starch gelatinizes, and crust forms as the dough cooks, producing a stable result [7]. Gelatinization is caused by a sequence of overlapping molecular processes: (i) swelling, which happens when starch granules take in water and swell; (ii) hydration-facilitated melting, which happens when their crystalline structure is irreversibly lost; and (iii) disruption, which is caused by the hydration of disordered polymer chains and partial dispersion of amylose and amylopectin. The process of starch gelatinization is endothermic, and the number of transitions—four in most cases—depends on the composition of the dough, primarily on its water content. The dissociation of previously produced amylose-lipid complexes is related to the third and fourth transitions, whereas the first and second transitions show the disorder (melting) of starch crystallites

To maximize the quality of the finished product, it is imperative to comprehend how gelatinization operates [7]. Peak gelatinization temperatures, in example, can be delayed by even low amounts of sodium chloride, sucrose, fat, and emulsifiers Along with partially solubilized starch, swollen granules lose their crystallinity [7] and are crucial structural components of bread [7]. Lucas, pp. 335–354, 2014.shown that the pace of heating has an impact on swelling; specifically, a higher heating rate broadens the temperature range over which the granules enlarge. Lastly, Mondal and Datta[6].point out that bread products with gelatinized starch have an amorphous structure.

Lucas [13] notes that the process of dough hardening and turning into bread is aided by both protein aggregation and gelatinization. These processes, which begin at about 60 °C for starch gelatinization and 65–70 °C for protein denaturation, are frequently linked to a slow rise in gas pressure and cell expansion. This method emphasizes how rising temperatures enhance gas production while restricting dough extension, favouring the influence of gases with low vaporization temperatures on cell expansion. Still, The latter author points out that there is no correlation between temperature and pressure build-up during baking, according to recent tests using miniature sensors.

Above 60 °C heating also progressively changes the protein network, most notably by polymerizing glutenins and converting gluten gel to coagel According to Yin and Walker [5], proteins become insolubilized as a result of the creation of new covalent connections, or disulfide bonds, and gluten compressibility reduces while hardness increases. Gluten films are ruptured by heat denaturation, and this process intensifies when the internal bread temperature goes above 75 °C. Starch granule swelling and soluble amylose exudate commence at the same time (60 °C), and internal pressure from trapped gas and water vapour keeps the structure inflated[5]. Cell membrane gelatinization causes dough to become more viscous and less extensible, which raises internal pressure in sealed gas cells and causes the membranes to burst. The dough's ability to expand is likely limited by the interchange of gas molecules between neighbouring cells and their eventual transportation outside of it.
Lastly, the authors note that low molecular weight proteins tend to rise while polymeric proteins tend to decrease.

4.3. The dough-crumb transition and changes in structure

Variations in dough viscosity have been studied in relation to the dough-crumb transition. Lucas[13] provides a literature summary. Three unique stages of variation are identified, which correlate to transition temperatures of 55, 75, and 100 ◦C. During the initial stage, the viscosity gradually reduces until the dough reaches a temperature of approximately 55-60°C. All materials that undergo no transformation experience this drop. Protein coagulation and starch gelatinization cause a linear rise in viscosity during the second phase, which occurs between 55 and 75 °C. In the third stage. (In the range of 75 to 100 ◦C), viscosity falls roughly linearly. There are two possible explanations: either (i) the gas nuclei that were initially in the dough or those that developed during the mixing step expand when heated, or (ii) the grains of gelatinized starch soften and break. The more likely of the two theories is the latter.Protein denaturation and starch gelatinization both impact diffusion by releasing and absorbing water, according to Mondal and Datta [7]. These two occurrences, which are related to the dough-crumb transition, take place in the 60–85 °C temperature range. Two other physical changes that occur with baking are crust formation and crumb resilience, according to Yin and Walker [5]. Crumb robustness is correlated with the shape of minuscule gas cells encased in the protein-starch-lipid matrix and becomes noticeable above 80 °C. Cell structure is greatly impacted by baking conditions, thus achieving the right balance between internal and external dough changes is essential to producing a crust that looks good. Ultimately, the formation of crust colour and flavour is attributed to two heat reactions: caramelization and the Maillard reaction.

According to Walker[7], under all baking circumstances, the local temperature recorded in the crumb centre does not rise above 100 °C.[16], on the other hand, observe that the crust temperature quickly rises above this degree due to the fact that water evaporates from the surface more quickly than it can be moved from the centre. Generally speaking, the crust or crumb will

https://doi.org/10.52202/074664-0008

heat up more quickly in an oven at a higher temperature. Steam baking, on the other hand, produces distinct behaviour. Bredariol et al.'s summary of the variations between bread made with and without steam[20].Condensed water on the bread's surface reduces heat transfer during steam baking, which lowers the surface's heating rate. The heat flow to the centre of the dough increases as the majority of the dew evaporates, speeding up the crumb's heating process. The scientists observed that crumb pores grew larger and more numerous during the baking process under all baking parameters. Lastly, they observed that, whether steam is present or not, the pore size increases with baking temperature.

4.4. Crust formation and browning

An evaporation front that advances towards the core forms the bread crust[15]. The product's exterior is distinguished by a steadily declining water content, reaching 5–10% (wet basis), and a sharply rising temperature, surpassing 100 °C. The crust's structural features are derived from these changes in temperature and moisture content, which also prevent the product's interior from being dehydrated by limiting the diffusion of water vapour to the surrounding air. Although the internal transport processes of thin items, like biscuits, vary, surface modifications yield comparable features. According to Manhiça[17]., browning is influenced by the dough's composition, which includes sugars, amino compounds, and leavening agents, as well as the temperature and water activity.

Numerous variables, including water evaporation, the dough's initial porosity, the degree of drying at the dough's surface, the length of baking, and the temperature at which cells collapse, affect crust thickness . Compared to crumb, crust has a lower water content and is denser due to its smaller, longer, or fewer cells. In bulk terms, the ratio of crust to crumb is 1-2. The high temperature causes faster evaporation at the dough top, which prevents the crust's starch from fully gelatinizing. However, the increased viscosity of surface layers during baking is mostly caused by the decrease in water content rather than partial starch gelatinization. Another essential component that keeps the crust's alveolar structure intact is rigidity. Ultimately, the Maillard reaction is accelerated by low water activity and high temperatures in external dough layers, producing a variety of desired flavors and aromas, including methyl-propanal, 4-hydroxy-2, 5-dimethyl-3(2H)-furanone, and 2-acetyl-1-pyrroline [21]

A thorough analysis of the impact of drying rate on dough viscosity during the conversion of dough into crust and crumb was put forth[19]. The experiments were conducted using dynamic mechanical thermal analysis (DMTA) in an environment that was as similar to conventional baking as possible. Water content and temperature were tracked during the tests. The authors noted that even though the water content reduced, the dough's viscosity behaved as though the water content remained constant at high heating rates (corresponding to a typical top crust, 15–30 ◦C/min) [19]. Furthermore, the authors contended that this behaviour consisted of a non-changing viscosity phase at first, followed by a rise at roughly 60 ◦C, which was associated with the start of starch gelatinization [19]. A thorough analysis of the impact of drying rate on dough viscosity during the conversion of dough into crust and crumb was put forth[19]. The experiments were conducted using dynamic mechanical thermal analysis (DMTA) in an environment that was as similar to conventional baking as possible. Water content and temperature were tracked during the tests. The authors noted that even though the water content reduced, the dough's viscosity behaved as though the water content remained constant at high heating rates (corresponding to a typical top crust, 15–30 ◦C/min) [9]. Furthermore, the authors contended that this behaviour consisted of a non-changing viscosity phase at first, followed by a rise at roughly 60 ◦C, which was associated with the start of starch gelatinization [9].

The primary causes of bread flavour seem to be crust development and browning. Rather than caramelization, a Maillard-type process is primarily responsible for browning. It happens when the temperature is higher than 110 °C, and weight loss during baking and oven temperature have been experimentally correlated [7]. One crucial feature of baked goods is their surface colour, which can even be regarded as a crucial index. It is a particularly significant feature for cookies and biscuits, as it is directly linked to the desired texture, aroma, and visual qualities[17]

Intermediate chemicals are responsible for browning; these compounds eventually polymerize to generate brown pigments [1]. Therefore, during baking, colourful compounds are produced as a result of non-enzymatic chemical interactions. The Maillard reaction occurs when reducing sugars are heated in the presence of amino acids, proteins, and/or other nitrogen-containing compounds, whereas caramelization refers to a complex set of reactions that happen when carbohydrates, specifically sucrose and reducing sugars, are heated directly. Purlis [15]

Not advantageous to health. Purlis[15] and Isleroglu [17] both point out that the production of acrylamide is a detrimental effect of the Maillard process. Acrylamide is a neurotoxin and has been identified by the International Agency for Research on Cancer as a potential human carcinogen. Food matrix and operational parameters, including moisture content, pH, and time-temperature ratios, affect the synthesis and build-up of acrylamide. Therefore, close monitoring of this parameter is necessary to ensure the safety of the final product.

5. Main findings

5.1. Results of the systematic literature review

The first dataset had 4200 texts in it. After using the selection criteria listed in section 2 (search technique), this was whittled down to 63. After duplicates were eliminated, there were 61 articles remaining: 49 research papers, 8 reviews, and 4 book chapters. Using flow charts, Fig. 1 presents an overview of the selection procedure that adhered to the PRISMA statement [12]. Results for ScienceDirect, PubMed, and the Web of Science are displayed in Figures 1(a), (b), and (c), respectively.

5.2. Ovens

A crucial step in creating bread is baking. Dough is changed into a light, porous, easily digestible, and delicious product by heat[5]. Ovens of several kinds are employed. According [9], industrial ovens are mainly gas-powered and categorized as direct (forced convection), indirect, microwave, radiation, or infrared systems. Woodchip burners, electricity, and oil are examples of additional energy sources. Sub-paragraphs that follow provide an overview of the most popular systems.

5.2.1. Traditional furnace

According to Manhiça[1], fixed grate burning of biomass is a common practice in small batch furnaces. For instance, there are two types of furnaces in Mozambique: semi-direct and indirect. The combustion chamber and the baking chamber are the two separate chambers that define the indirect oven. The combustion chamber is forced to have air circulation. Three tiers, each with three or four drawers, make up the baking chamber. It is intended to transfer heat from a coil to the product, mostly by radiation and natural convection. The combustion chamber heats the circulating fluid inside the coil. Spraying water vapour within the oven has a big impact on the colour of the crust. A combustion chamber is another component of the semi-direct oven. In addition, the semi-direct oven has a baking chamber and a combustion chamber that are heated by burning biomass. Both chambers are composed of refractory ceramic material. The combustion chamber has a funnel-shaped top and is parallelepiped in design.

https://doi.org/10.52202/074664-0008

Even though indirect furnaces use 0.55 kg of wood as opposed to 0.90 kg of wood per kg of baked wheat flour, they are still less economical to operate on an industrial scale due to maintenance and electricity expenses[1]. Furthermore, because of the way dough is delivered into a semi-direct oven and its geometry, the temperature profile within is not consistent. Consequently, there is variation in the quality of the bread. However, design optimization would raise the calibre of the finished product and baking considerably.

5.3. Effects of the use of steam during baking

The methods used in the production of steamed and oven-baked bread differ greatly, and the resulting goods have distinct flavours, textures, nutritional values, and shelf lives[16]. Compared to traditional baking, the processing methods for Chinese steamed bread may minimize nutrient loss and need less heat. Lysine and vitamin B1 are lost when conventional bread is baked at 200 °C [20]. According to[16]., steamed bread has a greater protein efficiency ratio than baked bread. It is thought to be a healthier meal and is more suited for diabetic diets since it does not undergo the Maillard reaction, which would produce harmful by products like furan and acrylamide. It is also low in sodium and oil [16].A steamer set over boiling water is used to cook Chinese steamed bread. A roll-sized product with a smooth white skin and no crust is produced by this approach[15]. The physic-chemical characteristics of bread, especially its crust and crumb, can be greatly impacted by steaming[1]. In order to enhance crust brightness during the early baking phases, steam is frequently used [17]. Thus, the amount of steaming has a big impact on crust browning. Fortoul and Altamirano[5].We discovered that cakes prepared in extremely humid conditions had lighter crusts and more shine. The quantity of gelatinized starch granules on the crust could be the cause of these effects. Water vapour from steaming may be responsible for these events, since it facilitates the production of a starch paste that eventually gelatinizes, creates dextrin, and caramelizes to impart colour and gloss.

Furthermore, Altamirano-[1].report that more Steam increases the stiffness, distortion, and force of failure. A soft crust is indicated by a low failure force, whereas a high value denotes a more stiff and brittle crust. Failure deformation may be brought on by variations in the crust's moisture level, which can alter the size and distribution of starch granules and impact the mechanical characteristics of cereal-based foods' cell walls. Changes in the starch on the crust surface brought on by the steam are most likely the cause of the increased failure firmness. The creation of cross-links between partially soluble starch and gluten proteins is primarily responsible for firmness. The latter authors' microstructure research revealed that the amount of steam considerably changed the crust's structure, influencing the size and shape of the bubbles. The creation of cross-links between partially soluble starch and gluten proteins is primarily responsible for firmness. The amount of steam considerably altered the crust structure, influencing bubble size and shape, the protein-starch network, and surface density, according to the latter authors' microstructure analysis. The characteristics of the crust, specifically its colour, gloss, and thickness, significantly impact how fresh bread tastes. Once more, the amount of steam has a big impact on them. Thicker crusts with high vapour transmission rates and permeability are produced by low steam levels. Porosity, which functions as a barrier, can be directly linked to lower crust permeability. The authors come to the conclusion that in order to get a hard and brittle crust, the proper amount of steam must be applied.

6. Discussion:

6.1. Strategies to improve baking

The creation of cross-links between partially soluble starch and gluten proteins is primarily responsible for firmness. The amount of steam considerably altered the crust structure, influencing bubble size and shape, the protein-starch network, and surface density, according to the latter authors' microstructure analysis. The characteristics of the crust, specifically its colour, gloss, and thickness, significantly impact how fresh bread tastes. Once more, the amount of steam has a big impact on them. Thicker crusts with high vapour transmission rates and permeability are produced by low steam levels. Porosity, which functions as a barrier, can be directly linked to lower crust permeability. The authors come to the conclusion that in order to get a hard and brittle crust, the proper amount of steam must be applied. It could enhance baking and energy efficiency in addition to oven design and operating conditions. By measuring the energy savings in an industrial oven and highlighting areas for additional savings (heat recovery from flue gases, enhanced heat transfer, and oven wall insulation), they confirmed the effectiveness of their method.

6.1.2. Progress in microwave baking

Microwave technology could be used in conjunction with various heating methods. Combining with forced hot air is one option [7]. A normal microwave oven has very little airflow, which results in extremely low rates of heat and moisture transmission. As a result, impingement jet technology was introduced by the latter writers. By impinging the food surface with high-velocity air jets, this method increases mass transfer and heat rates.

Sumnu [9].conducted a comparative study between two commercially available microwaves and three heating modes: microwave plus infrared, microwave plus impingement jets, and impingement jets alone. The study revealed temporary alterations in multiple bread qualities. Compared to breads prepared using the other two methods, breads baked in a microwave plus IR oven had much reduced specific volume, porosity, and moisture content. Furthermore, they emphasized that there was no discernible difference between the three heating modes in terms of variations in thermal conductivity and dielectric characteristics. Yolacaner[9].corroborated these intriguing findings, emphasizing that the microwave plus impingement jet combination was able to lessen crust evaporation that is often reported in microwave baking alone.

Steam and microwaves are two more recent innovations. In this instance, one alternative mode of operation involves feeding steam into the ovencavity[9]. The latter authors' setup comprises of a water tank, a food container with many openings in the

Bottom, and a substance in the wall parts that shields microwaves. The food within the container is protected from direct microwave radiation. However, more study is required because there isn't a unit that can be purchased commercially. Another intriguing concept is put forth by [17]. The authors examined the impact of using a microwave and halogen lamp in combination with various baking techniques, with a focus on bread quality and energy usage. According to their findings, bread baked under a halogen lamp had a lower specific volume and resulted in greater weight reduction than bread prepared in a microwave. They did point out that the combination had two benefits: the halogen lamp enhanced crust browning, and microwave baking reduced time. All things considered The combination produced a larger weight while producing an appropriate specific volume and crust colour that were comparable to traditionally baked goods. reduction and heightened rigidity. However, in the case of gluten-free bread, where the lack of the gluten network poses a significant issue, increased stiffness can be intriguing[9].

6.2. Strategies to improve ovens

Compared to other methods of food manufacturing and preservation, conventional baking uses a lot of energy, hence it's crucial for production chains to have ovens designed optimally[13].In

https://doi.org/10.52202/074664-0008

addition to having a favourable impact on the environment, new innovations and improved management could increase the quality of bread and bakery products while lowering energy use[7].Some of the most intriguing oven enhancement techniques are described in the subparagraphs that follow.

6.2.1. Solar energy ovens

Growing energy consumption, especially in developing nations, puts our planet's fragile balance at further risk. Furthermore, substantial production capacity is required to meet the increased demand for power, as noted by [15].Businesses are badly impacted by this, especially bakeries, which rely entirely on electricity. One choice is to use renewable energy sources, like solar power. A hybrid solar-powered oven would lessen, but not totally eliminate, losses from power outages and load shedding.

The efficiency of a solar energy oven was calculated by[15].using the energy required to bake three loaves of bread. They determined that 2.182 kW of power would be needed, and that solar energy could provide this. They determined that 28 evacuated cube collectors were needed after selecting them as the power source. An intriguing solution to lessen environmental pressures would be a solar-powered oven, though further research is necessary, especially for large-scale operations.

7. Conclusions and future trends

The significance of choosing the best baking method in relation to the intended product attributes has been emphasized by this review. Heat transfer modalities are well understood and have been extensively documented in the literature; nevertheless, more investigation is required into novel hybrid approaches including vacuum–conventional and microwave–IR baking. The ideal time-temperature ratio during baking is a significant subject that requires more research.

This review made clear that combining steam and vacuum technology with traditional ovens may make them better. Particularly for the latter, the results show promise in terms of shorter baking times, less energy usage, and less of an impact on the environment—all without having a discernible detrimental influence on bakery goods.

Blending microwave ovens with other baking methods could also greatly improve them. The benefit of microwave baking is that radiation is easily absorbed by the wet interior and goes through a dry crust. But a major drawback for many bakery goods is the inability to achieve a brown crust, especially when it comes to bread. The advantages of IR heating for browning and crisping combined with the time-saving benefits of microwave heating could be combined in this situation when microwave and IR, as well as IR-visible heating, are used together. Once more, when compared to traditional baking, this hybrid method minimizes baking time, energy use, and environmental effects.

This assessment concluded by pointing out that advances in science and technology have greatly advanced baking technique. A plethora of cutting-edge methods are becoming available as viable substitutes for traditional oven baking. Still, more work needs to be done to enhance the plants, machinery, and ovens. The culinary sector is always looking for fresh concepts. Both in industrialized and emerging nations, there is a pressing need for efficient baking techniques and ovens that can satisfy the demands of large-scale factories while using renewable energy sources. The food business, and the baking sector in particular, would benefit from such ideas as they tackle the enormous problem of environmental sustainability.

8. Funding

There was no funding for this research from any public, private, or non-profit organizations. A statement of conflicting interests. The authors state that none of the work described in this publication appears to have been influenced by any known competing financial interests or personal ties.

9. References

[1] Altamirano-Fortoul, R., Le-Bail, A., Chevallier, S., & Rosell, C. M.: Effect of the amount of steam during baking on bread crust features and water diffusion. Journal of Food Engineering. 108(1), 128-134 (2012).
[2] Arepally, D., Reddy, R. S., Goswami, T. K., & Datta, A. K.: Biscuit baking: A review. LWT–Food Science and Technology, Article 109726 (2020). Bock, J. E., Wrigley, C. W., & Walker, C. E.: Bakeries: The source of our unique wheat-based food, bread. Encyclopedia of Food Grains. 3(2), 335-342 (2016).
[3] Bredariol, P., de Carvalho, R. A., & Vanin, F. M.: The effect of baking conditions on protein digestibility, mineral and oxalate content of wheat breads. Food Chemistry. 332, 127399 (2020). Bredariol, P., Spatti, M., & Vanin, F. M. (2019).
[4] Bredariol, P., Spatti, M., & Vanin, F. M.: Different baking conditions may produce breads with similar physical qualities but unique starch gelatinization behavior. LWT – Food Science and Technology. 111, 737-743 (2020).
[5] Campbell, G. M., Sharp, C., Wall, K., Mateos-Salvador, F., Gubatz, S., Huttly, A., et al.: Modelling wheat breakage during roller milling using the double normalized Kumaraswamy breakage function: Effects of kernel shape and hardness. Journal of Cereal Science. 55(3), 415-425 (2012).
[6] Cappelli, A., Bettaccini, L., & Cini, E.: The kneading process: A systematic review of the effects on dough rheology and resulting bread characteristics, including improvement strategies. Trends in Food Science & Technology. 104, 91-101 (2020a). Cappelli, A., Canessa, J., & Cini, E. (2020b).
[7] Cappelli, A., & Cini, E.: Effects of CO2 snow addition during kneading on thermoregulation, dough rheological properties, and bread characteristics: A focus on ancient and modern wheat cultivars. International Journal of Refrigeration. 117, 52-60 (2020).
[8] Cappelli, A., & Cini, E.: Challenges and opportunities in wheat flour, pasta, bread, and bakery product production chains: A systematic review of innovations and improvement strategies to increase sustainability, productivity, and product quality. Sustainability. 13(5), 2608 (2020).
[9] Cappelli, A., Cini, E., Lorini, C., Oliva, N., & Bonaccorsi, G.: Insects as food: A review on risks assessments of Tenebrionidae and Gryllidae in relation to a first machines and plants development. Food Control. 108, 106877 (2020c). Cappelli, A., Guerrini, L., Cini, E., & Parenti, A. (2019a).
[10] Cappelli, A., Guerrini, L., Parenti, A., Palladino, G., & Cini, E.: Improving whole wheat dough tenacity and extensibility: A new kneading process. Journal of Cereal Science. 90, 102852 (2020d). Cappelli, A., Mugnaini, M., & Cini, E. (2020e).
[11] Cappelli, A., Oliva, N., & Cini, E.: Improving roller milling technology using the break, sizing, and reduction systems for flour differentiation. LWT–Food Science and Technology. 133, 110067 (2020f).
[12] Cappelli, A., Oliva, N., & Cini, E.: Stone milling versus roller milling: A systematic review of the effects on wheat flour quality, dough rheology, and bread characteristics. Trends in Food Science &Technology. 97, 147-155 (2020g).
[13] Cappelli, A., Parretti, C., Cini, E., & Citti, P.: Development of a new washing machine in olive oil extraction plant: A first application of usability-based approach. Journal of Agricultural Engineering. 50(3), 134-142 (2019b).
[14] Di Silvestro, R., Di Loreto, A., Marotti, I., Bosi, S., Bregola, V., Gianotti, A., Quinn, R., & Dinelli, G.: Effects of flour storage and heat generated during milling on starch, dietary fibre, and polyphenols in stoneground flours from two durum-type wheats. International Journal of Food Science and Technology. 49, 2230-2236 (2014).
[15] Forsythe, H. D. M., & Madyira, D. M.: Experimental performance assessment of a solar-powered baking oven. Procedia Manufacturing. 35, 535-540 (2019). Grenier, D., Le Ray, D., & Lucas, T. (2010).
[16] He, Y., Guo, J., Ren, G., Cui, G., Han, S., & Liu, J.: Effects of konjac glucomannan on the water distribution of frozen dough and corresponding steamed bread quality. Food Chemistry. 330, 127243 (2020).
[17] Isleroglu, H., Kemerli, T., Sakin-Yilmazer, M., Guven, G., Ozdestan, O., Uren, A., & Kaymak-Ertekin, F.: Effect of steam baking on acrylamide formation and browning kinetics of cookies. Journal of Food Science. 77(10), E257-E263 (2012).

[18] Keskin, S. O., Sumnu, G., & Sahin, S.: Bread baking in halogen lamp–microwave combination oven. Food Research International. 37(5), 489-495 (2004).
[19] Kowalczewski, P.Ł., Gumienna, M., Rybicka, I., Gorna, B., Sarbak, P., Dziedzic, K., & Kmiecik, D.: Nutritional value and biological activity of gluten-free bread enriched with cricket powder. Molecules. 26(4), 1184 (2021a).
[20] Kowalczewski, P.Ł., Walkowiak, K., Masewicz, Ł., Smarzyński, K., Thanh-Blicharz, J. L., Kačaniová, M., & Baranowska, H. M.: LF NMR spectroscopy analysis of water dynamics and texture of Gluten-Free bread with cricket powder during storage. Food Science and Technology International. Article 1082013220987914 (2021b).
[21] Li, X., Wang, C., & Krishnan, P. G.: Effects of corn distillers dried grains on dough properties and quality of Chinese steamed bread. Food Sciences and Nutrition. 8(8), 3999-4008 (2020).

A Review of Research on Social Media's Effects on Travel and Hospitality

Shugufta Pandith[1] and Dr Kirti Sharma[2]

[1,2] Department of Tourism and Hospitality Management, Chandigarh University Mohali India

Abstract:

This study examines the part of social media in the tourism industry and explores its effect on various aspects, including customer decision-making, promotional activities, and customer interaction. It highlights the significant capability of social media as a merchandising strategy for tourism products. However, it also acknowledges that the field of social media in tourism is unmovingly in its nonage, and comprehensive research is required to understand its influence on the industry, society, and the economic contributions it brings. The paper emphasizes the need for further investigation into the social and economic implications of social media in the real sphere. Additionally, it discusses the challenges faced in studying online resources and the experiences of populations or communities affected by unequal power relationships. The paper also delves into the impact of the internet community on youth tourism, electronic word-of-mouth dissemination, and the importance of online review sites like Trip Advisor. It explores bibliometric analyses, the accuracy of information in online evaluations, and the effects of social media evaluations on restaurant consummation and business value. The study also addresses the mediating role of marketing skills, social media communication strategies, and the relationship between social media and business reputation. Furthermore, it explores the impression of social media on music festivals, the travel and hospitality sectors, and Vienna hotels.

Keywords: Social Media, Tourism, Competitiveness, Digital Environment

1. Introduction:

The tourism and hospitality industries have been significantly and widely impacted by social networking.The following are some significant ways that social media has affected and changed various sectors: [1]

Greater Exposure and Reach: Social media podium have provided tourism and hospitality businesses with an opportunity to reach a global audience at a relatively low cost. Through effective use of social media marketing, businesses can showcase their offerings, engage with potential customers, and increase brand awareness. [2] Influencer Marketing: The rise of travel bellwether's on social media has designed new avenues for tourism and hospitality businesses to promote their products and services. Collaborating with influencers who have a large following

https://doi.org/10.52202/074664-0009

and a niche focus on travel allows businesses to tap into their audience and gain exposure to a highly targeted market. [3]

User-Generated Content: SM manifestos have set off a treasure trove of user-generated content related to travel and hospitality. Travelers share their experiences, photos, and reviews, providing authentic and real-time information to others. This user-generated content serves as valuable social proof and effects the decision-making process of prospective travelers. [4]

Consumer Engagement and Relationship Building: Social media platforms enable direct interaction between businesses and customers. Hotels, airlines, and tour operators can engage with travelers, respond to inquiries and feedback, provide personalized recommendations, and address customer concerns. This fosters a sense of trust and connection, enhancing the overall customer experience. [5]

Online Booking and Reservations: These days, a lot of social media sites have integrated booking tools that let users reserve travel, lodging, and experiences right on the site. This makes passenger bookings more convenient and 6] seamless by streamlining the process.. [7]

Fig.1. Synchronicity of Social Media. Destination Promotion and Tourism Marketing

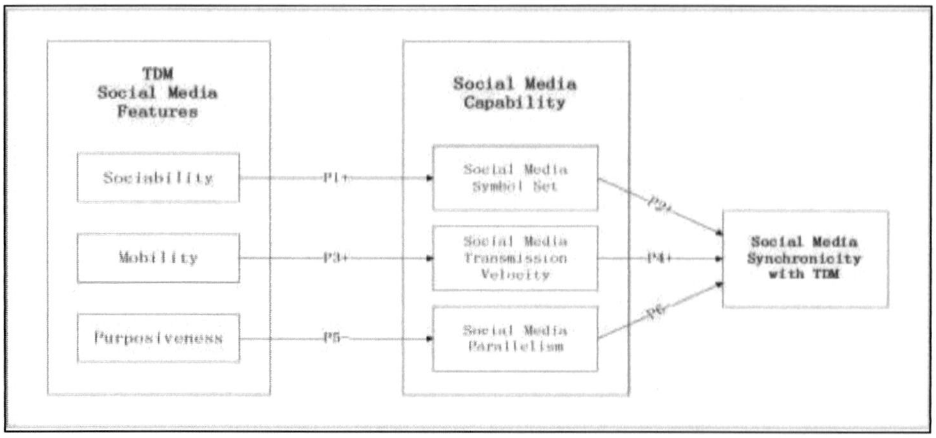

Tourism boards and destination marketing organizations utilize social media platforms to promote their locations, attractions, and events. They can target specific demographics, showcase unique experiences, and create engaging Campaigns to attract visitors. [8]
Crisis Management and Communication: In times of crises, such as natural disasters or pandemics, social media plays a pivotal part in disseminating statistics, updating tourers on safety measures, and managing crisis communication. It provides a real-time channel for businesses and authorities to keep travelers informed and address concerns. [9]

Market Research and Insights: Social media podiums offer valuable data and insights for businesses in the tourism and hospitality fabrication. Through communal listening and analytics tools, businesses can gather information about customer preferences, trends, and sentiments, enabling them to tailor their offerings and marketing strategies accordingly. [10]

But it's crucial to remember that social media also has its drawbacks.. Negative reviews and feedback can quickly spread and damage a business's reputation. Maintaining an active presence and managing online reputation becomes crucial. Additionally, the fast-paced nature of internet community needs businesses to be agile and responsive, constantly adapting to changing trends and customer expectations. [11]

In general, social media has transformed the travel and hospitality sector by providing new opportunities for promotion, customer engagement, and brand building. It has reshaped the way traveler's research, plan, and experience their trips, while also empowering businesses to connect with a wider audience and stay relevant in a digital age.

1.1 Related Work

This manuscript lists the social media's use in tourism that is currently known and makes recommendations for further investigation. SM plays fantastic role in many facets of tourism, particularly in the way that customers decide what to buy, how to engage them in promotional activities, and how to communicate with them. Utilizing social media to its fullest potential has shown to be a successful marketing technique for travel-related goods. Additionally, this article contends that SM in the tourism industry continues to be in its nonage. It is challenging to promote thorough examination of SM's effects on all facets of the tourism industry, including society, and to demonstrate how much money SM brings in to the sector. The financial and social fabric of reality has been strengthened by SM.[12]

Through the analytical application of the idea of communicative action, developing internet cultures and socio-technical practices have been explored. It demonstrates how social media and travel are connected. It investigates how correlative technologies have influenced the creation and alteration of this business in contrast to what the dictionary says. This paper depicts the equivocal capability of tourism social media as information and communications technology for liberation but also a medium for prioritization, control and misuse of social media. New methods of inquiry and theoretical paths need to be adopted by academic community to examine the multiplex reciprocity between tourism and technology development. This paper also provokes Habermas' disunity between broadcasting and administered media. The hurdles between the two types are nowadays highly unfocused. While assessing the review, depicts that technology research in tourism is basically influenced by Pollyannaism and technical interests and condemnatory eyeshot on the ill treatment of electronic sources and the study of population or communities that experience discrimination and exclusion because of unequal power relationships, their voices are few and far between. [13]

The main focus is on the on the youth tourism. Youth represents the highest percentage of technology users and takes an active part in various high yielding processes of tourist operations, as development of designs and dispersal of novel products. The major goal of this research is to analyze the effects of social media and ICTS upon young travelers as well as the characteristics of young travelers who actively engage in SM and the implications of the information they receive from SM and ICTS on their travel choices. [14]

Electronic word of mouth, or EWOM, is spread through travel review websites, which are most frequently utilized in the hotel sector. It gives potential visitors the chance to carefully consider the views, remarks, experiences, and contributions. The most popular website is Trip Advisor, which was established in 2000 and is constantly expanding, and it has already had a significant impact on tourist demand and supply. According to a World Travel Market industry analysis from 2011, over one- third of the UK travel sector chooses its hotels using review websites like Trip Advisor, and nearly three-fourths of all travelers consult them before making a hotel reservation (Phocus-wright, 2014). The aim of the research is to define online Whiner's profile, to characterize online objections and to find out which factors (e.g. appraisements, demographic)

https://doi.org/10.52202/074664-0009

can affect online objections and the Trip Advisor which is considered to be the most acceptable review site in the hospitality industry was chosen. [15]

This study used two bibliometric analytic techniques to offer a thorough and organized overview of academic literature on social media. Between 2007 and 2016, 16 corporate and hospitality/tourism magazines yielded an entire collection of 406 publications on social media. Cocitation research. The subject of tourism and hospitality showed a diversified conceptual framework, whereas Word-of-Mouth was designated as the primary conceptual basis for social media study in business. The researchers then used analysis to track how research themes in both categories changed over time. The comparison of the two domains' social media studies revealed four areas of overlap: the development of the field over time; the rise in use of term "social media"; new phenomenon of social networking sites; and the research concentration on managerial applications. The report also recommended developing a future research agenda for digital media studies concerning the hotel and tourist industries.[16][17]

This highly limits the ability to generalize and consequences of SM analytics studies. The review compares and contrasts Trip Advisor, Expedia, and Yelp three of the most well-known online review sites in terms of the accuracy of the data contained in online evaluations of all the hotels. There are notable disparities between how the hotel industry is depicted on various networks. Particularly, there are significant differences across online evaluations in terms of language qualities, cognitive facets, attitude, rating, effectiveness, and connections between these features. The study identifies a number of themes for future studies in SM monitoring in the tourism and hotel sectors as well as a framework for understanding
Methodological challenges.[18][19]

The purpose of this study is to examine the moderating role that an excellence certificate plays in the relationship between operating efficiency measures, social media reviews of restaurants, and financial performance. The quantity of online evaluations that customers provide has a considerable positive impact on the restaurants' performance. The quality of the food being provided, the number of customers served each hour of work, and customer satisfaction ratings all have an impact on how well a restaurant performs. It has been demonstrated that a culinary excellence certificate dramatically modifies the correlation between the volume of evaluations and the overall quality of restaurants. In terms of net revenue, guest numbers, and average bill, an eatery with an excellent certificate gains better from positive feedback or an improved overall score than restaurants without a top-notch certificate. Therefore, in order to boost their bottom line, restaurateurs having an elite certificate need maintain a high degree of favorable client feedback.[20][21]

The connection between social media activity at a restaurant company and business value, although social media has grown in importance for restaurant owners' interactions with both current and future customers, its actual financial effects are uncertain and there is scant empirical support. This study indicates the beneficial and progressive impact of restaurant companies' actions. After adjusting for a series of business factors, the impact of SM on company worth. [22][23]

The aim of the study is to investigate the mediating impact that marketing skills have in the relationship between social media and profitability in the hotel sector. For the data analysis in this study, a structural equation modeling approach was adopted. Data from postal questionnaires were collected, with 384 lodgings in the UK serving as the sample. The data analysis's conclusions demonstrate a robust and positive relationship between social media use and corporate success. The association between social media usability and corporate output is, however, strongly and favorably moderated by marketing skills such as branding and revolution, according to the research..[24][25]

The global hotel chain gave the information on hotel performance and online reviews. The hotel chain snap up the information from online reviews that had been compiled by a well-known social media organization for the hospitality industry. The results demonstrate that, after overall ratings, replies to negative comments are the second most essential determinant of hotel success. Higher overall ratings and a higher number of complaint responses both improve the hotel's performance. Therefore, it is crucial for hotel marketing to manage social media reviews, especially their overall score and reaction to negative comments. [26][27]

Despite the fact that it is widely acknowledged that corporate communication directly affects the image of a company, little is known about the connections between a company's social media communication strategy and how that company's reputation develops in an online setting. This piece adds to the corpus of information by examining how SM communication methods affect the reputations of businesses. The hotel sector serves as the study's context. The findings shed light on the difficulties in creating social networking strategies that impact business reputation. [28][29]

The management of music events and the effects of online social networks on relationships with clients are the paper's two main research areas. In the modern digital era of commercial media, less is understood about how different travel company customer opinions change on interpersonal platforms and, consequently, if those perceptions affect intended marketing goals like reputation. The research was used to develop and validate a conceptual model using structural equation modelling. The results show that relationships formed on social media can lead to positive PR and other desired benefits, and that digital network does, in fact, have a significant footprint on festival-goers' sentiments and loyalties to determined companies.[30][31]

This study looks into how social networking sites have affected the travel and hospitality sectors. A sort of social media called a website for social networking gives users a place to get fasten with one another. Because social media is so flexible, it may be utilized for both private and career reasons. Social media platforms are one among the essential instruments that serve a significant and advantageous purpose in the hotel and tourism industries. Social networking site content may have both beneficial and negative effects on marketing. Consumer behavior, relationships with consumers, cost- and timeeffectiveness, brand awareness and corporate image development, company promotion, consumer targeting, and lastly drawbacks will all be covered in this study paper.[32][33]

Business as well as academia have paid close attention to the potential impact of online communities on the travel and hospitality industries. The vast bulk of research, however, has been conducted from the standpoint of the customer instead of the service provider. The methods hotels employ for assessing the effects social networking has on their unique goals and the numerous aspects, notably both financial and non-monetary ones that are taken into account are not well studied. By conducting informal conversations with holidaymakers in Vienna, Austria, the current study investigates how Vienna hotels evaluate the influence of online channels on their business goals.The results show that when using social media marketing in Vienna, advertisers prioritize customer satisfaction, engagement, and brand awareness over financial measures as their primary return on investment (ROI).[34]

How social media influences traveler making choices along with tourist administration and operations is a hotly contested topic. Strategically, social media is crucial to the sustainability of the travel and tourist sector. The industry is reacting slowly to the business opportunities that social media presents. Due to the pressure to go digital, some firms may accept new technology

 https://doi.org/10.52202/074664-0009

without taking into account how it will affect the business in a digital setting. The internet community is always growing. The term "increasing significance and interest of social networking sites in tourism" was theoretically coined as more studies were found during the preceding several years. Creating and disseminating social media sources with data control and proprietary rights raises some problems. The development of social media—web-based, portable gadgets that turn textual interactions into dynamic dialogue—and online social networks have been the most notable technological advances in recent years. Individuals that tweet their opinions and observations for the world to see are referred to as citizen journalists, and they should be aware that their acts could result in legal action.[35]

The current study evaluated the social media usage patterns of upscale accommodations and provides recommendations to enhance the efficacy of organic posts. Most five star harbors functioning in Turkey has been included in this study, and during the busiest travel month of August, 441 of 612 institutions had their freely accessible post data on Facebook profiles evaluated. 2,012 Facebook posts' worth of information were painstakingly collected and reduced to 1,847. Following the collection of the post data, a content evaluation form was used to code it, and ordinal regression was used to analyses it. The findings emphasized the importance of the kind of content in obtaining better follower involvement in addition to the advantages of interactive and colorful material. Location and facility type (hotel, vacation village, and thermal hotel) were also found to have a substantial impact on follower interactions.[36]

The results of this study demonstrate that the volume of internet-based reviews that customers provide significantly improves restaurants' performance. Additionally, the way eateries execute is affected by client happiness scores, the amount of patrons served for each hour of labor, and the use of managerial indicators of efficiency and social media evaluations. In terms of net income, guest matters, and average bill, restaurants with an excellence certification profit more than those without one from additional reviews or an improved overall rating. As a result, proprietors of restaurants with an excellent certificate ought to keep a high degree of favorable client feedback with the goal to boost their bottom line.[37]

The study paper demonstrates how clever travel marketers are benefiting from social media involvement and discusses the novel consumer decision route via examples from all areas of the travel industry. There are explorations of the theoretical as well as practical repercussions and suggestions for further research in this rapidly changing digital environment.[38]

Using a theoretical framework incorporating the Technology-Organization-Environment (TOE) framework and the theory of diffusion of Innovation Theory (DIT), the current study seeks to empirically explore the key variables that impact a hospitality firm's decision to use media. Seven factors that affect the adoption of corporate media channels have been identified using the technique of structural equation modelling (SEM) and poll findings from 106 Greek hospitality firms.. Three categories—technological, organizational, and environmental—are used to classify these variables. In respect to their ability to predict outcomes, technological traits are thought to be the most significant of the three kinds of influences. Our findings' theoretical and practical ramifications are discussed. [39]

In addition to the cumulative effect of social media use and advertising spending on firm value, this study hypothesizes and experimentally investigates the correlation between using online media and the qualitative worth of tourist and hospitality firms. This study indicates that social media engagement has a strong and positive association with the firm value of the tourism and hospitality industries by using a continuous, long-term data collection from companies that have issued domestic A-shares in China. But because the two types of investments complement each other, spending more on advertising lessens the beneficial effect that networking has on a company's worth.[40]

2. Objective of the studies

(I) To provide a comprehensive overview of the existing research on how social media impacts various aspects of the tourism and hospitality industries.

(II) To identify key factors and variables that influence the relationship between social media and tourism/hospitality outcomes.

3. Research Methodology

The researchers referred to the secondary sources which are enlisted in Scopus during the period of 2005 till 2023. The papers were thoroughly governed in the light of social media's impact on tourism and hospitality industry.

4. Bibliometric Analysis

3.1 The Transformative effect of Social Media on Tourism

Social media has undeniably revolutionized the world of tourism, leaving an indelible mark on how we explore, plan, and share our travel experiences. Its impact has been nothing short of transformative, ushering in a new era of connectivity, inspiration, and engagement for travelers worldwide.

Amongst the most remarkable features of social media's influence on tourism is its ability to inspire wanderlust. Platforms like Instagram, Facebook, and YouTube showcase breathtaking landscapes, awe-inspiring destinations, and culturally immersive experiences through captivating visuals and engaging narratives. As we scroll through our feeds, we are transported to distant lands, exploring the world through the eyes of fellow travelers and content creators. Social media has become a wellspring of inspiration, igniting our curiosity and motivating us to venture beyond our comfort zones.

Moreover, social media has become an invaluable resource for travelers. Through user-generated content, we gain access to authentic and real-time information, tips, and recommendations from those who have already embarked on similar journeys. We can uncover hidden gems, discover off-the beaten-path experiences, and make more informed decisions when planning our trips. Travelers' reviews, photos, and videos provide valuable insights that enrich our travel experiences and help us create memories that resonate long after the trip concludes.

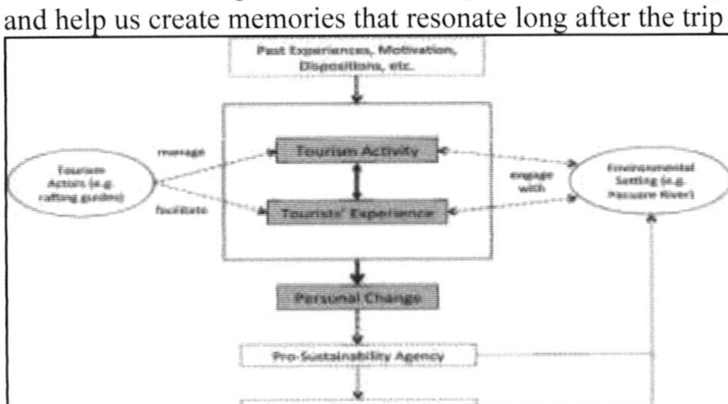

Fig 2. Conceptual Model of transformative tourism [41]

https://doi.org/10.52202/074664-0009

The impression of SM on the tourism industry goes beyond mere inspiration and information. It has fostered direct engagement between travelers and businesses, eliminating traditional barriers and creating seamless connections. Hotels, airlines, and tourism agencies now actively engage with their customers, responding to inquiries, providing personalized recommendations, and offering exclusive deals through social media platforms. This direct communication not only enhances the customer experience but also allows businesses to understand and cater to the evolving target audience's requirements and preferences.

Social media has also empowered a new generation of travel influencers and content creators. These digital storytellers curate captivating narratives and share their unique perspectives, shaping travel trends and influencing consumer behavior. Their firsthand experiences and recommendations hold significant sway, making them powerful catalysts for the tourism industry. Collaborations between businesses and influencers have become increasingly prevalent, amplifying the exposure of destinations, accommodations, and experiences to a wider audience.

While social media's impact on tourism has been overwhelmingly positive, it is important to acknowledge a few challenges. The authenticity of user-generated content can sometimes be called into question, as some individuals may prioritize aesthetics over genuine experiences. Additionally, the pressure to capture picture-perfect moments for social media sharing can occasionally detract from

In conclusion, the effect of SM on tourism obsolete nothing short of transformative. It has shaped the way we discover, plan, and share our travel experiences, connecting us with the world like never before. Social media's ability to inspire, inform, and engage has opened up a realm of

the raw and authentic travel experience.

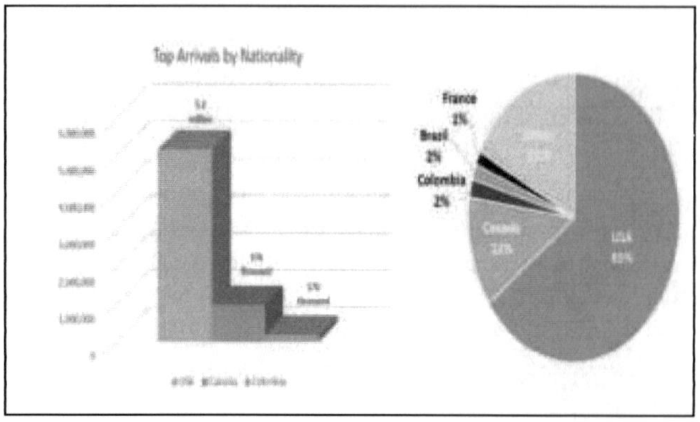

Fig.3. International Tourists arrived in Mexico by air in 2020 [42]

possibilities for travelers and businesses in the tourism business. As we navigate the ever-evolving landscape of social media, its influence on tourism is poised to continue shaping the way we explore and create lifelong memories.

3.2 Analysis of the Impression of Social Media on Tourism

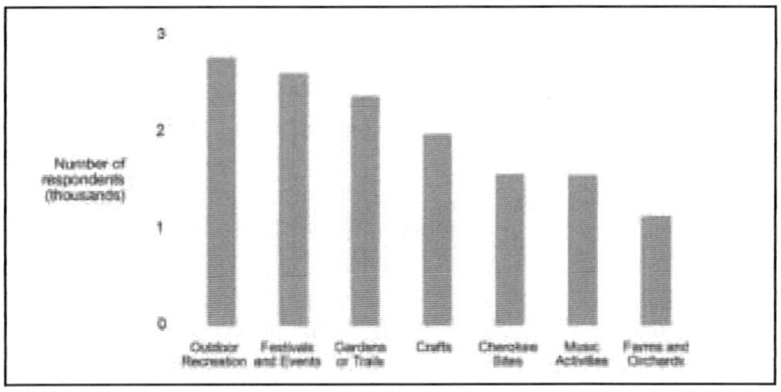

Fig.4.Tourisim Activity preferences in the Blue Ridge National

Heritage Area. [43]

Social media has revolutionized the tourism industry, bringing about significant changes in how people engage with travel-related information, make decisions, and share their experiences. In this analysis, we will delve deeper into the impression of social media on tourism, exploring its implications on various aspects of the industry.

3.2.1 Accessible Inspiration and Discovery:

Social media platforms have become powerful sources of travel inspiration, enabling users to discover new destinations, attractions, and experiences. The visually rich content shared by influencers, travelers, and tourism organizations exposes users to a myriad of possibilities, enticing them to explore beyond their usual preferences and embark on new adventures.

3.2.2 User-Generated Content and Authenticity:

User-generated content plays a crucial role in influencing travel decisions. Travelers now have access to authentic reviews, personal experiences, and recommendations shared by their peers. The transparency of user-generated content on social media fosters trust and gives a more genuine perspective, helping travelers make informed choices about accommodations, attractions, and activities.

3.2.3 Direct Engagement between Travelers and Businesses:

Social media platforms enable direct interaction between travelers and tourism businesses.Through direct interaction, companies may respond to consumer questions, offer prompt assistance, and make tailored recommendations. Travelers can express their concerns, seek information, and share feedback, leading to enhanced customer satisfaction and improved services.

3.2.4 Amplification through Influencer Marketing:

https://doi.org/10.52202/074664-0009

The rise of travel influencers has a profound impression on the tourism industry. Influencers with a large following and niche expertise can effectively promote destinations, accommodations, and experiences to their engaged audience. Collaborations between influencers and tourism businesses amplify brand exposure and generate interest, driving tourism and fostering destination marketing.

3.2.5 Viral Campaigns and Trendsetting:

Social media has the power to make content go viral, creating trends and shaping travel behaviors. Viral campaigns, challenges, and hashtags generate buzz, attracting attention to specific destinations or unique experiences. This viral nature of social media drives engagement, raises awareness, and positions certain locations as trending hotspots, resulting in increased visitor numbers.

3.2.6 Enhanced Customer Experience and Service Personalization:

Social media networks offer a channel for businesses to offer personalized experiences and tailor their services to individual customer needs. By leveraging data and insights from social media analytics, businesses can procure a deeper understanding of customer predisposition and deliver customized offers, improving customer satisfaction and loyalty.

3.2.7 Crisis Management and Real-Time Communication:

During times of crises or unforeseen events, social media serves as a crucial communication tool for the tourism industry. It enables businesses and authorities to disseminate information, address concerns, and provide updates in real-time. This timely and transparent communication fosters trust and helps manage crises effectively.

The impression of social media on tourism is far reaching and transformative. From inspiring travelers to providing access to authentic information, facilitating direct engagement, and shaping travel trends, social media has become an integral part of the tourism landscape. The ability to inspire, connect, and influence has made social media an indispensable tool for businesses and travelers alike. However, it is important to navigate the disagreement associated with social media, such as maintaining authenticity, managing reputational risks, and adapting to the evolving digital landscape. Overall, the impact of SM on tourism is poised to continue shaping the industry, driving innovation, and transforming the way we explore the world.

Tourism activities from 2010 to 2022 From 2010 to 2022, tourism activities experienced significant growth and underwent various changes due to global trends, technological advancements, and external factors. Here is an overview of some key developments during this period:

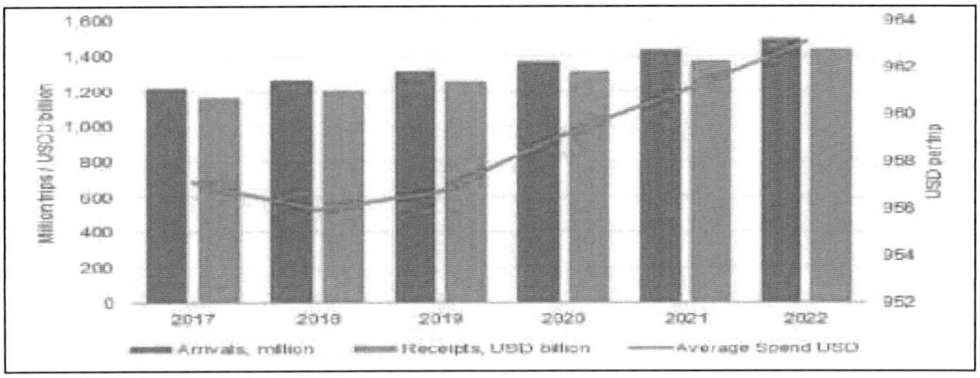

Fig.5. World Inbound Arrivals Recipients and Average Spend 2017-2022 [44]

Rise of Digital Travel: The widespread adoption of smartphones and the availability of high-speed internet access fueled the growth of digital travel. Online travel agencies (OTAs) and travel booking platforms gained popularity, allowing travelers to research, plan, and book their trips conveniently. Online reviews, user-generated content, and social media became influential factors in shaping travel decisions.

Emergence of Sharing Economy: The sharing economy disrupted the tourism industry with platforms like Airbnb and Uber. Accommodationsharing platforms offered alternative options for travelers, allowing them to stay in local homes and experience a more authentic and affordable stay. Similarly, ride-sharing services provided convenient and cost-effective transportation options.
Sustainable Tourism and Responsible Travel: Increasing awareness of environmental issues and social responsibility led to a growing emphasis on sustainable tourism practices. Travelers and destinations alike began prioritizing eco-friendly initiatives, responsible wildlife tourism, and community engagement. Sustainable certifications and initiatives gained traction as travelers sought more responsible and ethical experiences.

https://doi.org/10.52202/074664-0009

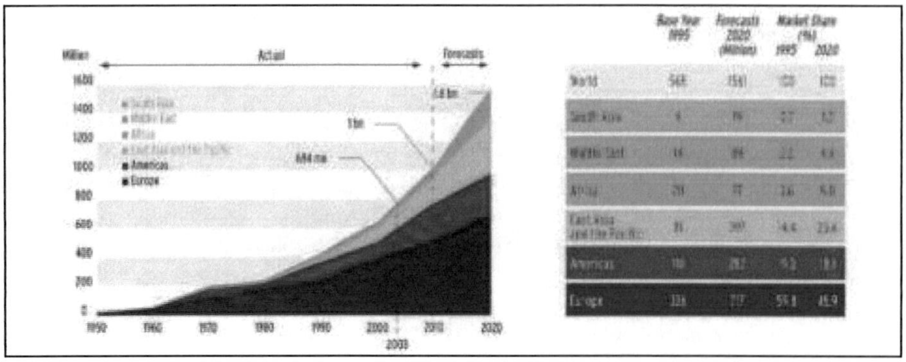

Fig.6. International Tourist arrivals by regions since 1950 and forecasts for 2020 [45]

Growth of Experiential Travel: Travelers shifted from passive sightseeing to seeking immersive and experiential travel. Authentic and immersive experiences such as cultural tours, adventure activities, food and wine experiences, and local interactions gained popularity. Travelers sought unique and personalized experiences that offered a deeper connection with the destination.

Social media and influencers' impact: Social media tenents became high-octane tools for travel inspiration, planning, and sharing experiences. Influencers with large followings started playing a significant role in shaping travel trends and destination choices. Influencer collaborations and sponsored content became common as businesses recognized the impact of SM in reaching and engaging with travelers.

Impact of Global Events and Disruptions: Various global events and disruptions influenced tourism activities during this period. Economic recessions, natural disasters, terrorist attacks, and the COVID19 pandemic significantly impacted travel patterns, leading to fluctuations in tourist arrivals, changes in travel behavior, and a heightened focus on health and safety measures.

Rise of Wellness and Health Tourism: The importance of health and well-being gained prominence, leading to a rise in wellness and health tourism. Travelers sought destinations and experiences that offered relaxation, rejuvenation, and holistic wellness activities such as spa retreats, yoga and meditation retreats, and wellness-focused resorts.

Technology and Travel Experience: Travelers could access digital guides, navigation tools, and immersive virtual tours to plan and enhance their trips. Digital payment systems and contactless solutions also facilitated smoother transactions.

Shift in Destination Preferences: Travelers' destination preferences evolved over time. Emerging destinations, off-the-beaten-path locations, and cultural immersion gained popularity as travelers sought unique and less crowded experiences. The focus shifted from mainstream tourist hotspots to more authentic and sustainable destinations.

Recovery and Adaptation during COVID 19: Tourism was negatively harmed by the COVID-19 epidemic globally, resulting in travel restrictions, lockdowns, and a decline in international travel. The industry faced unprecedented challenges, but also showcased resilience and adaptability. Health and safety protocols, domestic and local tourism, and digital innovations played key roles in the recovery process.

Overall, from 2010 to 2022, tourism activities experienced transformative changes driven by technology, sustainability, experiential travel, and shifting traveler preferences. The industry faced challenges and disruptions but also embraced opportunities for innovation, digitalization, and a renewed focus on responsible and authentic travel experiences.

5. Conclusion and Scope Of The Study

Based on the provided content, it is concluded that internet community has a significant influence on the tourism and hospitality industry. It entertains a crucial part in various aspects such as customer decision-making, promotional activities, interaction with customers, and information dissemination. Social media platforms, including review sites like Trip Advisor, have a significant influence on tourism demand and supply. The use of social media ICTs by young travelers is also highlighted.

However, there are challenges and limitations associated with SM in the touristy. The accuracy and reliability of information on online review sites can vary, and there are discrepancies in how the hotel industry is depicted on different platforms. The management of social media reviews and customer feedback is crucial for hotel marketing and performance. Additionally, there is a concern for further study on the impressions of social media on business reputation, customer relationships, and financial achievements in the tourism and hospitality sectors.

Overall, social media has transformed the way tourism and hospitality businesses operate and interact with customers. It offers opportunities for marketing, engagement, and brand recognition. However, it also presents challenges in terms of information accuracy, customer satisfaction, and managing online reputation. Future studies should concentrate on overcoming these issues and investigating the potential possibilities in the sector.

6. Discussion and Limitations

Social media has completely changed how customers communicate with travel and hospitality companies. Travelers now rely heavily on online media sites like YouTube and Instagram to find activities, lodgings, and places to visit. These platforms' emphasis on visuals enables companies to directly present their products to prospective clients, influencing their travel choices.A notable consequence of social media usage is the increase of user-generated content (UGC). Travelers use a variety of venues to share their experiences, opinions, and suggestions, which has a big impact on other users' choices. Offering genuine insights into their goods and services, this organic content is a potent marketing tool for companies.Social media sites function as avenues for feedback and consumer support. Travelers frequently utilize online sites like Facebook and Twitter to directly communicate their thoughts, grievances, and praises to businesses. Potential customers' perceptions of an organization's credibility and corporate image can be significantly impacted by how they handle these kinds of interactions.Social media networks offer organizations useful data insights for focused marketing strategies. Businesses can maximize the reach and efficacy of their marketing campaigns by customizing them to target consumer segments based on demographic and behavioral insights.

7. Limitations

https://doi.org/10.52202/074664-0009

Sample Bias: Sample bias is a problem that many research that look at how social media affects travel and hospitality may have. Studies frequently concentrate on users of particular platforms or demographics, which might not be entirely representative of the general public. This may restrict how broadly applicable the results are.

Rapid Technological Changes: New platforms, features, and trends are frequently introduced, resulting in a continually changing social media scene. Because of this, study findings could be out of date very soon, particularly in sectors with high turnover rates like tourism and hospitality.

Measurement Difficulties: It might be difficult to put a number on how social media affects travel and hospitality. The complete impact of SM on consumer behavior and industry trends may not always be fully captured by metrics like engagement, reach, and sentiment analysis.

Ethical Concerns: Privacy, data use, and online manipulation are some ethical issues that social media brings up. These ethical issues should be covered in research on its effects to guarantee the responsible use of SM data in the travel and hospitality sector.

https://doi.org/10.52202/074664-0009

8. References

[1] Zeng, B., Gerritsen, R.: What do we know about social media in tourism? A review. Tourism Management Perspectives. 10, 27-36 (2014).

[2] Munar, A.M.: Paradoxical digital worlds. In Tourism Social Media: Transformations in Identity, Community and Culture (Vol. 18, pp. 35-53). Emerald Group Publishing Limited (2013).

[3] Bizirgianni, I., Dionysopoulou, P.: The influence of tourist trends of youth tourism through social media (SM) & information and communication technologies (ICTs). Procedia-Social and Behavioral Sciences. 73, 652-660 (2013).

[4] Miguéns, J., Baggio, R., Costa, C.: Social media and tourism destinations: TripAdvisor case study. Advances in Tourism Research. 26(28), 16 (2008).

[5] Leung, X.Y., Sun, J., Bai, B.: Bibliometrics of social media research: A co-citation and co-word analysis. International Journal of Hospitality Management. 66, 35-45 (2017).

[6] Xiang, Z., Du, Q., Ma, Y., Fan, W.: A comparative analysis of major online review platforms: Implications for social media analytics in hospitality and tourism. Tourism Management. 58, 51-65 (2017).

[7] Kim, W.G., Li, J.J., Brymer, R.A.: The impact of social media reviews on restaurant performance: The moderating role of excellence certificate. International Journal of Hospitality Management. 55, 41-51 (2016).

[8] Kim, S., Koh, Y., Cha, J., Lee, S.: Effects of social media on firm value for US restaurant companies. International Journal of Hospitality Management. 49, 40-46 (2015).

[9] Tajvidi, R., Karami, A.: The effect of social media on firm performance. Computers in Human Behavior. 115, p.105174 (2021).

[10] Kim, W.G., Lim, H., Brymer, R.A.: The effectiveness of managing social media on hotel performance. International Journal of Hospitality Management. 44, 165-171 (2015).

[11] Floreddu, P.B., Cabiddu, F., Evaristo, R.: Inside your social media ring: How to optimize online corporate reputation. Business Horizons. 57(6), 737-745 (2014).

[12] Hudson, S., Roth, M.S., Madden, T.J., Hudson, R.: The effects of social media on emotions, brand relationship quality, and word of mouth: An empirical study of music festival attendees. Tourism Management. 47, 68-76 (2015).

[13] Yazdanifard, R., Yee, L.T.: Impact of social networking sites on hospitality and tourism industries. Global Journal of Human-Social Science: Economics. 14(8), 1-5 (2014).

[14] Benea, I.A.: Influences of social media on the tourism and hospitality industry. Modul Vienna University, Vienna (2014).

[15] Van der Bank, C.M., van der Bank, M.: The impact of social media: advantages or disadvantages. African Journal of Hospitality, Tourism and Leisure. 4(2), 1-9 (2014).

[16] Aydin, G.: Social media engagement and organic post effectiveness: A roadmap for increasing the effectiveness of social media use in hospitality industry. Journal of Hospitality Marketing & Management. 29(1), 1-21 (2020).

[17] Kim, W.G., Li, J.J., Brymer, R.A.: The impact of social media reviews on restaurant performance: The moderating role of excellence certificate. International Journal of Hospitality Management. 55, 41-51 (2016).

[18] Hudson, S., Thal, K.: The impact of social media on the consumer decision process: Implications for tourism marketing. Journal of Travel & Tourism Marketing. 30(1-2), 156-160 (2013).

[19] Pateli, A., Mylonas, N., Spyrou, A.: Organizational adoption of social media in the hospitality industry: An integrated approach based on DIT and TOE frameworks. Sustainability. 12(17), p.7132 (2020).

[20] Yang, S., Liu, Y., Wu, X.: 1+ 1< 2! Effects of social media engagement and advertising on firm value of tourism and hospitality companies. Journal of Hospitality & Tourism Research. 45(8), 1417-1439 (2021).

[21] Sakas, D.P., Simos, T.E.: A fifth algebraic order trigonometrically-fitted modified Runge-KuttaZonneveld method for the numerical solution of orbital problems. Mathematical and Computer Modelling. 42 (7-8), 903-920 (2005).

[22] Konstantopoulos, N., Sakas, D.P., Triantafyllopoulos, Y.: The strategy of stakeholder briefing during merger negotiation in the bank market. Journal of Management Development. 28 (7), 622-632 (2009).

https://doi.org/10.52202/074664-0009

[23] Konstantopoulos, N., Sakas, D.P., Triantafyllopoulos, Y.: The dimension of communication in the merger: Simulation with dynamic model. AIP Conference Proceedings. 963 (2), 1062-1065 (2007).

[24] Triantafyllopoulos, Y., Konstantopoulos, N., Sakas, D.P.: The role of leadership in high tech manufacturing companies in a changing environment. Key Engineering Materials. 495, 176-180 (2012).

[25] Triantafyllopoulos, Y., Konstantopoulos, N., Sakas, D.P.: The performance management after mergers and acquisitions in high technology manufacturing business. Key Engineering Materials. 495, 171-175 (2012).

[26] Vaxevanou, A.Z., Konstantopoulos, N., Sakas, D.P.: Outsourcing or insourcing in the high technology systems sector in a maritime company. Key Engineering Materials. 495, 163-166 (2012).

[27] Terzi, M.C., Sakas, D.P., Vlachos, D.: Marketing dynamic simulation modelling in high tech laboratories. Key Engineering Materials. 495, 23-27 (2012).

[28] Terzi, M.C., Sakas, D.P., Seimenis, I.: Pricing strategy dynamic simulation modelling within the high-tech sector. Key Engineering Materials. 495, 167-170 (2012).

[29] Markaki, E.N., Sakas, D.P., Chadjipantelis, T.: Selecting the project teams' members. A challenging human resources management process for laboratory research. Key Engineering Materials. 495, 159-162 (2012).

[30] Sakas, D.P., Vlachos, D.S., Simos, T.E.: Adaptive neural networks for automatic negotiation. AIP Conference Proceedings. 963 (2), 1355-1358 (2007).

[31] Sakas, D.P., Vlachos, D.S., Simos, T.E.: Fuzzy constraint-based model for efficient management of dynamic purchasing environments. AIP Conference Proceedings. 963 (2), 1351-1354 (2007).

[32] Sakas, D.P., Vlachos, D.S., Simos, T.E.: Adaptive techniques for online auctions. AIP Conference Proceedings. 963 (2), 1359-1362 (2007).

[33] Sakas, D.P., Konstantopoulos, N., Triantafyllopoulos, Y.: Contribution of the executives in bank sector mergers: Application with a simulation model. AIP Conference Proceedings. 963 (2), 1054-1057 (2007).

[34] Vlachos D.S., Simos T.E.: Partitioned Linear Multistep Method for Long Term Integration of the NBody Problem. Applied Numerical Analysis & Computational Mathematics. 1 (2), 540-546 (2004).

[35] Kosmas O.T., Vlachos D.S.: Simulated annealing for optimal ship routing. Computers & Operations Research. 39 (3), 576-581 (2012).

[36] WYSE Travel Confederation: Climate Change & Youth Travel, a Youth Travel Industry Guide. Retrieved from http://www.wysetc.org/resource/resmgr/research_reports/climate_change_industry_guide.pdf (2007).

[37] ELTRUN: Tell us your opinion on Facebook. Retrieved from http://www.epistimonikomarketing.gr/article_show.php?article_id=3113 (2009).

[38] World Tourism Organisation (WTO): Youth Travel Matters-Understanding the global phenomenon of youth travel. Retrieved from http://www.siimt.com/work/sites/siimt/resources/LocalContent/1172/3/YouthTravel_protec.pdf (2008).

[39] Boyd, D. M., Ellison, N.B.: Social Network Sites: Definition, History & Scholarship. Retrieved from http://onlinelibrary.wiley.com/doi/10.1111/j.1083-6101.2007.00393.x/full#b73 (2007).

[40] Westbrook, R.A.: Product/consumption-based affective responses and post-purchase processes. Journal of Marketing Research. 24 (3), 258-270 (1987).

[41] Litvin, S.W., Goldsmith, R.E., Pan, B.: Electronic word-of-mouth in hospitality and tourism management. Tourism Management. 29(3), 458-468 (2008).

[42] Yoo, K.H., Gretzel, U.: Influence of personality on travel-related consumer-generated media creation. Retrieved from http://www.sciencedirect.com/science/article/pii/S0747563210001226#b0450 (2011).

[43] Line T., Jain, J., Lyons, G.: The role of ICTs in everyday mobile lives. Retrieved from http://www.sciencedirect.com/science/article/pii/S096669231000102X#bib6 (2010).

[44] Xiang, Z., Gretzel, U.: Role of social media in online travel information search. Tourism Management. 31(2), 179-188 (2009).

[45] Qualman, E.: Socialnomics: How Social Media Transforms the Way We Live and Do Business. Hoboken, New Jersey: John Wiley & Sons, Inc. (2009).

[46] Kaplan, A.M., Haenlein, M.: Users of the world, unite! The challenges and opportunities of Social Media. Retrieved from http://www.scribd.com/doc/63799736/Kaplan-andHaenlein-2010-Social-Media (2009).

[47] Rerres, K., Kafeza, E.: Tourism in digital economy. Research Institute for Tourism, Athens GR. (2009).

[48] Richards, G., Wilson, J.: New Horizons in Independent Youth and Student Travel–Today's Youth Travellers: Tomorrow's Global Nomads. ISTC-ATLAS. Retrieved from http://www.openingdoorsopeningminds.org/d/FINAL_Industry_Report.pdf (2003).

[49] Carlson, N.: At Last – The Full Story of How Facebook Was Founded. Retrieved from http://www.businessinsider.com/how-facebook-was-founded-2010-3?op=1 (2010).

[50] Ellemers, N., Spears, R., Doosje, B.: Social Identity: Context, Commitment, Content. Oxford: Blackwell (1999).

 https://doi.org/10.52202/074664-0009

The Role of Automated Controls and Streamlined Compliance in Managing Risks in Digital Finance.

SHIBAHATHULLA TK[1] and Prof. (Dr) MOHAMMED ASHRAF ALI[2]
[1.] Research scholar, Research and PG Department of Commerce Aligarh Muslim University,
[2.] Professor, Research and PG Department of Commerce Aligarh Muslim University

Abstract
Digital finance leverages emerging technologies to provide financial services through digital channels. While expanding access and efficiency, digital finance also introduces risks related to cyber threats, market volatility, unfair outcomes, and compliance burdens. This paper examines two complementary approaches to governing these risks - automated controls embedded in business processes combined with streamlined, collaborative compliance programs prioritizing key objectives. Intelligently automated controls like behaviour analytics and transaction guardrails help embed oversight and monitoring within real-time digital finance activities. Making controls an automated compliance feature enhances transparency, efficiency and protection at scale. Meanwhile, streamlined frameworks enabled by regulatory technology focus on high-risk priorities like financial integrity and consumer fairness while minimizing complexity. Employed together, automated controls and optimized compliance offer a proactive governance model attuned to digital finance's rapid pace of change and disruption across banking, investing and insurance. Constructing core technological capabilities and oversight protocols ex-ante also improves system-wide preparedness when risks inevitably emerge. Though challenges around hidden biases and new threats exist, thoughtfully embedding controls and compliance within digital finance's DNA carries promise. Automated, transparent systems can promote accountability and accessibility while unlocking immense productivity and inclusion gains when governance evolves in parallel with innovation.

Keywords: Digital Finance, Trends Driving Digitization of Finance, Emerging Risk Landscape in Digital Finance, Role of Automated Controls and streamlined Compliance as Complements

Introduction

The financial industry has experienced rapid digitization over the past decade through exponential growth in mobile banking, marketplace lending, and advising capabilities[50]. While promising greater accessibility, efficiency and customization, digital finance also raises risks related to cyber threats, market volatility from electronification, and unfair outcomes driven by biased algorithms. Balancing innovation opportunities with emerging vulnerabilities remains an overriding priority for private and public sector oversight[18].Digital finance refers to the use of new technologies like cloud computing, big data, artificial intelligence, blockchain, and advanced analytics to provide financial services through digital channels. It enables innovative financial models like marketplace lending, enhanced credit scoring, and streamlined compliance that can expand access to financing and financial inclusion [53]. However, digital finance also introduces new risks around adoption lags, infrastructure gaps, and potential payment disruption that require mitigating. In short, digital finance leverages emerging technologies to facilitate financial services through digital means, promising benefits but also requiring effective governance[92].Digital risk refers to the use of digital technologies to enhance risk management effectiveness and efficiency. This includes leveraging process and decision automation, digitized tracking and early alert systems, workflow optimization, optical character recognition, advanced analytics like machine learning and AI, and applying robotics to procedures and interfaces[89]. Essentially, digital risk management entails systematically aligning processes, data, analytics, IT systems, organizational structures, talent, and culture to take advantage of the capabilities of digital technologies. The goal is to improve risk operations through increased automation, better data, more sophisticated analytics, and an organizational commitment to optimizing risk through digital means[38].Digital currencies like electronic money serve as alternative online payment options. The growth of e-commerce and financial crises in places like Cyprus and Greece have increased demand for digital payment alternatives as investors seek out new mechanisms. A key goal in developing digital currencies has been boosting e-commerce productivity by cutting transaction times and costs[54]. Two main types of digital currencies have emerged: cryptocurrencies and non-cryptocurrencies. Cryptocurrencies are decentralized, peer-to-peer digital monies that rely on cryptography for security, like Bitcoin, Litecoin, Zerocoin and Peercoin. Non-cryptocurrencies represent a promise to pay a set amount of precious metal, including e-cash, e-gold and e-bullion. While the lack of mainstream acceptance has hindered the development of non-cryptocurrencies, some network-based systems like Digicash, CyberCoin, and Netcash have served as currency proxies[37].Digital investment platforms form ecosystems with high compliance risks that threaten sustainability. The 27% risk of investors losing expected returns shows the need to embed compliance controls. Conducting investment compliance procedures as assurance engagements within these ecosystems can enhance reliability, trust, and risk mitigation. This supports the methodological development of compliance and stronger regulation of digital investment activities[39].Digital finance offers inclusive worth by supplying companies and individuals with sustainable, affordable financial services [107]. However, it can also restrict access to these services and alter traditional economic qualities. Though the financially disadvantaged may resent the wealthy, digital finance can also benefit them, for instance, by decreasing information imbalance. It can enhance capital financing efficiency, identify and manage risks better, reduce transaction expenses, and lessen municipal debt[106]. This paper specifically examines two interrelated approaches to governing risks in digital finance - automated monitoring and controls embedded in business processes combined with streamlined and collaborative compliance programs focused on principal risk objectives. Automated controls calibrated to transactions, portfolios and customer analytics provide macroprudential supervision[90]. Meanwhile, streamlined frameworks enhance regulatory coordination on priorities like financial integrity, consumer fairness and business resilience while minimizing unnecessary complexity burdens. Employed in a comprehensive and complementary fashion, embedded controls and refined compliance offer a proactive governance model for the rapid pace of technology change across banking, investing and insurance[84]. Constructing core technological capabilities and oversight protocols ex-ante also improves system-wide preparedness when risks inevitably materialize from new innovations or volatile operating environments. The following analysis offers concepts and evidence for policymakers and industry leaders working to align promise and prudence amid the digital finance transformation[44].

Objective

1. Assess the effectiveness of automated controls in detecting and preventing fraud in digital finance activities.
2. Evaluate whether streamlined, risk-based compliance frameworks reduce compliance costs for financial institutions compared to rules-based approaches.
3. Identify best practices for integrating automated controls and streamlined compliance to balance innovation and oversight in digital finance.
4. Analyze the role of regulatory technology (RegTech) in supporting streamlined compliance and scalable oversight in digital finance ecosystems.
5. To study the Role of Automated Controls and Streamlined Compliance in Managing Risks in Digital Finance.

LITERATURE REVIEW

DIGITAL FINANCE

https://doi.org/10.52202/074664-0010

As financial services digitize, new technologies like cloud computing, blockchain, and artificial intelligence are enabling more efficient yet complex operations. These innovations promise benefits but also pose risks related to cybersecurity, money laundering, privacy breaches, and algorithmic biases[51].Digital finance inherently facilitates solving information asymmetry by identifying and incorporating industry and individual conduct as market components, enhancing financial institutions' data-gathering capacity. Though the government regulates the economy, it must adhere to financial market regulations [109]. As digital financial technologies progress, municipal financial data like fixed assets, liabilities and capital flows become more transparent. Enhancing financial transparency importantly reflects government credibility, impacting the effectiveness of governmental bond ratings and local government bond financing[82]. Financial systems in the age of big data should use data to gain business insights while balancing efficiency and control to support sustainable, profitable growth. Finance should research and utilize data to provide company-wide business insights and ensure growth via enhanced business models, resource distribution, and structure. Finance's ultimate purpose is to improve efficiency to drive company expansion. Large firms need automation, IT and digital tech for smooth operations. Accounting should actively pilot automation and AI, delegating standard accounting to machines. This summarizes how Huawei Finance is becoming digitalized using quality data [104]. Digital finance refers to the influence of emerging technologies on financial services, encompassing an array of products, applications, processes, and business models that are transforming traditional banking. While financial innovation isn't new, investment in new technologies has substantially grown recently, quickening innovation [35]. We now use mobile tech for payments, transfers, and investments via tools that were non-existent years ago. Artificial intelligence, machine learning, distributed ledgers, cloud computing and big data analytics have enabled established institutions and new entrants to offer novel services and models, benefiting consumers and companies through greater access, choice and operational efficiency[2]. Digital financial servicesthat rely on digital technologies for delivery and consumer use. Fintech refers to digital technologies with the potential to transform financial services provision by developing new or modifying existing business models, applications, processes and products. In practice, "fintech" broadly denotes the ongoing wave of new digital financial services like web, mobile, cloud, machine learning, digital ID and APIs [78]. Fintech firms are new financial sector entrants specializing in digital financial offerings such as digital payments, insurance, banks, and peer lending. Big techs are large tech companies like search engines, social media, and e-commerce platforms that now also provide digital financial services [15]. Digital finance refers to financial services like lending, investing, and bill payment that is delivered through digital channels such as mobile phones, mobile wallets, personal computers, the internet, or debit and credit cards linked to a reliable digital payment system. There are two broad categories of literature on these technology-enabled financial services [29].Digital technology has enabled the rise of digital finance, a new set of financial service delivery methods for activities like lending, investing, and bill payment. There are two broad categories of literature on digital finance [48]. Digital finance has been a source of global concern for many years. There are diverse definitions of digital finance from both academics and industry practitioners. Some governments and monetary authorities also define digital finance differently, including how payment data is classified, with some still integrating it with electronic transactions [19]. Altogether, there is a lack of consensus on what constitutes digital finance, whether it includes activities like lending, investing and bill payment delivered through channels like mobile devices, the internet, and card networks linked to digital payment systems [77]. As a result, the literature on these technology-enabled financial services can be categorized into two broad perspectives, even as the scope of what is considered "digital finance" continues to evolve.

Trends Driving Digitization of Finance

Digitalization impacts finance, noting its role in jobs, economic growth, and financial inclusion. Rapid digitalization is transforming financial services provision and employment. The changing landscape raises issues around work conditions, business models, regulation, and managing the effects of digitalization on workers [55]. The pandemic demonstrated finance's support role and drove digital adoption. Constituents should constructively dialogue to find agreeable solutions focused on skills, conditions, health and safety, work-life balance, privacy, and data protection. In addition, the digitisation of finance promoted financial inclusion globally, which is a vital driver of economic advancement. It was accelerating the pace, safety, and transparency of transactions while enabling the creation of sustainable financial products customized to people's needs [27]. Technology eliminated obstacles to offering financial services, such as lack of identification or formal income and geographical remoteness [47]. Digitization was also imperative for businesses on a macro scale as a foundation for enriched productivity and seizing new market prospects.Several technological and competitive factors underpin the accelerating digitization of financial services over the past 10-15 years[32].

Rising consumer expectations for digital services:

Customers increasingly expect to be able to manage their finances through digital channels like mobile apps and websites. Banks and other financial service providers are digitizing to meet these expectations [11].The rise of digital technology has fundamentally impacted how consumers behave and make purchasing decisions. The digital age has completely transformed buyers by shifting them away from traditional approaches and towards digital platforms and devices [28]. Today's digitally-savvy consumers have moved beyond merely seeing TV ads to consuming marketing messages across their smartphones, tablets, laptops, and various social media channels. Marketers must identify, reach, and engage these new species of nimble, actively connected customers across multiple devices and touchpoints in this new digital ecosystem [95]. These digitally empowered consumers present marketers with both significant challenges and incredible rewards compared to more traditional shoppers. Modern customers have indispensable expectations around convenience, availability, and value when they shop. They want their product and service needs met on their own time, through their channel of choice, and at a reasonable price [101]. This new but fundamental consumer expectation is a key reason why digital channels are being favoured for purchases [56]. The digital experience has armed customers with quick access to extensive information for brand comparisons, the ability to transact through diverse gadgets, and opportunities to share post-purchase feedback and commentary. As technology progresses, consumer shopping habits and behaviours evolve, and companies must continue adapting their strategies to remain relevant[102].

Moving to the cloud

Cloud computing allows financial institutions to become more agile, launch digital offerings faster, scale better, and reduce reliance on legacy systems [14].Cloud computingis transforming many sectors, including finance. Cloud banking utilizes internet-based, pay-as-use services to expand real-time computing capabilities. This aims to optimize supply chains for greater efficiency and lower costs. While cloud computing began in the 1950s, it developed in the 1990s with the internet and advanced in 2006 when Amazon enabled small businesses to rent computing power [25]. Since then, technology has quickly changed banking. Mobile banking, digital payments, and new technologies are shaping the future. The widespread adoption of digital wallets like Apple Pay reinforces the disruptive technological revolution that is challenging traditional banking [99]. Cloud computing stores data on virtual rather than physical drives, enabling easier sharing and access. This reduces the need for physical infrastructure, with average cost savings of 23% in recent studies. A survey found that 60% of banks now allocate 20-39% of their budget to cloud applications for associated cost savings and flexibility from usage-based billing [60].Cloud banking allows firms to respond quickly to shifting consumer needs and technology changes. Speed and agility will be vital competitive strengths from now on. While steady progress continues, traditional banks may struggle to maintain dominance as demographics and behaviors evolve alongside technology. Security fears exist, but examples show that cloud migration can increase stability, efficiency, and risk reduction [70]. Still, challenges like data encryption and vendor lock-in must be addressed. Flexible databases and integrated extensive data systems are also needed.

 https://doi.org/10.52202/074664-0010

Widespread cloud computing adoption could significantly boost economic growth and job creation by enabling greater innovation and productivity[9]. It could also expand offerings like cash management and business analytics where transaction banking currently falls short. While cloud computing progresses steadily, traditional banks may struggle to maintain dominance as demographics and behaviors shift alongside technology. Security fears exist, but examples show that cloud migration can increase stability, efficiency, and risk reduction. Still, obstacles like data encryption and vendor lock-in must be addressed [73]. Flexible databases and integrated extensive data systems are also needed. If adopted at scale, cloud computing could boost GDP and create millions of jobs in Europe by enhancing innovation and productivity. It could also expand offerings like cash management, payments, and business analytics where transaction banking has fallen short[71].

Use of data andanalytics

Vast amounts of customer data combined with advanced analytics, AI, and machine learning allow hyper-personalization of offerings and improved risk modelling. Targeted cross-selling through digital channels is increasing [3]. Large customer data sets and advanced analytics help financial institutions achieve greater customer understanding, enhance risk management, and tailor products. Digital channels also supply additional data for analysis[74]. How big data has transformed every industry, including business, as well as many institutions and individuals. It has changed how people think about data analysis and decision-making. With the enormous volumes of data generated daily, the ability to effectively manage, store, examine, and interpret data has become critical. Big data offers advantages like enhanced decision-making, increased efficiency, and new opportunities for companies [32]. As a result, businesses now have superior insight into customer behavior, market trends, and operational inefficiencies. Companies can leverage big data analytics to improve decision-making processes, optimize operations, and develop new products and services [26]. Big data has revolutionized how organizations capture, analyze and act on data. This has led to data-driven decision-making, greater operational efficiency and innovative new offerings. Big data has had a profound and transformational impact across industries, including altering mindsets around data and analytics [65]. While definitions vary, an excessively narrow vision of big data is impractical and only useful on the cutting edge of business analytics. Past claims that big data requires data exceeding traditional tools' processing abilities are outdated, given advancing capabilities in storage, processing, structuring, and analysis that make big data feasibility steadily resolvable [58]. The statistics analysis community provides a more useful conceptualization: Big data describes the huge volumes of structured and unstructured data captured and processed daily[24]. Big data analytics becomes practicable largely following computing advances under Moore's Law, tracking gains in bandwidth, software, storage, and processing. Propelling big data analytics is decreasing data storage costs, increasing processing capabilities, demand for big data analysis, and the usefulness of predictions made[5].

Competition from fintech disruptors

The rise of financial technology companies offering innovative digital banking, lending, insurance and other services has forced incumbent players to accelerate their digitization efforts to compete. Start-ups are disrupting parts of financial services with innovative business models and digital-first, mobile experiences [12]. Incumbents must accelerate digitization to compete [22].FinTech should be conceptualized more broadly in two key ways than current definitions allow. Embracing this expanded conceptualization is likely a prerequisite to recognizing that big data analytics provides essential insight into FinTech. First, advancements in information, communication, and computing technologies have underpinned most financial market innovations for centuries. Second, financial innovation built on technological progress is far from new; rather, it has driven financial market evolution for much of history. FinTech innovation has accelerated enormously over the past three decades with the advent of financial engineering, a mathematically-based discipline largely focused on arbitrage, derivatives and innovations in financial instruments, tools and strategies [58]. Today, FinTech takes many forms with impact across financial sectors, reaching well beyond financial engineering. A FinTech taxonomy is proposed to enable rapid regulatory response when FinTech fails, a readiness approach allowing swift action by regulators and victimized private parties. Key historical examples of both successful and failed FinTech regulation are reviewed, highlighting the current opportunity to transform vast historical financial databases into FinTech big data [42].FinTechhas increased competitive pressure on incumbents by directly targeting consumers. The battle to "own the customer" will be hard-fought since value chain links closest to end users typically earn the highest margins. Some fintechhasskilfully captured high-margin incumbent offerings like international transfers, diminishing the profitability of incumbents' remaining product mix. However, many fintech struggle to scale [34]. Re-bundling helps some digital banks broaden their scope and compete at scale, albeit with thinner margins. Disruption may be limited where fintech relies on incumbents for operations [77]. Many still leverage banks to store funds and process payments, with user transactions originating from bank accounts. Services like Wise also utilize bank rails for cash-in/out and compliance. In some markets, fintech has built internal infrastructure, replacing incumbents for such functions[54]. New innovations like stable coins and CBDCs may further reduce fintech dependence on legacy institutions, especially for payments. Fintech disruption may be limited in areas where fintechdepends on incumbent institutions for key operations. As mentioned, many FinTechs continue to leverage banks to store funds and make payments. Fintech transactions often originate from users' existing bank accounts[27].

Mobile Banking Adoptionand Open Banking

With increasing smartphone penetration, customers expect to be able to bank, invest, apply for loans, and manage finances digitally via mobile apps. Financial institutions are prioritizing mobile delivery. Policy changes like open banking regulations are promoting open API ecosystems that enable easier exchange of financial data between institutions and third-party developers. This spurs greater connectivity and financial service innovation [67]. Open banking represents the first practical application of the combined concepts of open data and open finance [6].It promised to promote innovative data-driven financial services, increase consumer choice, reduce costs, and stimulate competition [81].Unlike traditional banking, which operates on closed, siloed systems controlled by individual financial institutions, open banking is premised on open and interconnected systems. In traditional banking, consumer financial data is effectively locked within individual providers and is unable to be easily shared. In contrast, open banking's open architecture based on APIs enables consumer financial data to be securely shared between financial institutions with consumer consent. This shift from closed to open systems marks a major difference between status quo banking models and the emerging open banking paradigm [69].Mobile banking began being adopted in 1999. Today, mobile banking has achieved widespread usage, with many consumers leveraging it to make transactions, send or receive payments, and purchase products. Mobile banking can be defined as a type of mobile commerce application which allows financial institutions to enable customers to perform banking activities through mobile devices [76]. Mobile banking allows financial transactions to be conducted on mobile devices, a service offered by financial institutions, particularly banks. It enables clients to perform various activities depending on the institution, which is made easier nowadays through mobile apps. Users can check balances, view statements online, transfer funds, and make prepaid purchases directly on their phones. Mobile banking facilitates managing finances digitally without visiting physical branches. The growing adoption of mobile applications has allowed easier access and usage across banking and financial sectors. Overall, mobile banking enables the transition to digital financial services, transforming how institutions and customers manage money [10]. Mobile banking involves using a mobile device to conduct financial transactions. This service is offered by financial institutions, particularly banks. There are several types of mobile banking services: Account information access - clients can view balances, statements, transaction history, etc.Transactions - make payments, transfers, deposits, withdrawals, etc [92]. Investments - manage investment accounts, trade stocks and funds. Support services - contact the bank, report issues, chat with representatives. Content

 https://doi.org/10.52202/074664-0010

and news - institutions provide tips, notifications, promotions and other updates. Many banks utilize SMS text messages and mobile applications to regularly update clients on account activity, send fraud alerts, and notify them of service maintenance, among other matters. This keeps customers informed about their accounts and potential security issues [84].

Emerging Risk Landscape in Digital Finance

Risk management allows companies to thoroughly evaluate all relevant business and regulatory risks and control and track actions in an organized way. With the recent rise in regulatory requirements and increasingly vocal shareholders, many companies have become more aware of the need to identify and manage risk areas across their operations, including operational, financial, information technology, brand, and other risks [100]. In addition, risks are no longer viewed as only the responsibility of executive specialists; company boards want transparency into risk exposures so they can provide effective long-term strategic leadership. As a result, companies are taking a systematic approach to pinpointing, prioritizing, quantifying, and responding to all kinds of business risks, with the goal of managing any associated exposures [75]. Digital risk refers to any digital capabilities that enhance risk management effectiveness and efficiency, particularly through process and decision automation, as well as digitized tracking and early warning systems [90]. The digital risk approach utilizes workflow automation, optical character recognition, advanced analytics like machine learning and AI, new data sources, and robotic process automation to streamline processes and interfaces. At its core, digital risk means strategically adapting processes, data, analytics, IT infrastructure, overall organizational design, talent, and culture [97]. Essentially, it is a coordinated overhaul of all these elements to take advantage of automation, analytics, and other digital technologies to transform risk management.Myriad uncertainties for both service providers and authorities accompany the digital transition in finance across cyber, market, compliance and liability domains[31].

Cyber Risks

As finance becomes more digitized, companies are relying more on online platforms, digital payment systems, interconnected systems, and extensive data. This digital transformation exposes the financial system to heightened cybersecurity threats and risks [17]. Some key cyber risks include:- Data breaches that expose highly sensitive customer financial information and transactions- Cyber attacks that disrupt key financial infrastructure like payment systems, stock exchanges, and other market platforms- Fraud through techniques like phishing attacks, ransomware, and identity theft targeting customers and financial institutions- Attacks on core banking systems and back-end infrastructure that could cripple critical finance functions- Use of cryptocurrencies and hidden online marketplaces to hide illicit financial activity[101]. These threats underline the crucial need for robust cyber resilience and defence in the finance industry as it adopts more innovative digitized services. Financial institutions need to assess cyber risks as part of their overall risk management strategy. This includes investments in security infrastructure, strict cyber protocols, emergency response plans, regular testing, updates on new threats, and staff training. Maintaining trust and stability in global digital finance relies heavily on the ability to create a safe digital environment resilient to cyber-attacks [88].Financial services are highly exposed to cyber threats today owing to reliance on digital innovation. As financial firms digitize critical operations via cloud computing, interconnected systems, and external partnerships, vulnerabilities increase. Successful cyber-attacks could severely disrupt banking, compromise customer data, undermine financial stability and trigger economic impacts.Thus, robust cyber risk management is crucial for the resilience of the financial sector [1].Cyber risks jeopardize the confidentiality, availability, and integrity of information systems and data. By exploiting vulnerabilities in critical infrastructure and assets, cyber threats can severely impact operations, compromise sensitive information, and endanger organizations. Building cyber resilience through robust defences and safeguarding critical systems is key to securely navigating intensifying cyber threats. Cyber threats are growing more advanced and costlier, necessitating greater clarity around the potential impact to inform coherent, adaptable, global cybersecurity capacity building

[4].Cyber incidents incur higher costs for larger, interconnected firms, but the financial sector sees lower average costs due to greater IT security investment. While cloud dependence reduces small-scale cyber costs, systemic cloud provider risks likely raise tail risks. As digital adoption spreads globally, cyber threats can emerge from anywhere worldwide. It is inaccurate to presume attacks originate from a few restricted nations subject to control. With no true geographic boundaries in cyberspace, malicious actors have limitless space to instigate cyberattacks unconstrained by physical location. This global interconnectivity enables threats to strike from practically any origin, making the whole digitally connected planet a potential launch point for cyberattacks [36].

Market Risks

Market risk refers to the risk of losses in financial markets due to movements in market factors like asset prices, interest rates, foreign exchange rates, and commodities prices[4]. Here are some of the key roles market risk plays in digital finance: Digital assets price volatility - Cryptocurrencies and other digital financial assets tend to demonstrate significant price volatility driven by factors like hype cycles, regulations, security issues, etc.[86]. This introduces major market risk. Algorithmic/high-frequency trading (HFT) risks - HFT relies on complex algorithms and amplified speed/turnover. Coding errors or flash crashes can trigger massive, unexpected losses. Interest rate risks in lending apps/platforms - Many fintech lenders rely on short-term variable rate funding exposed to interest rate moves. Rate hikes can create financial viability issues[23]. Foreign exchange fluctuations risks in cross border transactions - Digital payment services enabling instant global peer-to-peer transfers lead to forex risks from exchange rate volatility. Managing these market risks requires appropriate governance, risk capabilities,and buffers tailored to the digital dynamics, as traditional methods may not suffice [16]. This is an evolving challenge as digital finance innovations continue to outpace regulations. Electronification leads to automated, accelerated volatility across trading, lending and personal wealth platforms[36].

Compliance Risks

As digital finance innovations aim for incredible speed and efficiency, they can also increase compliance risks related to consumer protection, data privacy, financial crimes, and regulatory requirements[68]. Fintech lenders, digital wallet apps, robo-advisors, and cryptocurrencies operate in a complex, fragmented regulatory landscape with heightened risks of non-compliance [62]. Keeping up with vast volumes of consumer data and surveillance for illegal transactions is challenging. Non-compliance can result in litigation, regulatory punishments, and penalties, undermining consumer trust[63]. Managing compliance proactively by design through governance frameworks, automated systems, and oversight controls helps mitigate these risks in an era of accelerated digital finance growth and disruption.The rising complexity of supervising real-time transactions, algorithmic decisions, and cross-border data flows[79].

Liability Risk

The shift towards self-directed and automated digital financial services is amplifying liability risks associated with consumer losses from transaction errors, investment losses from poor recommendations, or privacy breaches due to inadequate cybersecurity[89]. Technology failures exposing consumer data or disrupting core banking operations also pose expanded liability threats for digital finance[7]. With blurred lines on accountability, consumers can pursue litigation against various entities like financial institutions, fintechs, data aggregators, analytics providers, and cloud computing platforms that power digital finance[104]. Managing these interconnected and hard-to-define liability risks thus becomes crucial through upfront contracts, adequate disclosure, class action waivers, creating reserves, securing appropriate insurance, and utilizing technologies like blockchain for transparency[20].

Automated Controls & Streamlined Compliance as Complements

As digital finance innovations intensify speed, complexity, and interconnected ecosystems, manual compliance processes struggle with regulation, oversight, fraud monitoring, and reporting[109].

https://doi.org/10.52202/074664-0010

Streamlined compliance enabled by automated controls offers risk-based prioritization through behavior analytics, predictive models and in-process red flagging. By embedding compliant-by-design practices within digital finance building blocks, smart contracts, and straight-through processing, automated controls scale protection across firms and consumers while minimizing friction[82]. Testing metrics also strengthen governance and strategic decision-making on regulatory technology. Though digitized controls pose their own risks from coding errors or bias, thoughtful automation and compliance integration create transparent, accountable systems attuned to both innovation and protection across entities and changing regulations in increasingly dematerialized, decentralized finance [47].

Automated Controls

As digital finance introduces greater speed, complexity, data volumes, and interconnectedness, manual compliance processes struggle to efficiently track transactions, identify suspicious activities, and meet reporting obligations[106]. Intelligently automated controls are becoming integral to managing risks by embedding guardrails within digitized banking, insurance, trading, and payment activities[42].Controls like transaction pattern analysis, geofencing, biometrics, cybersecurity safeguards, and account restriction rules help pre-empt fraud and defaults within real-time systems at scale. Though digitized controls carry coding and bias risks, their risk-calibration, testing and oversight, when thoughtfully implemented,enhance transparency, efficiency and oversight across digital finance processes and changing regulations [61], Deutsche Bank's "Operations 4.0" exemplifies the digital transformation of financial services through artificial intelligence, robotics, the Internet of Things, and cloud computing [21]. By automating inefficient manual back-office functions like trade settlement, this smart automation promises immense productivity gains but also risks eliminating thousands of repetitive clerical roles. As the industry digitizes back-end processes, the resultant workforce and operational shifts constitute a revolutionary fourth industrial revolution for finance97].Digitizing workflows is essential for deploying automation or artificial intelligence. Robotic Process Automation uses software "bots" to mimic human tasks like filling forms, allowing machines to complete manual processes. While robots and automation have long been integrated into manufacturing, evolving RPA capabilities and lower costs now enable similar productivity improvements in services [95]. Given rising labour costs but continually decreasing robot prices, the cost-benefit tilt toward automation adoption appears inexorable. The increasing digitization and automation of operations opens new possibilities for more intelligent decision-making. As self-driving cars use sensors and advanced algorithms to steer autonomously, operations may become more autonomous and smarter[43]. By integrating artificial intelligence-powered cognitive abilities into robotic process automation software, we get cognitive RPA bots that are much more intelligent than basic RPA. While regular RPA automates repetitive, rules-based computer tasks done by people, cognitive RPA can automate perceptual and judgment-based jobs needing abilities like language processing, machine learning, and speech recognition[6].

Streamlined Compliance

As digital finance continues to grow, ensuring compliance with regulations is crucial yet challenging due to the speed of transactions [80]. Streamlined compliance utilizes artificial intelligence and blockchain to seamlessly integrate regulatory requirements into digital financial services. By digitizing compliance protocols and using tools like regulatory technology (RegTech) and smart contracts, financial processes can remain fast, innovative and legally compliant[66]. Streamlined compliance enables the benefits of digital finance, including improved financial access and efficiency, while still providing necessary consumer protections. In an era of digital disruption in finance, a compliant yet streamlined approach will allow digital innovation in finance to flourish responsibly [73]. Compliance means organizations implement controls and procedures to satisfy governmental, regulatory, industrial, and internal policy demands. Since attaining compliance is not a one-time event, organizations must establish repeatable compliance processes to adhere to regulations that are affordable over time [44]. Making compliance repeatable allows organizations to sustain compliance at lower costs instead of expensively re-establishing initial levels each time rules change. By embedding efficient compliance processes into operations, businesses can cost-effectively maintain compliant status continuously [98].The Compliance Framework rests on three pillars: inform staff on obligations and legislative changes affecting their business unit activities; compel annual compliance declarations from top administration, including Vice Chancellors; and build assurance through audits and reviews for continuous improvement. The Framework receives regular monitoring and evaluation for efficacy. Like auditing, validating business process compliance may require repetitive tests against requirements, especially for IT operations impacting operational, managerial and supporting processes [112]. For example, financial reporting standards could mandate the building and maintenance of secure networks, which PCI compliance also requires. Firms must provide regular compliance updates, shifting from periodic to continuous verification. Researchers propose automating compliance management through IT infrastructure that handles data collection, control implementation, and analytics[112]. Third-party software solutions now offer ready-to-use compliance automation, but hard-coded measures make changes costly. Further, hardcoded compliance rules risk incompleteness and inaccuracy [64]. Moreover, risk control systems themselves introduce risks by enabling violations. Although research often isolates risk management, governance, and compliance, some frameworks, such as OCEG's GRC Capability Model, integrate them, as governing bodies now promote them. The optimal approach combines the advantages of practical and theoretical methodologies while minimizing limitations[13].

Research Methodology

This paper focused on The Role of Automated Controls and Streamlined Compliance in Managing Risks in Digital Finance. The data for this study was obtained mainly from secondary sources, such as world-famous journals, reports, executive opinions, and literaturereviews. This paper adopted a qualitative research methodology based on a thorough review and analysis of existing literature on digital finance, automated controls, and streamlined compliance. The data was obtained primarily from secondary sources, including reputable academic journals, industry reports, publications by international organizations, and expert opinions. An extensive literature review was conducted to gather relevant information on the key topics, including the definition and scope of digital finance, emerging technologies driving its growth, the associated risk landscape, and the roles of automated controls and streamlined compliance frameworks in governing these risks. Published materials were carefully selected based on their credibility, relevance, and recentness. The gathered data was then synthesized and critically analyzed to identify patterns, trends, best practices, and insights about the research objectives. The analysis involved examining the existing knowledge, contrasting different viewpoints, and drawing logical conclusions supported by evidence from the literature. Particular emphasis was placed on evaluating the effectiveness of automated controls and streamlined compliance as complementary approaches to managing risks in the rapidly evolving digital finance ecosystem. The research methodology aimed to provide a comprehensive understanding of the subject matter by consolidating and interpreting the available knowledge from multiple authoritative sources. The findings and conclusions were formulated based on the critical analysis of the reviewed literature.

Findings and Conclusion

This paper argues that automated controls and streamlined compliance offer complementary oversight models appropriate for the increasing digitization of finance. Embedding monitoring directly within processes allows dynamic detection of risks as innovations develop and deploy. Meanwhile, refined regulatory principles concentrated on key risk priorities can manage compliance burdens amid growing complexity. Digital finance leverages emerging technologies to provide financial services through digital channels, transforming traditional models. While promising expanded access, efficiency, and customization, digital finance also introduces new vulnerabilities that require effective governance. Two complementary approaches are proposed: automated controls embedded within digitized finance activities and streamlined, collaborative

 https://doi.org/10.52202/074664-0010

compliance frameworks focused on key risk priorities. Intelligently automated controls like behaviour analytics, predictive models, and transaction guardrails help embed compliance, oversight and fraud monitoring within real-time digital finance systems. Making controls an automated feature of digitized workflows enhances transparency, efficiency and protection at scale. Streamlined compliance then utilizes regulatory technology and smart contracts to seamlessly integrate oversight with innovation, balancing speed and safety. Adopted in a comprehensive, complementary fashion, automated controls and optimized compliance offer a proactive governance model attuned to digital finance's rapid pace of change and disruption. By constructing core technological capabilities and regulatory protocols ex-ante, the financial system also improves preparedness when inevitable risks emerge from innovations or volatile environments. While challenges around patching coding gaps, hidden biases, and new threats persist, thoughtfully embedding controls and compliance within digital finance's DNA carries immense promise. Automated, hyperconnected systems can become accountable, transparent and accessible while unlocking immense productivity and inclusion gains when governance evolves in parallel. As finance progresses towards dematerialized, decentralized operations, solutions grounded in protection, ethics, and shared prosperity will help secure a responsible digital future. This paper adopted a qualitative research methodology based on a thorough review and analysis of existing literature on digital finance, automated controls, and streamlined compliance. The data was obtained primarily from secondary sources, including reputable academic journals, industry reports, publications by international organizations, and expert opinions. An extensive literature review was conducted to gather relevant information on the key topics, including the definition and scope of digital finance, emerging technologies driving its growth, the associated risk landscape, and the roles of automated controls and streamlined compliance frameworks in governing these risks. Published materials were carefully selected based on their credibility, relevance, and recentness. The gathered data was then synthesized and critically analyzed to identify patterns, trends, best practices, and insights about the research objectives. The analysis involved examining the existing knowledge, contrasting different viewpoints, and drawing logical conclusions supported by evidence from the literature. Particular emphasis was placed on evaluating the effectiveness of automated controls and streamlined compliance as complementary approaches to managing risks in the rapidly evolving digital finance ecosystem. The research methodology aimed to provide a comprehensive understanding of the subject matter by consolidating and interpreting the available knowledge from multiple authoritative sources. The findings and conclusions were formulated based on the critical analysis of the reviewed literature.

REFERENCE

1. Aldasoro, I., Gambacorta, L., Giudici, P. and Leach, T.:The drivers of cyber risk. *Journal of Financial Stability*, *60*, 100989(2022).
2. Aben, J. and Etti, P.:European Digital Finance. In *Digital Development of the European Union: An Interdisciplinary Perspective* (67-84). Cham: Springer International Publishing(2023).
3. Akyüz, A. and Mavnacıoğlu, K.:Marketing and financial services in the age of artificial intelligence. *Financial Strategies in Competitive Markets: Multidimensional Approaches to Financial Policies for Local Companies*, 327-340(2021).
4. Alexander, C. :*Market risk analysis, value at risk models* (Vol. 4). John Wiley & Sons (2009).

5. Al-Qirim, N., Tarhini, A. and Rouibah, K.:August. Determinants of big data adoption and success. In *Proceedings of the 1st International Conference on Algorithms, Computing and Systems*, 88-92(2017).
6. Arner, D.W., Barberis, J. and Buckey, R.P.:FinTech, RegTech, and the reconceptualization of financial regulation. *Nw. J. Int'l L. & Bus.*, *37*, 371(2016).
7. Arner, D.W., Barberis, J. and Buckley, R.P.:The evolution of Fintech: A new post-crisis paradigm. *Geo. J. Int'l L.*, *47*, 1271(2015).
8. Arner, D.W., Buckley, R.P. and Zetzsche, D.A.:Open banking, open data and open finance: lessons from the European union. *Forthcoming in Linda Jeng (ed), Open Banking (Oxford University Press, 2021)*, 21-69(2021).
9. Atlan, T.: 'Cloud Migration Data Security: 5 Challenges and Ways to manage,' *Atlan*, 10 October. https://atlan.com/cloud-migration-data-security/,(2023).
10. Atobishi, T., Bahna, M. and Fogarassy, C.:The main influencing factors of mobile banking adoption in the open innovation business environment (case study). *HUNGARIAN AGRICULTURAL ENGINEERING*, (40), 28-38(2021).
11. Balkan, B.:Impacts of digitalization on banks and banking. *The Impact of Artificial Intelligence on Governance, Economics and Finance, Volume I*, 33-50(2021).
12. Barroso, M. and Laborda, J.:Digital transformation and the emergence of the Fintech sector: Systematic literature review. *Digital Business*, *2*(2), 100028, (2022).
13. Black, J., Hopper, M. and Band, C.:Making a success of principles-based regulation. *Law and financial markets review*, *1*(3), 191-206(2007).
14. Bommadevara, N., Del Miglio, A. and Jansen, S.:Cloud adoption to accelerate IT modernization. *McKinsey Digital,* (2018).
15. Boot, A., Hoffmann, P., Laeven, L. and Ratnovski, L.:Fintech: what's old, what's new?. *Journal of financial stability*, *53*,100836, (2021).
16. Brad, S.A.,Jones. :*SPEECH Emerging Threats to financial stability – New challenges for the next decade*. https://www.rba.gov.au/speeches/2023/sp-ag-2023-10-31.html, (2023).
17. Buckley, R.P., Arner, D.W., Zetzsche, D.A. and Selga, E.: The dark side of digital financial transformation: The new risks of fintech and the rise of techrisk. *UNSW Law Research Paper*, (19-89), (2019).
18. Cangiano, M., Gelb, A. and Goodwin-Groen, R.:Public financial management and the digitalization of payments. In *Public financial management and the digitalization of payments: Cangiano, Marco| uGelb, Alan| uGoodwin-Groen, Ruth*. Washington DC: Center for Global Development,(2019).
19. Catapano, F., Di Iorio, S., Sementa, P. and Vaglieco, B.M.:Analysis of energy efficiency of methane and hydrogen-methane blends in a PFI/DI SI research engine. *Energy*, *117*, 378-387(2016).
20. Claessens, S., Frost, J., Turner, G. and Zhu, F.:Fintech credit markets around the world: size, drivers and policy issues. *BIS Quarterly Review September,* (2018).
21. Clarke, P.:'Deutsche Bank turns to 'robots' to meet 18,000 job cut target,' *https://www.fnlondon.com/articles/deutsche-bank-turns-to-robots-to-meet-18000-job-cut-target-20191118*, 18 November, (2019).
22. Crittenden, A.B., Crittenden, V.L. and Crittenden, W.F.:The digitalization triumvirate: How incumbents survive. *Business Horizons*, *62*(2), 259-266(2019).
23. Cullen, J.:Exchange-traded funds (ETFs) and FinTech: Market efficiency and systemic risk. In *Routledge Handbook of Financial Technology and Law* 227-244(2021). Routledge.
24. Dagilienė, L. and Klovienė, L.:Motivation to use big data and big data analytics in external auditing. *Managerial Auditing Journal*, *34*(7),750-782(2019).
25. Dandapani, K.:Electronic finance – recent developments,' *Managerial Finance*, 43(5), 614–626(2017). https://doi.org/10.1108/mf-02-2017-0028.

 https://doi.org/10.52202/074664-0010

26. Dash, S., Shakyawar, S.K., Sharma, M. and Kaushik, S.: Big data in healthcare: management, analysis and future prospects. *Journal of big data*, *6*(1), 1-25(2019).
27. Dhar, V. and Stein, R.M.:FinTech platforms and strategy. *Communications of the ACM*, *60*(10), 32-35(2017).
28. Dluhopolskyi, O., Pakhnenko, O., Lyeonov, S., Semenog, A., Artyukhova, N., Cholewa-Wiktor, M. and Jastrzębski, W.:Digital financial inclusion: COVID-19 impacts and opportunities. *Sustainability*, *15*(3), 2383 (2023).
29. Dr Sunnatha Vincent Jaiwant, S.:Changing Structure of Consumer Buying Behaviour and Expectation in the Digital Era. In *Industry 4.0 and the Digital Transformation of International Business* , 207-215(2023). Singapore: Springer Nature Singapore.
30. Durai, T. and Stella, G.:Digital finance and its impact on financial inclusion. *Journal of Emerging Technologies and Innovative Research*, *6*(1), 122-127(2019).
31. Ebert, C. and Duarte, C.H.C:Digital transformation,' *IEEE Software*, 35(4), 16–21(2018). https://doi.org/10.1109/ms.2018.2801537.
32. Elgendy, N. and Elragal, A.:Big data analytics: a literature review paper. In *Advances in Data Mining. Applications and Theoretical Aspects: 14th Industrial Conference, ICDM 2014, St. Petersburg, Russia, July 16-20, 2014. Proceedings 14* , 214-227(2014). Springer International Publishing..
33. El-Zoghbi, M., Chehade, N., McConaghy, P. and Soursourian, M.:The role of financial services in humanitarian crises, (2017).
34. Erick Feyen, E., Frost, J., Gambacorta, L., Natarajan, H. and Saal, M.:Fintech and the digital transformation of financial services: implications for market structure and public policy. *BIS papers,* (2021).
35. European Union *Overview of digital finance.:*https://finance.ec.europa.eu/digital-finance/overview-digital-finance_en, (2022).
36. Fabris, N.:Challenges for modern monetary policy. *Journal of Central Banking Theory and Practice*, *7*(2),5-24 (2018).
37. Fenwick, M. *et al.*:'A sustainable platform economy & the future of Corporate governance,' *ECGI Working Paper Series in Law* [Preprint], (2019).
38. Frankenfield, J.:Cryptocurrency explained with pros and cons for investment. *Dostupné*, *4*, 2023 (2023).
39. Ganguly, S., Harreis, H., Margolis, B. and Rowshankish, K.:Digital risk: Transforming risk management for the 2020 s. *McKinsey & Company,* .(2017).
40. Giudici, P.:Fintech risk management: A research challenge for artificial intelligence in finance. *Frontiers in Artificial Intelligence*, *1*, 1, (2018).
41. Gomber, P., Kauffman, R.J., Parker, C. and Weber, B.W.:On the fintech revolution: Interpreting the forces of innovation, disruption, and transformation in financial services. *Journal of management information systems*, *35*(1), 220-26(2018).
42. Gomber, P., Koch, J.A. and Siering, M.:Digital Finance and FinTech: current research and future research directions. *Journal of Business Economics*, *87*, 537-580 (2017).
43. Gozman, D., Liebenau, J. and Mangan, J.:The innovation mechanisms of fintech start-ups: insights from SWIFT's innotribe competition. *Journal of Management Information Systems*, *35*(1), 145-179(2018).
44. Graham, L.:*Internal control audit and compliance: documentation and testing under the new COSO framework*. John Wiley & Sons, (2015).
45. Guide, P.: Digitisation and Informality–Harnessing Digital Financial Inclusion for Individuals and MSMEs in the Informal Economy. *G20 Argentina Presidency,* (2018).
46. Gupta, H. and Chaudhary, R .:'An Empirical study of volatility in cryptocurrency market,' *Journal of Risk and Financial Management*, 15(11), 513(2022). https://doi.org/10.3390/jrfm15110513.
47. Harreis, H.:The future of risk management in the digital era. *Mckynsey& Company,* (2017).

48. Hassan, A.: *Challenges to Implementing Financial Inclusion in Sub-Rural Areas in Bauchi State* (Doctoral dissertation, American University of Nigeria), (2022).
49. Hess, T., Matt, C., Benlian, A. and Wiesböck, F.:Options for formulating a digital transformation strategy. *MIS Quarterly Executive*, *15*(2), (2016).
50. International Labour Organization *Technical meeting on the impact of digitalization in the finance sector.*:https://www.ilo.org/sector/activities/sectoral-meetings/WCMS_824051/lang--en/index.htm, (2021).
51. Jameaba, M.S.: Digitization revolution, FinTech disruption, and financial stability: Using the case of Indonesian banking ecosystem to highlight wide-ranging digitization opportunities and major challenges. *FinTech Disruption, and Financial stability: Using the Case of Indonesian Banking Ecosystem to highlight wide-ranging digitization opportunities and major challenges (July 16 2, 2020),* (2020).
52. Javaid, M., Haleem, A., Singh, R.P., Suman, R. and Khan, S.: A review of Blockchain Technology applications for financial services. *BenchCouncil Transactions on Benchmarks, Standards and Evaluations*, *2*(3), 100073, (2022).
53. Juan Antonio Ketterer Ketterer, J.A.:Digital finance: New times, new challenges, new opportunities, (2017).
54. Jumena, E.:Fintech Lending Jangan Jadi Digital Renteni,(2018)..
55. K Dandapani, K.:Electronic finance–recent developments. *Managerial Finance*, *43*(5), pp.614-626(2017).
56. Kagermann, H.:Change through digitization—Value creation in the age of Industry 4.0. In *Management of permanent change* ,pp. 23-45(2014). Wiesbaden: Springer Fachmedien Wiesbaden.
57. Kazancoglu, I. and Aydin, H.:An investigation of consumers' purchase intentions towards omni-channel shopping: A qualitative exploratory study. *International Journal of Retail & Distribution Management*, *46*(10), 959-976(2018).
58. Khatri, A., Singh, N.P. and Gupta, N.:Big data analytics: direction and impact on financial technology. *Journal of Management Marketing and Logistics*, *8*(4), pp.218-234(2021).
59. J, P.K. :*What is Big Data? / Uses of Big Data / Types Of Big Data / Big Data Analytics Interview Questions*. https://www.linkedin.com/pulse/what-big-data-uses-types-analytics-interview-questions-jha,(2023).
60. Kiiru, E.W.:*A survey on cloud computing adoption in Kenya's banking industry* (Doctoral dissertation), (2011).
61. Kumar.:*AI-Specific Cybersecurity Risks & Detections Methodology: Let's secure the Financial Sector.* https://www.linkedin.com/pulse/ai-specific-cybersecurity-risks-detections-lets-secure-yogendar-kumar-z2smf,(2024).
62. Kurz, H.D., Schütz, M., Strohmaier, R. and Zilian, S.S.:*Routledge Handbook of Smart Technologies*. Routledge, (2022).
63. Lancieri, F.:Narrowing data protection's enforcement gap. *Me. L. Rev.*, *74*, p.15 (2022).
64. Laura Freire.:*Reduce Compliance Risks with Compliance Automation | Apiax*. https://www.apiax.com/compliance-automation/,(2022).
65. Liu, Y.:Big data and predictive business analytics. *The Journal of Business Forecasting*, *33*(4), p.40 (2014).
66. Lobanov, L.:'Regtech: How Regulatory Technology is Streamlining Compliance in Financial Services,' *Medium*, 9 November. https://medium.com/@lea.lobanov/regtech-how-regulatory-technology-is-streamlining-compliance-in-financial-services-d8fce194402f, (2023).
67. Ltd, V.P.:*Fintech and the Future of Work: How Technology is Changing the Way We Bank*. https://www.linkedin.com/pulse/fintech-future-work-how-technology-changing-way-we-bank-vibidsoft, (2023).
68. Magnuson, W. :Financial regulation in the Bitcoin era. *Stan. JL Bus. & Fin.*, *23*, 159(2018).

 https://doi.org/10.52202/074664-0010

69. Mansfield-Devine, S.:Open banking: opportunity and danger. *Computer Fraud & Security*, *2016*(10), pp.8-13(2016).
70. McAfee, A.: What every CEO needs to know about the cloud. *Harvard business review*, *89*(11), 124-132(2011).
71. *Measuring the economic impact of cloud computing in Europe.*:https://digital-strategy.ec.europa.eu/en/library/measuring-economic-impact-cloud-computing-europe, (2017).
72. Mohammadzai, G., Pashtoon, K. and Aini, I.: Compliance Requirement for Dealing with Risks, Governance and IT Compliance. *Integrated Journal for Research in Arts and Humanities*, *3*(4), 150-156(2023).
73. Murinde, V., Rizopoulos, E. and Zachariadis, M.:The impact of the FinTech revolution on the future of banking: Opportunities and risks. *International review of financial analysis*, *81*, 102103 (2022).
74. Nann, S., Krauss, J. and Schoder, D.: Predictive analytics on public data-the case of stock markets, (2013).
75. Olaniyi, O., Olabanji, S.O. and Abalaka, A.:Navigating risk in the modern business landscape: Strategies and insights for enterprise risk management implementation. *Journal of Scientific Research and Reports*, *29*(9), 103-109(2023).
76. Oliveira, T., Faria, M., Thomas, M. A., Popovič, A,:Extending the understanding of mobile Banking adoption: When UTAUT meets TTF and ITM. International Journal of Information Management, 3(5), 689-703(2014).
77. Omarini, A.:FinTechs: Unbundling and re-bundling in the open industry of banking. In *The Routledge Handbook of FinTech*,pp. 193-215(2021). Routledge.
78. Ozili, P.K.:Impact of digital finance on financial inclusion and stability. *Borsa istanbul review*, *18*(4), 329-340(2018).
79. Papantoniou, A.A.:Regtech: steering the regulatory spaceship in the right direction?. *Journal of Banking and Financial Technology*, *6*(1),1-16(2022).
80. Pazarbasioglu, C., Mora, A.G., Uttamchandani, M., Natarajan, H., Feyen, E. and Saal, M.:Digital financial services. *World Bank*, *54*, (2020).
81. Perera, A. and Iqbal, K.: Big Data and Emerging Markets: Transforming Economies Through Data-Driven Innovation and Market Dynamics. *Journal of Computational Social Dynamics*, *6*(3), 1-18(2021).
82. Portilla, A., Vazquez, J., Harreis, H., Pancaldi, L., Rowshankish, K., Samandari, H. and Staples, M.: The future of risk management in the digital era. *Institute of International Finance & McKinsey & Company, Inc.*, *110*(11), 91-93(2017).
83. Quintyn, M.M. and Taylor, M.M.:*Should financial sector regulators be independent?*. International Monetary Fund, (2004).
84. Ramsey, J.:*9 Ways Banks Can Use SMS to Engage Their Customers | emfluence Marketing Platform*. https://emarketingplatform.com/blog/9-ways-banks-can-use-sms-to-serve-their-customers/,(2023).
85. Rasoulkhani, K., Brannen, L., Zhu, J., Mostafavi, A., Jaselskis, E., Stoa, R., Li, Q., Alsharef, A., Banerjee, S. and Chowdhury, S.:Establishing a future-proofing framework for infrastructure projects to proactively adapt to complex regulatory landscapes. *Journal of Management in Engineering*, *36*(4), 04020032, (2020).
86. Rau, R., Wardrop, R. and Zingales, L. eds.:*The Palgrave handbook of technological finance*. Palgrave Macmillan, (2021).
87. S, P.:*Digital Risk Management*. https://www.linkedin.com/pulse/digital-risk-management-paul-sankar, (2023).
88. Safitra, M.F., Lubis, M. and Fakhrurroja, H.:Counterattacking cyber threats: A framework for the future of cybersecurity. *Sustainability*, *15*(18), 13369(2023).
89. Sanad, Z. and Al Lawati, H.:Board gender diversity and firm performance: the moderating role of financial technology. *Competitiveness Review: An International Business Journal,* (2023).

90. Saptarshi Ganguly, S., Harreis, H., Margolis, B. and Rowshankish, K.: Digital risk: Transforming risk management for the 2020 s. *McKinsey &Company*,(2017).
91. Seth Bruhn, M. *et al.*:*ASSESSMENT OF THE SHORTFALLS AND OPPORTUNITIES IN FINANCING MICRO, SMALL AND MEDIUM ENTER*Bruhn,2017*PRISES IN EMERGING MARKETS, MSME FINANCE GAP*. INTERNATIONAL FINANCE CORPORATION, (2017).
92. Shaikh, A.A. and Karjaluoto, H.: Mobile banking adoption: A literature review. *Telematics and informatics*, *32*(1), 129-142(2015).
93. Shaikh, A.A., Alharthi, M.D. and Alamoudi, H.O.: Examining key drivers of consumer experience with (non-financial) digital services—An exploratory study. *Journal of Retailing and Consumer Services*, *55*,102073, (2020).
94. Shofawati, A. :The role of digital finance to strengthen financial inclusion and the growth of SME in Indonesia. *KnE Social Sciences*,389-407(2019).
95. Smit, S., Tacke, T., Lund, S., Manyika, J. and Thiel, L.:The future of work in Europe, (2020).
96. Sokolova, N.G. and Titova, O.V.: May. Digital marketing as a type: concept, tools and effects. In *1st International Scientific Conference" Modern Management Trends and the Digital Economy: from Regional Development to Global Economic Growth"(MTDE 2019)*, 508-512(2019). Atlantis Press.
97. SS&C Blue Prism.:*The benefits of Back-Office Automation*. https://www.blueprism.com/resources/blog/the-benefits-of-back-office-automation/,(2023).
98. Stöber, T., Kotzian, P. and Weißenberger, B.E.:Design matters: on the impact of compliance program design on corporate ethics. *Business research*, *12*, pp.383-424(2019).
99. Surbiryala, J. and Rong, C.: August. Cloud computing: History and overview. In *2019 IEEE Cloud Summit*, 1-7(2019). IEEE.
100. *Sustainable Development Disciplines for Humanity*.:*Sustainable development goals series*.
101. *The Global Cyber Threat to Financial Systems – IMF F&D.*, (2021). https://www.imf.org/external/pubs/ft/fandd/2021/03/global-cyber-threat-to-financial-systems-maurer.htm, (2023).
102. Tuten, T.L.:Principles of marketing for a digital age. *Principles of Marketing for a Digital Age*, pp.1-100(2022).
103. *USAID Digital Payments India Merchant and Consumer Research Resources* .:https://giw2017.org/resources/usaid-digital-payments-india-merchant-and-consumer-research-resources, (2016b).
104. Van Loo, R.:Rise of the digital regulator. *Duke LJ*, *66*, p.1267(2016).
105. Weiwei Huang, W.:Built on value: the Huawei philosophy of finance management (456). Springer Nature,(2019).
106. Wong, M.: *Analysis of Financial Crimes Compliance Shortcomings in Financial Technology Institutions: a Criminological Alternative to Outdated Compliance Models* (Master's thesis, Utica College), (2020).
107. Xiaoming Yuan, X. and Li, H.:Digital Finance Affects the Consumption Path of Urban Digital Finance and Residents: the Expansion of Digital Finance to Consumption in the Perspective of Space Spillover. *Journal of the Knowledge Economy*,1-26(2023).
108. Xu, G., Li, G., Sun, P. and Peng, D.:Inefficient investment and digital transformation: What is the role of financing constraints?. *Finance Research Letters*, *51*, 103429(2023).
109. Yang, D. and Li, M.:Evolutionary approaches and the construction of technology-driven regulations. *Emerging Markets Finance and Trade*, *54*(14), 3256-3271(2018).
110. Yang, G., Ding, Z., Wu, M., Gao, M., Yue, Z. and Wang, H.: Can digital finance reduce carbon emission intensity? A perspective based on factor allocation distortions: evidence from Chinese cities. *Environmental Science and Pollution Research*, *30*(13), 38832-38852(2023).
111. Zetzsche, D.A., Arner, D.W. and Buckley, R.P.:Decentralized finance (defi). *Journal of Financial Regulation*, *6*, 172-203(2020).

https://doi.org/10.52202/074664-0010

112.Z Marwane, W., Stein, S., Markovic, I., &Pulvermaller, S. :Compliance management for small and medium-sized enterprises. In Proceedings of the 15th Americas conference on information systems, (2009).

THE EVALUATION OF SUSTAINABLE INVESTING: A COMPREHENSIVE ANALYSIS

Mohammed Ashraf Ali[1], Mohammad Zulqar Nain*[2], Mohd Hamid[3], Mehfooz Zaki[4]
[1] Professor, Department of Commerce, Aligarh Muslim University, Aligarh, India
[2] Research Scholar, Department of Commerce, Aligarh Muslim University, Aligarh, India
[3] Research Scholar, Department of Commerce, Aligarh Muslim University, Aligarh, India
[4] Research Scholar, Department of Commerce, Aligarh Muslim University, Aligarh, India

Abstract

Sustainable investing has gain significant attention of the researchers as it is a crucial field of researchwhich have environmental, social, and governance (ESG) factors consequences. This study explores various aspects of sustainable investing, its approaches, development of sustainable finance, barriers, and potential outcomes. The study also focuses on the risk profile and the return of sustainable finance, and the performance of portfolio.

This longitudinal research reviews the literature between 2005-2023, where detailed analysis of the various strategies, ESG criterion,SRI,Impact investing techniques procedures used to evaluate and report sustainability measures.

The study offers guidance on how to get beyond these obstacles and raise the legitimacy of sustainable investing methods. With a forward-looking outlook, by unveiling new developments, innovations, and possible breakthroughs in sustainable investing. It highlights how technology, new legislative initiatives, and cooperative efforts will shape the direction of finance going forward by incorporating governance, social, and environmental factors.

Keywords: Sustainable Investing, Greenwashing, Environmental factors, Returns and RiskProfiles, Environmental, Social, and Governance (ESG).

1. Introduction

The financial landscape is changing rapidly with the rise of sustainable investing, which acknowledges the relationship between social and environmental responsibility and economic prosperity. This strategy aims to match financial objectives with beneficial societal and environmental effects. It is based on the recognition of global issues including social injustice and climate change. The fundamental tenets of environmental, social, and governance (ESG) criteria are becoming more and more integrated into the decision-making processes of investors.[1] Sustainable investment places a higher priority on long-term value generation and risk reduction than conventional profit-centric methods. Sustainable investing grows in importance as companies navigate a world where social welfare, environmental conservation, and ethical business practices are critical[1]. This influences corporate conduct, shapes financial markets, and encourages innovation.

Beyond moral issues, the importance of sustainable investing is becoming apparent as a motivating factor influencing corporate and financial market behavior. Companies must implement sustainable practices not only for moral reasons but also to draw in investment and stay competitive as global issues take center stage in public conversation.[2] This change encourages companies to see sustainability as a source of resilience and a driver of long-term financial performance, redefining the notion of value creation. As a result, sustainable investing acts as a catalyst for good, putting money in the hands of companies that uphold moral principles and build a more just and sustainable future.[3]Recognizing the material influence of environmental, social, and governance (ESG) aspects, institutional investors are actively incorporating sustainable investment strategies into their portfolios in a radical shift. This revolutionary movement is changing the way capital is allocated, giving businesses that are dedicated to sustainable business practices more weight.

https://doi.org/10.52202/074664-0011

Investors today examine non-financial performance indicators closely in addition to traditional financial measurements to obtain a comprehensive picture of a company's effects on the environment, society, and corporate governance. Sustainable investment is more than just a fad; it is a fundamental restructuring of the financial ecosystem, reflecting the understanding on the part of society that responsible stewardship is a prerequisite for financial success and that this has ramifications for the health of the world and its inhabitants.

The review paper aims to critically examine the landscape of sustainable investing, emphasizing a specific focus on the evaluation methodologies employed in assessing the effectiveness and impact of sustainable investment strategies. Through a comprehensive analysis of existing literature, the paper seeks to provide insights into the various frameworks, metrics, and challenges associated with the evaluation of sustainable investments, contributing to a deeper understanding of this evolving and impactful facet of the financial world.Understanding and assessing the effectiveness of sustainable investing strategies is crucial for several reasons. First, as sustainable investing gains prominence, investors need reliable methods to evaluate the impact of their financial choices on environmental, social, and governance (ESG) factors.Rigorous assessment ensures that investments align with ethical principles and contribute positively to sustainable development. Second, risk management depends on assessing how well sustainable investing techniques work. Investors can reduce long-term financial risks and make well-informed decisions by identifying potential hazards associated with ESG aspects.[4] Additionally, it supports the development of robust portfolios that can withstand the shifting global landscape of problems including social injustice, climate change, and corporate governance issues.

Moreover, the evaluation of sustainable investment approaches advances accountability and openness. The public at large, stakeholders, and investors all want more openness when it comes to the practical effects of investments.[2] Sturdy assessment techniques make it easier to communicate quantifiable results, allowing investors to show that they are committed to ethical and significant investing methods.[5] Additionally, encouraging innovation and advancement in the industry depends on having a solid understanding of how effective sustainable investing options are. It is possible to improve frameworks and measurements through ongoing assessment, which fosters the development of best practices. This in turn motivates companies to implement more environmentally friendly procedures, resulting in a positive feedback loop that is advantageous to investors as well as the larger world society.

In the end, evaluating how well sustainable investing techniques work comes down to how well they can match financial aims with more general societal and environmental goals. Investors may drive the shift to a more sustainable and equitable global economy, encourage responsible business behavior, and effect good change by closely examining the effects of their investments.

2. Literature Review

Sustainable investing, often referred to as socially responsible investing (SRI) or environmental, social, and governance (ESG) investing, is an investment approach that goes beyond traditional financial considerations to integrate environmental, social, and governance factors into the decision-making process. The key principles of sustainable investing revolve around achieving positive and sustainable outcomes in addition to financial returns. Here are the core principles:

S.No.	Key Principles	Sustainable Outcomes
1.	**Environmental Considerations**	Sustainable investing emphasizes the integration of environmental factors into investment decisions. This includes assessing a company's impact on natural resources, energy consumption, pollution, and overall environmental sustainability. Investors seek to support businesses with eco-friendly practices and a

		commitment to reducing their carbon footprint.
2.	**Social Responsibility**	Social considerations involve evaluating a company's impact on society, including its treatment of employees, relationships with communities, human rights practices, and efforts to promote diversity and inclusion. Sustainable investing aims to support businesses that contribute positively to social well-being and uphold ethical standards in their operations.
3.	**Governance Practices**	Governance principles focus on evaluating the quality of a company's leadership, internal controls, and overall corporate governance structure. Investors assess factors such as executive compensation, board independence, transparency, and adherence to ethical business practices. Sustainable investing encourages investment in companies with strong governance frameworks.
4.	**Long-Term Value Creation**	Sustainable investing is centered on the idea of generating long-term value for investors while contributing to positive societal and environmental outcomes. It emphasizes the integration of sustainability considerations into investment strategies to foster resilience and stability in the face of emerging global challenges.
5.	**Impact on Stakeholders**	Sustainable investing considers the broader impact that companies have on various stakeholders, including customers, employees, communities, and the environment. Investors seek to align their portfolios with businesses that prioritize the well-being of all stakeholders, aiming for a positive societal impact beyond financial returns.
6.	**Engagement and Advocacy**	Sustainable investors often engage with companies to encourage positive change. Through active ownership and advocacy, investors aim to influence corporate behavior, promote transparency, and contribute to the continuous improvement of sustainability practices within companies.

The historical development and evolution of sustainable investing trace a trajectory marked by shifting societal values, environmental awareness, and a growing recognition of the interconnectedness between financial success and broader global challenges. While the roots of ethical investing can be traced back centuries, the modern concept of sustainable investing has evolved over the past few decades.The 1960s and 1970s witnessed the emergence of socially responsible investing (SRI), driven by ethical considerations and a desire to avoid investments in industries perceived as harmful, such as tobacco or weapons manufacturing. During this period, investors began to incorporate moral and social criteria into their investment decisions, reflecting a nascent awareness of the impact of businesses on society. The 1980s saw the formalization of SRI with the establishment of socially responsible mutual funds. These funds applied exclusionary screens to avoid investments in companies with negative social or environmental impacts.

The scope of SRI remained relatively narrow, primarily focusing on negative screening rather than actively promoting positive societal and environmental outcomes. The 1990s marked a pivotal moment with the introduction of environmental, social, and governance (ESG) criteria. The ESG framework expanded the scope of ethical investing by considering positive criteria, such as

https://doi.org/10.52202/074664-0011

companies with strong corporate governance, ethical business practices, and environmental sustainability. Institutional investors increasingly began integrating ESG factors into their decision-making processes.

The early 2000s, the term "sustainable investing" gained prominence, reflecting a broader and more holistic approach that encompassed environmental, social, and governance considerations.[3]. The United Nations Principles for Responsible Investment (UN PRI), established in 2006, provided a global framework for institutional investors to integrate ESG principles into their investment practices.The post-2008 financial crisis era witnessed a surge in interest in sustainable investing as investors sought to align their portfolios with long-term value creation and risk mitigation. The development of impact investing, which explicitly seeks measurable positive social and environmental impacts alongside financial returns, further diversified the sustainable investing landscape.

In recent years, sustainable investing has reached new heights, with a proliferation of ESG funds, increased corporate disclosure on sustainability metrics, and growing mainstream acceptance. Regulatory developments and industry standards, such as the Task Force on Climate-related Financial Disclosures (TCFD), have further legitimized the importance of considering sustainability factors in financial decision-making.[6]Today, sustainable investing continues to evolve, driven by an increasing global awareness of urgent environmental challenges, social inequalities, and the need for responsible corporate governance. The historical journey of sustainable investing reflects a paradigm shift in how investors perceive the role of finance in addressing complex, interconnected issues on a global scale.[3]

Sustainable investing encompasses various approaches that integrate environmental, social, and governance (ESG) factors into investment decision-making. Each approach reflects a distinct set of principles and objectives. Here are discussions on three key approaches: ESG criteria, impact investing, and socially responsible investing (SRI).[1]

2.1 ESG Criteria

Overview: ESG criteria involve the systematic consideration of environmental, social, and governance factors in the evaluation of investment opportunities. These criteria provide a framework for assessing a company's performance in areas such as carbon emissions, labor practices, board diversity, and ethical business conduct.

Implementation: Investors employing ESG criteria analyze companies based on a set of predefined ESG metrics. Ratings agencies and ESG research providers assess and score companies on their ESG performance. Investors use these scores to make informed decisions, integrating ESG considerations into portfolio construction and risk management.

Objective: The primary goal of ESG criteria is to identify companies that demonstrate strong sustainability practices and responsible corporate behavior. ESG integration is often viewed to enhance long-term financial performance by considering non-financial factors that may impact a company's resilience and reputation.

Impact Investing

Overview: Impact investing goes beyond ESG considerations by explicitly seeking positive, measurable social or environmental impacts alongside financial returns. This approach aims to generate a tangible, beneficial effect on specific issues such as clean energy, healthcare, or poverty alleviation.

Implementation: Impact investors actively select and allocate capital to projects, companies, or funds that align with their targeted impact goals. Investments are evaluated based on their potential to create positive outcomes, and impact metrics are closely monitored to assess the success of the investments in achieving their intended goals.

Objective: The primary objective of impact investing is to drive positive change by channeling capital toward solutions for societal and environmental challenges. Investors seek not only financial returns but also to contribute to measurable improvements in areas aligned with their impact goals.

2.2 Socially Responsible Investing (SRI)

Overview: SRI represents an early form of ethical investing that screens out investments in industries or companies that do not align with specific moral or ethical values. SRI may involve both negative screening (avoiding certain sectors) and positive screening (actively seeking investments in socially responsible companies).

Implementation: Socially responsible investors apply exclusionary screens to filter out investments in areas deemed socially or environmentally harmful, such as tobacco, weapons, or companies with poor labor practices. Positive screening involves actively supporting companies that align with ethical principles.

Objective: The primary objective of SRI is to align investment portfolios with investors' ethical or moral values. SRI seeks to avoid supporting businesses perceived as contributing negatively to society while promoting investments in companies with positive social and environmental practices.

While these approaches share the overarching goal of integrating sustainability considerations into investment decisions, they differ in their methodologies, objectives, and the extent to which they explicitly seek positive impact. Investors often choose an approach based on their values, risk tolerance, and desired outcomes for both financial returns and societal or environmental impact.

Several key studies and findings related to sustainable investing evaluation provide insights into the effectiveness and impact of sustainable investment strategies. Here is a summary of some notable research in this area:

S. NO.	Author	Year	Title	Journals/Sources	Finding
1	Anita Handayani& RofikohRokhim	2023	A Comparative Study of Financial Performance between Sustainable and Conventional Investment	Journal of Entrepreneurship and Business	The study gives firms direction in creating sustainable investment products and gives investors useful information in assessing the performance of sustainable investments. Additionally, it can help governments and regulators create an environment that supports environmentally friendly and sustainable investment opportunities while providing investors with profitable returns.
2	Markus Koenigsmarck, Martin Geissdoerfer	2023	Shifting the Focus to Measurement: A Review of Socially Responsible Investing and SustainabilityIndicators	Sustainability -Vol. 15	The writers provided an overview of the SRI concept's past, talked about contemporary developments in the industry, and combined definitions of impact investment and SRI. They created the Cambridge SRI indicator framework, a trinomial sustainability indicator framework, and offered a summary of sustainability indicators for SRIs.

https://doi.org/10.52202/074664-0011

3	Edmans A, Kacperczyk M	2022	Sustainable Finance	Review of Finance (2022)	The authors discover that, although the correlations are not very strong, unfavorable social and environmental events are followed by divestitures. Even more potent is the danger of future departure if environmental and social performance are maintained. The research paper presents a model in which investors' decisions about their portfolios are significantly influenced by sustainability. The taste forces add an additional 2% per year to the reported risk premia, and the exclusion forces contribute around 2.7% annually.
4	Majid Abbasian Mirza, Truzaar Dordi, Pedro Ildemaro Alguindigue, Ryan G. Johnson, Olaf Weber	2023	Sustainability in Private Capital Investing: A Systematic Literature Review	Journal of management and sustainabilit	The purpose of the study is to address research gaps by identifying current themes in the existing literature and demonstrating the dearth of academic literature on sustainable investment and private capital markets.
5	Charney S. Akala, Taryn Neuhaus, Indrani O' Leary-	2022	A Systematic Review of Sustainable Investment Approaches	International journal of economics and finance	The study carried out a thorough analysis of research on sustainable investment strategies, with a focus on impact investing, environmental, social, and governance (ESG), and socially responsible investing (SRI).

https://doi.org/10.52202/074664-0011

	Govender				The results showed that impact investing, ESG, and SRI share conceptual frameworks. Compared to typical markets, SRI investments had advantages in terms of diversity despite having a mixed return performance.
6	Mark Anthony Camilleri	2021	The market for socially responsible investing: a review of the developments	Social Responsibility Journal (Emerald)- Vol. 17	A comprehensive review of the literature on the origins and expansion of socially responsible investing (SRI) in the financial markets is presented in this paper. It draws attention to the growing amount of money that financial capital providers are devoting to investments with a good impact and sustainability.
7	Terence C. Burnham1, Harry Gakidis, Jeffrey Wurgler2, Jeffrey Wurgler	2017	Investing in the Presence of Massive Flows: The Case of MSCI Country Reclassifications	Social Science Research Network	Significant price overshooting occurs when reclassifications from one index to another occur between the announcement and the effective dates. When a market switches from an index with greater benchmarked ownership to one with less, like Emerging to Frontier, prices often decline and vice versa. But after a year, these price swings typically level off.
8	Paraskevi Ovezikoglou, Dimitrios Aidonis, CharisiosAchi	2020	Sustainability Assessment of Investments Based on a Multiple Criteria Methodologi	Sustainability (Multidisciplinary Digital Publishing Institute)- Vol. 12	The study provides a thorough framework for assessing the sustainability of industrial investments as well as an innovative method for multiple criterion decision analysis.

 https://doi.org/10.52202/074664-0011

	llas, Christo s Vlacho kostas, Dionys isBoch tis		cal Framework		The framework considers planning, designing, and prevention while evaluating a total of eighteen potential indicators based on relevant material from the literature. The concepts of sustainable development serve as the foundation for the evaluation of four criteria: the environment, society, economy, and technology.
9	Xiao-Guang Yue, Yan Han, Deima nte Teresie ne, Justina Merkyt e, Wei Liu	2020	Sustainable Funds' Performance Evaluation	Sustainability (Multidiscipli nary Digital Publishing Institute)- Vol. 12	The study found that sustainable funds are less risky than traditional funds, based on the analysis of annual returns, standard deviations, Sharpe ratios, skewness, and kurtosis. However, there was no clear evidence to suggest that sustainable funds generate higher returns compared to traditional funds or benchmark indexes.

3. Methodology

The methodology employed for the comprehensive analysis of sustainable investing involves a meticulous and integrated approach.[7] Data collection spans financial and non-financial realms, where financial metrics, such as return on assets and equity, are extracted from reliable sources like financial databases and company reports. Non-financial performance metrics, including environmental, social, and governance (ESG) factors, are meticulously gathered from reputable providers like "Morgan Stanley Capital International" (MSCI is a leading provider of research-based indexes and analytics, and it has become a widely recognized name in the finance and investment industry.), Sustainalytics, and other specialized databases, enriching the analysis with a holistic perspective.[8]

The methodology used in the thorough examination of sustainable investing includes benchmarking against industry-specific indices to compare the performance of sustainable investment portfolios with traditional benchmarks. Sensitivity analysis increases the study's depth and dependability by measuring the effects of variable modifications. Incorporating case studies provides qualitative perspectives to enhance quantitative analysis. By tracking changes over time, longitudinal analysis offers a dynamic picture of patterns in sustainable investment.[9] Transparency is guaranteed by ethical concerns, and the approach is validated and given credibility by subject matter experts through peer review processes. This all-encompassing strategy guarantees careful investigation of the complex dynamics involved in sustainable investing, providing a strong basis for perceptive findings.

Ratings based on Environmental, Social, and Governance factors have become essential for assessing the effectiveness of sustainable investments. These ratings, which come from specialist agencies, evaluate a company's governance, social, and environmental actions in a methodical manner.[10] While social dimensions include diversity, labor practices, and community relations, environmental components consider things like carbon footprint and resource management. Ethical business practices and the composition of the board are evaluated as part of governance review.

ESG ratings provide investors with a consistent and measurable metric that helps them make well-informed decisions that are in line with their sustainability goals.[11] Companies use the scores or ratings as a critical standard, which affects their brand and draws in socially conscious investors. Dynamic and frequently updated, ESG ratings are essential for guiding investment strategies and encouraging corporate social responsibility. They also represent a larger trend in the financial sector toward sustainability.

4. Frameworks for Sustainable Investing Evaluation

Sustainable investing is guided by frameworks that systematically evaluate a company's environmental, social, and governance (ESG) performance. One notable comprehensive framework that places a strong emphasis on openness is the Global Reporting Initiative (GRI). GRI offers reporting criteria for sustainability that address a wide range of topics, including human rights and governance.[12] Businesses that follow the GRI guidelines improve disclosure uniformity and comparability while providing investors with comprehensive information about their sustainable operations.

The Sustainability Accounting Standards Board (SASB), which focuses on financially significant ESG concerns industry-specific, is another important framework. By requiring companies to publish information pertinent to their industry, the SASB's industry standards assist investors evaluate potential risks and possibilities. By ensuring that ESG data is customized for each industry, SASB's methodology offers a more insightful and financially viable view of a company's sustainability performance.[6]

Risks and possibilities associated to climate change are highlighted by the Task Force on Climate-related Financial Disclosures (TCFD). Companies are encouraged by TCFD recommendations to provide information regarding the financial effects of climate change on their business. For investors looking to assess how businesses are responding to climate-related issues and aligning with international climate goals, this approach is essential. Investors receive a complete and standardized toolset for assessing sustainable investments, encouraging transparency, and supporting ethical business practices by combining these principles.Various frameworks and models that aim to thoroughly evaluate a company's environmental, social, and governance (ESG) activities serve as a reference for sustainable investing.[7][7]As a fundamental framework, the Global Reporting Initiative (GRI) emphasizes transparency through standardized sustainability reporting. GRI helps businesses to disclose information regularly, giving investors a complete picture of their sustainable practices. It covers topics ranging from human rights to governance.[12] Beyond these, the United Nations Sustainable Development Goals (SDGs) offer a global framework that companies integrate into their sustainability initiatives. Investors evaluate a company's contribution to wider global sustainability goals by aligning with certain SDGs. Furthermore, impact investing concepts allocate funds to initiatives that produce quantifiable financial returns in addition to social and environmental benefits. When taken as a whole, these frameworks and models offer a comprehensive strategy that helps investors assess sustainable investments and promotes corporate responsibility in the global economy.

5. Performance of Sustainable Investments

The analysis of the financial performance of sustainable investments in comparison to traditional investments has gained substantial attention in recent years. Numerous studies have investigated this relationship, yielding mixed findings and complex considerations.

Positive Aspects	Overall Trends	Challenges and Considerations

https://doi.org/10.52202/074664-0011

Long-Term Outperformance: Some research suggests that sustainable investments, particularly those integrating with the ESG criteria, have demonstrated competitive or even superior financial performance over the long term. **Risk Mitigation:** Sustainable investments are often viewed as having lower exposure to certain risks, such as regulatory, reputational, and environmental risks, contributing to greater resilience during market fluctuations. **Market Recognition:** Increasing investor awareness and demand for sustainable investments have led to a positive correlation between strong ESG performance and favorable market recognition, potentially influencing stock prices.	**Growing Interest:** The increasing interest in sustainable investing has led to a shift in capital allocation, with more investors considering ESG factors in their decision-making processes. **Integration into Mainstream Finance:** Sustainable investing is becoming increasingly integrated into mainstream financial practices, with a notable rise in ESG funds and sustainable indices. **Divergence of Findings:** Despite a growing body of research, there is no unanimous agreement on the financial performance of sustainable investments, with studies showing diverse outcomes influenced by factors such as time horizon, geographic region, and methodology.	**Short-Term Volatility:** In the short term, sustainable investments may experience fluctuations and underperformance compared to traditional counterparts due to market dynamics, sector biases, or macroeconomic trends. **Data Limitations:** Challenges in obtaining accurate and standardized ESG data may affect the precision of performance assessments, leading to variations in findings across studies. **Diverse Strategies:** The broad spectrum of sustainable investing strategies, from exclusionary screens to impact investing, makes it challenging to generalize performance outcomes, as each strategy may yield different results. **Industry Variances:** The performance of sustainable investments can be influenced by

		industry-specific factors, and certain sectors traditionally associated with sustainability may exhibit varying financial dynamics.

Sustainable investments, particularly those incorporating Environmental, Social, and Governance (ESG) criteria, have shown promise in mitigating downside risks and enhancing long-term performance. Research suggests that companies with robust sustainability practices may demonstrate greater resilience to certain risks, contributing to a potential risk mitigation effect. Contrary to the traditional belief that sustainability compromises financial returns, studies indicate that sustainable investments can exhibit competitive or superior long-term performance. This challenges the notion that there is a trade-off between sustainability and returns.[13]It is acknowledged that sector-specific variances exist in the risk-return trade-offs within sustainable investing. Industries with high ESG performance may experience different dynamics compared to sectors facing sustainability challenges, emphasizing the importance of considering sector-specific factors in sustainable investment strategies.

Positive ESG performance is increasingly correlated with favorable market recognition, potentially leading to positive impacts on stock prices. Companies demonstrating robust ESG credentials are likely to attract investors and experience enhanced valuation over time.[14] However, the analysis

 https://doi.org/10.52202/074664-0011

of risk-return trade-offs in sustainable investing faces challenges due to difficulties in obtaining accurate and standardized ESG data. Inconsistencies in reporting practices and data quality across companies can hinder the precision of risk assessments. Additionally, the relationship between risk and return in sustainable investing may vary over different time horizons. While some studies focus on short-term fluctuations, others emphasize the potential long-term benefits of sustainable practices in terms of risk reduction and return enhancement, highlighting the importance of considering time horizon factors in evaluating sustainable investment strategies.

The influence of sustainable investing on long-term financial returns is gaining prominence, driven by various factors. Studies indicate that investments incorporating environmental, social, and governance (ESG) factors may exhibit competitive or superior long-term financial performance compared to traditional counterparts. Beyond potential financial gains, sustainable investing is associated with risk mitigation, as companies with robust sustainability practices often demonstrate resilience to certain risks, contributing to sustained financial performance. The growing interest in sustainable investments has elevated market recognition for companies with strong ESG credentials, potentially enhancing stock prices and valuations. This shift in capital allocation, where more investors integrate ESG considerations into decision-making, further underscores the potential impact on the financial performance of companiesaligning with sustainability goals. Nevertheless, industry-specific dynamics play a role, with certain sectors experiencing distinct financial outcomes based on their sustainability practices. As sustainable investing becomes more integrated into mainstream finance through ESG funds, sustainable indices, and regulatory developments, its influence on long-term financial strategies continues to grow.[15]

6. Challenges and Criticisms

Evaluating sustainable investments presents a set of challenges that stem from the complexity and evolving nature of environmental, social, and governance (ESG) factors. Here are some key challenges associated with assessing sustainable investments:

1. The reliability of sustainable investment decisions is restricted by limited data quality and availability in Environmental, Social, and Governance (ESG) information since companies' disclosure policies vary widely in terms of transparency.

2. Greenwashing is the practice of businesses inflating their sustainability assessments, endangering investor backing for weak or dishonest projects, and violating their ethical standards.

3. To avoid obsolete assessments and guarantee compatibility with current expectations and standards, the dynamic nature of the ESG landscape—characterized by changing standards, rules, and investor preferences—necessitates constant modification.

4. It is difficult to quantify social impact on sustainable investments because it is difficult to assign numerical values to qualitative and subjective features. This makes it difficult for investors to compare and evaluate societal contributions in financial assessments.

5. For investors trying to find a balance between short-term financial advantages and long-term benefits from sustainability, navigating this conflict between both is a constant difficulty.

6. Global differences in ESG priorities provide difficulties; with different material issues in different areas, businesses managing differing expectations need to modify assessment frameworks to guarantee a thorough and precise assessment of sustainability performance.

The challenges in sustainable investing stem from limited and inconsistent data in Environmental, Social, and Governance (ESG) information, leading to reliability issues. Investors are at risk from greenwashing, or overstatement of environmental initiatives. The ever-changing ESG landscape necessitates ongoing adaptation to new requirements. It is difficult to quantify social impact, which makes it difficult for investors to compare societal contributions fairly. A continuous challenge for investors is striking a balance between short-term financial advantages and long-term sustainability. Furthermore, assessments are made more difficult by regional differences in ESG priorities, which calls for modifications to meet different expectations and guarantee accurate sustainability evaluations.

7. Future Directions

The future of sustainable investing is marked by several pivotal directions that reflect a global commitment to environmental, social, and governance (ESG) considerations. Regulatory standardization is anticipated to take center stage as governments worldwide recognize the need for consistent ESG reporting. This move is expected to bring about uniform frameworks, enhancing transparency and comparability in sustainability disclosures across industries and regions.

1. It is expected that impact measurement techniques will advance, satisfying the growing need for sophisticated instruments that let investors evaluate and convey the concrete social returns on their sustainable investments—benefits that go beyond conventional financial measurements.

2. The integration of technology, AI, and data analytics is going to transform ESG assessment. It will do this by using sophisticated algorithms and automation to make assessing a company's sustainability performance easier and more accurate for investors.

3. The growth of sustainable indices and benchmarks is indicative of a more general trend in the direction of market accountability and transparency. Investors may use these dependable benchmarks to follow and evaluate companies according to their ESG indicators, which helps them make well-informed investment decisions.

4. A global commitment to addressing environmental concerns through sustainable and innovative solutions is highlighted by a rise in investments towards renewable energy, carbon capture technologies, and other climate solutions.

5. It is anticipated that investor activism would increase, with investors adopting a more outspoken position to support strong ESG policies and moral corporate governance. This indicates a proactive dedication to promoting change and guaranteeing that companies take responsibility for their effects on the environment and society.

These future directions collectively underscore the evolution of sustainable investing towards a more standardized, technologically driven, and impactful landscape. Investors, companies, and regulators alike are gearing towards a future where sustainability considerations are integral to decision-making, reflecting a broader shift towards responsible and conscientious financial practices.

8. Conclusion

The literature review has revealed the multifaceted landscape of sustainable investing evaluation, emphasizing its growing significance in the financial world. Key findings highlight the diverse methodologies, frameworks, and metrics employed in assessing the sustainability performance of investments. While successful case studies underscore the potential for financial success aligned with environmental and social responsibility, challenges such as data quality, subjectivity, and short-term volatility have been identified. Ongoing research and development are crucial for addressing these challenges and advancing the field.

The literature points toward potential future directions, including enhanced impact measurement frameworks, the integration of artificial intelligence, dynamic risk-return models, and increased focus on biodiversity and social impact metrics. As sustainable investing continues to evolve, the need for standardized approaches, transparency, and a deeper understanding of interconnected factors becomes paramount. The broader implications of sustainable investing on global finance, regulatory landscapes, and corporate practices underscore the urgency of ongoing research to refine evaluation methods and ensure a robust foundation for sustainable finance in the future. In navigating the complex intersection of financial returns and sustainability goals, continuous exploration and innovation will be essential for shaping a resilient and impactful future for sustainable investment.

https://doi.org/10.52202/074664-0011

References

[1] D. N. Sharma.:The Ethics of Investment: Balancing Profit with Social Responsibility," *Interantional Journal of Scientific Research in Engineering and Management*, 07, 03, 3–7(2023).

[2] K. Sustainable, T. Chapter, and L. Goals.:*Corporate sustainability*, vol. 189, no. 11,(2021).

[3] M. Koenigsmarck and M. Geissdoerfer.:Shifting the Focus to Measurement: A Review of Socially Responsible Investing and Sustainability Indicators," *Sustainability*, vol. 15, no. 2, 984(2023).

[4] A. Edmans and M. Kacperczyk.:Sustainable Finance," *Rev Financ*, vol. 26, no. 6, 1309–1313(2022).

[5] A. Handayani and R. Rokhim.:A Comparative Study of Financial Performance between Sustainable and Conventional Investment," *Journal of Entrepreneurship & Business*, vol. 4, no. 2, 114–124(2023).

[6] C. S. Akala, T. Neuhaus, and I. O. Leary-Govender.:A Systematic Review of Sustainable Investment Approaches," *Int J Econ Finance*, 14, 12, 72(2022).

[7] P. Ovezikoglou, D. Aidonis, C. Achillas, C. Vlachokostas, and D. Bochtis.:Sustainability assessment of investments based on a multiple criteria methodological framework. *Sustainability (Switzerland)*, 12, 17, 1–13(2020).

[8] T. C. Burnham, H. Gakidis, and J. Wurgler.:Investing in tLeunghe presence of massive flows: The case of MSCI country reclassifications," *Financial Analysts Journal*, vol. 74, 1, 77–87(2018).

[9] P. Bansal.(2005). Evolving sustainably: A longitudinal study of corporate sustainable development," *Strategic Management Journal*, vol. 26, 3, 197–218.

[10] T. Krappel, A. Bogun, and D. Borth.:*Heterogeneous Ensemble for ESG Ratings Prediction*, vol. 1, no. 1. Association for Computing Machiner, (2021).

[11] M. A. Camilleri.:The market for socially responsible investments: A review and evaluation," *CSR and Socially Responsible Investing Strategies in Transitioning and Emerging Economies*, 171–188(2020).

[12] C. de Villiers, M. La Torre, and M. Molinari.: The Global Reporting Initiative's (GRI) past, present and future: critical reflections and a research agenda on sustainability reporting (standard-setting)," *Pacific Accounting Review,* (2022)..

13] X. G. Yue, Y. Han, D. Teresiene, J. Merkyte, and W. Liu.: Sustainable funds' performance evaluation," *Sustainability (Switzerland)*, vol. 12, no. 19, 1–20().

[14] C. E. Bannier, Y. Bofinger, and B. Rock.:The risk-return tradeoff: are sustainable investors compensated adequately?. *Journal of Asset Management*, vol. 24, no. 3, 165–172(2022).

[15] S. R. M. Vasconcelos, J. O. Dos Santos, J. C. Marion, and D. R. Bergmann.:Análise da rentabilidade dos fundos de investimentossustentáveisbrasileiros no período 2010-2016.*RevistaContemporânea de Contabilidade*, vol. 16, no. 39, 89–112(2019).

Exploring the Dynamics of Retail Investor Behavior: An Empirical Analysis of Determinants and the Impact of Self-perceived Confidence on Investment Decision

Mohammed Ashraf Ali[1], Md Shamim Gaddi[2] and Mohd Zafeer[3]

[1] Department of Commerce, Aligarh Muslim University,Aligarh, UP, India

[2] Department of Commerce, Aligarh Muslim University, Aligarh, UP, India.

[3] Department of Commerce, Aligarh Muslim University, Aligarh, UP, India.

Abstract:

This research delves into the intricate web of retail investor behavior by conducting an empirical analysis that examines the determinants influencing investment decisions and the significant role of self-perceived confidence in shaping these decisions. In a rapidly evolving financial landscape, understanding the factors driving retail investors' decision-making processes is crucial for market stakeholders, policymakers, and investors. Key determinants such as financial literacy, risk tolerance, market knowledge, social influences, demographic variables, and economic conditions are examined to elucidate their impact on investment decisions. Furthermore, the research investigates the nuanced relationship between self-perceived confidence and investment behavior.This study examined using a structured questionnaire to gather information from 360 retail investors located in four districts in India. PLS-SEM and SPSS were taken into consideration for testing the research hypotheses. The findings enlighten that determinants of retail investors, i.e., Self- perceived confidence, Risk tolerance, financial literacy, and social influences, are significantly associated with investment decisions. In the realm of behavioral finance, it is anticipated that the findings of this study will add substantially to the body of existing knowledge. Understanding the determinants of retail investor behavior and the role of self-perceived confidence can inform policymakers, financial educators, and investment professionals in designing targeted interventions to enhance investor decision-making skills and promote financial well-being.In a time when more people are investing in the financial markets on their own, this study is pertinent and important in building a community of knowledgeable and resilient investors.

Keywords: Retail Investor Behavior, Investment Decisions, Self-perceived Confidence, Behavioral Finance.

https://doi.org/10.52202/074664-0012

1. Introduction

Retail investors are essential players in the financial markets who enhance market efficiency and liquidity. Regulators, policymakers, and market participants must comprehend the conduct of these investors. Retail investors are an essential component of the financial market, contributing significantly to the overall trading volume. Understanding their investment behavior is crucial for policymakers and market participants alike. This research paper explores the dynamics of retail investors' behavior by analyzing the determinants of their investment decisions. Specifically, we will examine the key factors that influence their investment decisions.Retail investors' decisions are impacted by a variety of factors, such as psychological biases, market sentiment, and individual traits; institutional investors frequently adhere to systematic investment techniques. To better understand the dynamics of retail investor behavior, this study will concentrate on the factors that influence their investment decisions and how self-perceived confidence affects them.

The finance literature frequently discusses the variety of individual investors' financial behavior, particularly asset allocation and investing decisions. Investigating fundamental questions like what influences small and retail investor activity in the financial markets is crucial for improving our understanding of retail investor behavior. Do social environments and contextual learning over time, such as peer and educational instruction and learning about investing decisions, have an impact on the behavior of retail investors? Do their financial factors, such as income and investment levels, or demographic characteristics, like gender, age, and occupation, influence their behavior more?By elucidating the impact of behavior determinants in shaping retail investors' behavior, this research aims to contribute to the body of literature on behavioral finance while investigating these topics in an empirical setup.This research looks at three possible domains of determinants, including fundamental change, Unforeseen market occurrence, personal factors, and self-perceived confidence level of retail investors, in order to measure the behavior of these investors in the stock market. This study looks at the variables that could affect retail investors' trading patternsutilizing information obtained from a survey of over 360 retail investors selected from a group of investors who use a financial brokerage client with a solid reputation in the financial services industry.

The current empirical study of the factors influencing the trading behavior of retail investors reveals that financial and demographic characteristics like age, gender, education level, and investment levels influence their behavior across three domains; on the other hand, characteristics like marital status and occupation do not significantly influence their trading behavior. Additionally, it is found that a significant element influencing retail investors' trading behavior is their self-perceived confidence.Self-reported portfolio size and self-perceived confidence level are positively correlated, suggesting that overconfident retail investors tend to think that because they have better investing skills, their portfolios will inevitably perform better, and, as a result, they typically have larger investment portfolios than average.

By offering empirical data on how retail investors interpret shifts in the fundamentals, unforeseen developments in the market, and their own level of confidence about financial and investment decisions, as well as how these issues differ amongst people with different economic and demographic characteristics, this study adds to the growing body of literature on behavioral and personal finance. Retail investors' confidence in their ability to make trading and investment decisions is influenced by various factors, including age, gender, income, and stock market exposure.From the perspective of personal financial management, our analysis characterizes these variations fundamentally and explains their relevance.

2. Literature Review

There are various reasons why the topic of factors influencing investor behavior in the stock market is significant. Behavioral finance and economics is a newly popularized field of academic research that it is associated with. This domain is equally important for professionals in the personal finance and financial markets.Secondly, understanding investor behaviour helps to decipher the asset allocation problem's cross-sectional mechanism and, as a result, helps investors understand asset pricing. In terms of methodological contributions, this type of explanation is unquestionably thought to aid in advancing the case for the behavioral asset pricing approach. Last but not least, an abundance of empirical studies on investor behavior have been conducted in lots of countries and economies, primarily utilizing financial market data.Few primary research approaches, including surveys, psychometric tests, and experiments, have been used to try to understand investor behavior.This attempt adds to the vast amount of literature in behavioral finance and economics by presenting empirical evidence using these unique approaches that are often recognized as superior for understanding behavioral concerns.

Regarding different asset classes, the behavior dynamics of investors and the performance of investment portfolios are examined. [10] For instance, looks at the behavior of investors and mutual fund performance. He claims that if current investors stop reacting as strongly to the fund's lower-than-expected performance as opposed to its performance meeting their expectations, the fund manager will have more incentive to shift risk. However, the perspective of the financial market and financial services sector informs this relationship. Understanding whether different traits of individual and retail investors influence their investment behaviour is not very helpful.

Analyzing every aspect of investors' behavior, such as choosing how much money to invest in the stock market, takes a lot of work. However, studying every decision they make is possible. Small choices, for the most part, are always context-dependent: the investor might have decided to invest the money after learning that the company had made a significant announcement in the news, or he might have just realized that he would not be able to take a vacation and that the agent had recommended this particular stock [12].

Regarding investor behavior, thoroughly examining the significance of financial and demographic factors could be two of the most important aspects. The results could have implications for professionals in wealth management, retail and small investors, and other users. Academic research indicates that various demographic factors, including age, gender, income, marital status, and education, have an impact on an individual investor's investment decisions and risk tolerance [9], [11], [13] and [1], among others.These factors also influence an individual's aversion to realize losses and their selection of suitable instruments for investing [7], [14], [8], [6].

By analyzing the patterns of gross flows in the mutual fund sector, [10] give empirical evidence on investor behavior and imply that a certain percentage of investors assess and react to performance during periods noticeably shorter than a year. They contend that knowing how investors behave may affect fund advisors' motivations.

Much other research uses novel approaches to study investor behavior, including econometric analysis of transaction history, neural data analysis utilizing realization theory framework, and experimental methods examining genetic traits. [4] Show that, within a traditional utility maximization framework, variation in background risk exposure can account for part, but not all, of the observed cross-sectional variation in portfolio holdings. This background risk exposure can originate from real estate holdings, labor and entrepreneurial income, transaction costs, borrowing restrictions, and life cycle considerations.

Aversion to risk and mental aptitude are two of the most significant personality type elements that have been used to explain decisions regarding the percentage of risky assets in an investor's overall portfolio. Other factors that have been considered include socioeconomic and demographic data.

Overconfident investors base their decisions only on their own free will and think that other people's decisions regarding their investments could be impacted by their thoughts, emotions, or experiences. Their arrogant attitude completely supports their idea, and they consider the opinions and suggestions of others to be dull and illogical. Overconfident traders trade excessively and with overconfidence, disregarding the risk level. This may or may not be a good thing. They believe that taking on risks is a crucial part of their financial plan. Nonetheless, detractors argue that because chronic overconfidence tends to decrease market efficiency, it actually encourages more trading activity.

3. Research Methodology

Utilizing fundamental univariate statistical tests, such as the t-test (paired), we examine whether the influence of domain-specific factors influencing investor behavior differs among sample retail

https://doi.org/10.52202/074664-0012

investors' financial and demographic profiles. Any observation in one sample with a natural counterpart observation in another is compared using the paired t-test. In order to generate a t-statistic, we use this test to determine the difference between paired values of two or more samples (in our case, classified by domain and demographic profiles), taking into account the variation of values within each sample.Since our retail investor survey data has matched samples, the paired t-test can be applied here as a more potent alternative to a two-sample approach like the two-sample t-test.Furthermore, the linear relationship between the investor's opinion of the market and himself/herself and the financial and demographic factors is assumed to be driven mainly by investor behavior.

We also want to investigate further the variables that affect the amount of portfolios retail investors hold. In our survey, we asked retail investors to report the approximate size of their stock market investment portfolios. Based on their reporting, we classify investors into two groups of portfolio holders, small and big portfolio holders. We use a binary regression approach to analyze the impact of demographic, financial, and perceptual determinants of self-reported portfolios. The self-reported portfolio size is our dependent variable, with a value of 0 (Small portfolio investors) or 1(Large portfolio investors). The self-perceived confidence level and a collection of socioeconomic factors make up the explanatory variables.

Since self-perceived confidence level is directly correlated with portfolio size, we employ this measure as one of the explanatory factors.

Lastly, we are interested in determining whether there is a direct correlation between retail investors' self-perceived confidence in the stock market and their self-reported holdings. In order to investigate the association between self-confidence and portfolio size, we will be employing two stages least square (2SLS) regression estimation, as the problem of simultaneous equation bias is likely to arise when using the OLS regression estimation approach.

4. Results and discussion

Using survey data from a sample of retail investors who use financial brokerage house services, this study investigates the factors influencing the trading behavior of retail investors. A structured questionnaire was used to gather survey data, and it was disseminated via a variety of offline and online channels. The survey was confidential and optional to complete. The survey instrument includes information on the intended use of the data and the reason for data collection. The questionnaire had subsections on financial, psychological, and demographic items.A total of 670 retail investors received questionnaires.From this, about 53% (n = 360) of the participants answered the survey.

As expected, only 27% of the sample investors are women, with men making up 73%. According to [1], men usually reorganize their financial portfolios, whereas women take fewer risks in the stock market. Therefore, we don't mind if our sample has a skewed male-to-female ratio.The data we obtained from the sample respondents indicates that around 76% of them are married, 72% have a graduate degree, 11% are professionals, and 82% are salaried employees. We do not disclose information on the sample respondents' occupation, marital status, or level of education.

Several internal and external factors can influence the trading behavior of retail investors. Financial and psychological heuristics like self-control, caution, and prudence are examples of internal forces. Information asymmetry, financial addiction, and the state of the economy and market are examples of external factors that could influence their behavior [3]. A five-point Likert scale was used in this study to evaluate the trading behavior of retail investors.

We hypothesize that a variety of domains, such as fundamental change, unforeseen market occurrences, and personal factors, influence investors' self-perceived confidence levels. We used the three areas of fundamental change, sudden market occurrences, and personal factors to regress self-confidence to verify this hypothesis. The self-perceived confidence level of investors was adversely correlated with sudden market occurrences but positively correlated with fundamental change and individual variables (See Table 1). This supports our claim that retail investors feel more confident when making trading decisions and are able to deal with fundamental changes and personal circumstances that are predicted mainly and known to them beforehand. However, any sudden and unanticipated change in the market deflates their trust.

We were interested in finding out if these traits influence the reported size of retail investors' investment portfolios because we have researched how self-reported factors of their trading behavior vary across demographic characteristics. We arranged our sample of retail investors according to the size of their self-reported portfolios, and we looked at whether any demographic factors influenced these portfolios.

The statistically significant regression coefficient indicated a positive relationship between their portfolio size and key demographic attributes, including gender and their level of education. It suggests that male retail investors tend to have larger portfolios and that their education level influences retail investors' portfolio sizes. According to self-reported portfolio size, monthly income and investments as a proportion of income also have a favorable correlation (See Table 2). Not surprisingly, though, we discovered that larger portfolios are typically held by retail investors who have higher incomes and more stock market investment.It's interesting to note that sample

https://doi.org/10.52202/074664-0012

retail investors' portfolio sizes are independent of their age, marital status, and professional occupation. [2]Proposed that there exists a positive correlation between an investor's age and gender and the size of their portfolio. Age was not a significant factor in our findings, but gender did matter in our sample of retail investors.

Another important factor determining portfolio size is the confidence that retail investors believe in themselves. We discovered that the portfolio size of the sampled retail investors is positively and significantly correlated with self-confidence (See Table 2). The relationship between portfolio size and self-confidence suggests that when retail investors are confident in their ability to make money in the stock market, they tend to hold larger portfolios. For this possible causal association between portfolio size and self-confidence, we provide no empirical support in this paper. It is also possible that investors who have larger portfolios are more confident in their abilities and performance in the stock market.

Table 1: Regression results

Dependent variable	**Self-perceived confidence level**
Intercept	3.610 (0.691)***
Fundamental change	0.261 (0.054)***
Unforeseen market occurrence	−0.292 (−0.051)***
Personal factors	0.116 (0.043)**
Adjusted. R^2	0.681
F-stat.	148.987
Durbin-Watson	1.836

Notes: The dependent variable is self-confidence; the independent variables are fundamental change, unforeseen market occurrence, and personal factors. Standard errors are presented in parenthesis ( ). **, ***At the 5 and 1 percent significance levels, respectively, coefficients are statistically significant.

Table 2: Binary regression estimates

	Self-reported portfolio size
Intercept	−0.715***
	(0.121)
	[25.626]
Gender	0.614**
	(0.420)
	[1.730]
Age	0.321

https://doi.org/10.52202/074664-0012

	(0.162)
	[2.495]
Marital status	0.775
	(0.722)
	[1.653]
Occupation	−0.066
	(0.142)
	[0.145]
Education	0.823***
	(0.354)
	[3.141]
Income	0.876***
	(0.318)
	[7.806]
Investment	0.258***
	(0.162)
	[3.831]
Self Confidence	0.215***
	(0.051)
	[21.022]
Log likelihood	195.647
Cox and Snell R^2	0.235
Nagelkerke R^2	0.437

Notes: The dependent variable is the self-reported portfolio size; the explanatory factors are gender, age, marital status, education, occupation, monthly income, investments, and self-reported confidence level. The standard errors and Wald statistic are presented in parenthesis ( ) and [ ], respectively. **, *** coefficients are statistically significant at the 5 and 1 percent significance levels.

Using 2SLS estimate, we discovered that the sample investors' portfolio size is a growing function of their perceived confidence level. Since we anticipated that the dependent variable's error term, or the self-reported portfolio size, would be correlated with other variables in our estimation, we used a set of instrument variables to solve this issue by replacing the problematic variables(See Table 3). Our findings are consistent with the theory that an investor's trading behavior is primarily influenced by her self-perceived confidence and demographic traits, which are essentially products

https://doi.org/10.52202/074664-0012

of a number of social and economic factors. The results are strong and significant according to the Heckman test of endogeneity.

Table 3: Regression coefficient of 2SLS estimates

	Self-reported portfolio
Intercept	3.486***
	(0.145)
	[9.237]
Self-perceived confidence	0.481***
	(0.021)
	[4.031
Adjusted R^2	0.575
F-stat	24.328

Notes: The dependent variable is the portfolio size; the explanatory variable is the self-perceived confidence; the instrument variables are the income, investment, fundamental change, unforeseen market occurrence, and personal factors. The standard errors and t-statistic are reported in parenthesis () and [], respectively. ***At the 1 percent threshold, coefficients exhibit statistical significance.

5. Conclusions

This paper examined how retail investors' trading behaviors vary based on their financial and demographic variables. We assessed investor trading propensity using structured survey data in three areas: fundamental change, unforeseen market occurrences, and personal factors. We found that these domains varied across several financial and demographic criteria. Specifically, we empirically show that three areas of financial and demographic parameters, such as age, gender, and investment levels, determine their trading behavior.Retail investors' trading behavior is mostly dictated by their perceived level of confidence, which is influenced by both personal and predicted market fluctuations. However, characteristics like marital status and occupation have no significant impact on their trading behaviour. Rather portfolio size and level of self-perceived confidence are positively correlated, suggesting that overconfident investors often believe that higher returns are a direct outcome of their better investment skills. As a result, they hold larger investment portfolios. Retail investors' investment behaviour is predicted by their self-perceived confidence, which lends credence to the idea that investor education and evaluation are crucial for making prudent decisions regarding investments.

6. Limitations

In summary, the study's possible constraints include the fact that the majority of the retail investors are middle-aged, male, and well-educated. Such a feature implies that the sample would be more likely than the general investing population to exhibit skewed trading behavior and a riskier mindset [1].The degree of investment and size of a retail investor's portfolio could be a potential limitation of the study.

Fewer individual investors who invested a larger portion of their income in the stock market and, consequently, had larger investment portfolios made up our sample. These problems could show up in the self-perceived confidence level test.Because it used the psychometric technique to determine the potential influence of demographic factors on trading behavior, this study carried over the limitations associated with it, including measurement challenges and the reproducibility of the findings.

Acknowledgments: This study did not receive any fund.

Disclosure of interests: The authors declare that they do not have any competing financial interests or personal relationships that could have appeared to influence the work reported in this paper.

References

1. Barber, B. M., & Odean, T. :Boys will be boys: Gender, overconfidence, and common stock investment. The quarterly journal of economics, 116(1), 261-292 (2001).
2. Barber, B. M., Odean, T., & Zheng, L.: Out of sight, out of mind: The effects of expenses on mutual fund flows. The Journal of Business, 78(6), 2095-2120(2005).
3. Chandra, A., & Kumar, R. : Factors influencing Indian individual investor behaviour: survey evidence(2012).
4. Curcuru, S. E., Thomas, C. P., & Warnock, F. E.:On returns differentials. Journal of International Money and Finance, 36, 1-25 (2013).
5. Curcuru, S. E., Thomas, C. P., & Warnock, F. E. : On returns differentials. Journal of International Money and Finance, 36, 1-25(2013).
6. Daniel, K., Hirshleifer, D., & Subrahmanyam, A.:Investor psychology and security market under-and overreactions. the Journal of Finance, 53(6), 1839-1885(1998).
7. Fama, E. F., & French, K. R.:Testing trade-off and pecking order predictions about dividends and debt. Review of financial studies, 1-33(2002).
8. Filbeck, G., Hatfield, P., & Horvath, P.:Risk aversion and personality type. The Journal of Behavioral Finance, 6(4), 170-180 (2005)..
9. Hershey, J. C., & Schoemaker, P. J. (1980). Risk-taking and problem context in the domain of losses: An expected utility analysis. Journal of Risk and Insurance, 111-132.
10. Johnson, W. T. :Who monitors the mutual fund manager, as well as new or old shareholders. Unpublished working paper (2007)..
11. Riley Jr, W. B., & Chow, K. V.:Asset allocation and individual risk aversion. Financial analysts journal, 48(6), 32-37 (1992).
12. Salimov, D.:Determinants of Individual Investors' Behavior in Portfolio Decisions (Doctoral dissertation) (2012).
13. Schooley, D. K., & Worden, D. D.:Investors' asset allocations versus life-cycle funds. Financial Analysts Journal, 55(5), 37-43(1999).
14. Shefrin, H., & Statman, M.:Making sense of beta, size, and book-to-market. Journal of Portfolio Management,

https://doi.org/10.52202/074664-0012

Globalisation, Economic Growth and Employment Opportunities

Reetika Dhawan[1] and Dr. TrilochanKumar[2]
[1]Research Scholar, Department of Economics, Jammu University
[2]Assistant professor, Department of Tourism, UITHM, Chandigarh University

Abstract: Globalisation leads to the integration of economy of a country to the world economy and helps in expanding the wings. It allows free movement of trade, goods, people, services, technology, and capital. It refers to the integration of world economies together. The term sometimes also refers to the movement and mobilization of goods, services, people , knowledge across borders. The term has been widely used in 1980s which marks an era of technological advancement which makes it easier to boost growth and employment. Earlier the world economies were not connected to share knowledge resources and human resources but with the introduction of globalization. As a result of globalisation and free trade practices, there has been an expansion of market frontiers and productive activities.This Paper explains how Globalization has an impact on economic growth and employment opportunities across India.

"There is a consensus in the world today that economic development is not at all and the GDP is not necessarily a measure of progress of society"

K.R. Narayanan

Keywords: Globalization, IMF, Integration, Trade, Debt, WTO etc.

1. Introduction:

Globalisation is the integration of countries through foreign trade and investment. The economic crisis of 1991 in which India was trapped in economic debt and was not able to make repayments for what it borrowed from the neighboring foreign countries and it resulted in Inflation. This situation influenced the country to make certain developmental policies in order to draw out the country from debt. The policy formation for the structural development of the country in turn pushed it in debt and thus India took shelter under IBRD i.e. International Bank for Reconstruction and Development also known as World Bank and the IMF i.e. International Monetary Fund. The institution stated that India should open up its Trade and removes restrictions and open its economy and let it integrate with the other economies of the world. This gives birth to NEP i.e. New Economic Policy which has Liberalisation, Privatisation and Globalisation as its main components.

Liberalisation and Privatisation were the strategies of development and Globalisation was the outcome of the strategies. Globalisation leads to the integration of economy of a country to the world economy and helps in expanding the wings. It allows free movement of trade, goods, people, services, technology, and capital. It not only increases cross-border movement of capital, technology, goods and services but it also increases such economic activities that will straddle national boundaries. Globalisation can be economic, political, cultural or environmental. Economic globalization refers to the integration of financial markets and MNCs. Political globalization refers to the policies that boost trade. It deals with IMF, WTO etc. Cultural globalization refers to the communication, transportation that is extended with the help of technology.India has been undergoing a rapid developmental process from a very long time (11, 15). The Process was speeded up after the independence era. Through FDI i.e. foreign direct investment, technological advancement, Interdependence on other economies and integration with other economies leads to economic growth of the economy. India gained highly from the LPG model as its GDP increased to 9.7% in 2014-2015 (7). Globalisation has helped in removing major trade barriers.

2. Objectives of the Study

1. To study the impact of Globalisation on Economic Growth of a country.
2. To study the impact of Globalisation on Employment Opportunities.

In the Globalisation era, i.e. from 1990-91 to 1996-97 the GDP increased from 1.3% to 7.8%. In 1997-98, the GDP fall to 4.8% and it continued falling to 2002-02 to 4.0%. In 2003-04, it increased to 7.5% and showed increasing trends till 2006-07 and it was the highest i.e. 9.2% in 2006-07.The following table shows the impact of Globalisation on Economic Growth of country.

Table no. 1: Impact of Globalisation on Economic Growth of country.

Year	GDP in %
1990-91	1.3
1991-92	1.5
1992-93	5.3
1993-94	6.2
1994-95	7.3
1995-96	7.3
1996-97	7.8
1997-98	4.8
1998-99	4.3
1999-2000	5.1
2000-01	4.4
2001-02	5.8
2002-03	4.0

https://doi.org/10.52202/074664-0013

2003-04	7.5
2004-05	8.5
2005-06	9.0
2006-07	9.2
2007-08	8.2
2008-09	6.1
2009-10	7.8
2010-11	6.6
2011-12	5.1
2012-13	6.9
2013-14	6.7
2014-15	7.0

Source: Handbook of statistics on Indian economy, 2014-15

Globalisation allows FDI i.e. Foreign direct investment in the form of lands, capital, factories, MNCs which increases the growth of trade among countries and helps in the transfer of funds, technology etc. In today's highly competitive global environment, it is essential to make innovative and skilled use of information technology (6,10). Internationalisation has led to a growth in the number of multinational corporations operating in every region of the world. Prior to the phase of globalisation, the United States of America was the major player in multinational corporations. But once the trend of globalisation expanded, a great number of different countries entered the race to become multinational corporations (1). It allows technological advancement and stimulates new techniques of production which makes production cost effective and thus helps in the economic growth by increasing GDP. An international technological infrastructure has been built as a result of the development of technologies connected to telecommunications, such as the internet, telephones, wireless technologies, and undersea fibres. This has made it possible for information to be transferred across international borders in a more seamless manner. Regulations concerning intellectual property rights, patents, and international agreements are simple to implement. In areas where there has been an increase in both investment and trade, it is clear that the poverty rate has dropped for those regions. India is a nation that is experiencing a rapid increase in foreign direct investment (FDI), which has led to a decrease in the poverty rate. Mexico is in a stronger position than its neighbours to have prevailed over the macroeconomic crisis.

There are endless colossal components of globalization. For instance, as additional cash is poured into developing countries, there is a higher hazard for the in those global areas to monetarily succeed and increment their fundamental popular of living (5). Worldwide resistance supports inventiveness and development and proceeds with expenses for items/benefits in investigate. Developing nations can receive the rewards of front line age without present procedure a portion of the developing agonies related with progress of these advancements. Governments can all the more likely work on the whole toward regular wants now that there is an advantage in participation, an improved potential to associate what's more, facilitate, and a universal familiarity with issues. There is an extra get right of passage to abroad custom inside the state of motion pictures, tune, suppers, attire, and bounty more noteworthy. In fast, the division has more determination. Globalization has a compelling web sway on the monetary improvement of the world.

2.1. Globalisation and Employment:There are several facets and a dynamic relationship between globalisation and employment. In addition to this, they collaborate with one another (14, 8). On both of these topics, we have a considerable quantity of factual information at this point in time; however, there are virtually no factual indicators regarding the impact that globalisation of the banking sector has had on the labour market.Globalisation, when opens the home economy to world economy, it also opens various employment opportunities along with it (9). It is possible that globalisation will first have an effect on the number of jobs that are available in the economy, and as a result, it will have an effect on important macroeconomic parameters such as the unemployment rate and the ratio of businesses to the population (2). The fact that the influence might be different at the microscalefinancial level (foundation, venture, and monetary movement) and at the macroscale financial level (all the economy), just as it is in the short-term and long-term, is one of the factors that contributes to the complexity of the issue (3). Additionally, it has the potential to bring about job benefits for the nation as a whole as a result of increased profitability in the remaining projects, increased wages, and an increase in the demand for utilisation.

According to the findings of some of the most recent International Labour Organisation (ILO) surveys, the number of jobs that are currently available in the world is larger than it has ever been before. This optimistic viewpoint is supported by all reports(1). The true cause of these situations may in any case be attributed to factors other than the globalisation of the financial system, such as the expansion of segments. Globalisation may also have an effect on the structure of employment, such as the distribution of jobs across different financial activities (13; 14). The employment opportunities that are associated with particular financial activities may have a tendency to disappear, whilst employment opportunities that are associated with other activities, possibly new activities, are created as a result of shifting upper hands and examples of specialisation (12).

In this regard, the problem is made even more complicated by the fact that changes in the structure

 https://doi.org/10.52202/074664-0013

of jobs can be brought about by both the globalisation of the financial sector and the growth of technology. Job compensation may be affected in two different ways as a result of globalisation of the economy. To begin, increasing the total efficiency of the economy, also known as productivity, leads to a rise in real earnings that can be shared with salaries earned by members of the workforce. Secondly, by promoting the mobility of products and manufacturing factors, it is possible that it will eventually equalise pricing differences between countries. These price differences may include the cost of labour or the difference in remuneration for jobs. It would appear that both of these processes are taking place on a worldwide scale. (4) Recent studies conducted by the International Labour Organisation (ILO) indicate that the percentage of working citizens who are employed is falling. Furthermore, it appears that employment incomes in emerging nations are continually improving, which has resulted in a reduction of the gap between the salaries of people in different jobs around the world.While this is going on, it appears that the wage gap between those with the highest qualifications and those with the lowest qualifications is widening in industrialised nations.

The gaps in wages appear to be changing throughout the course of time. As a result of their connection to progress, professionals working in the field of research and development are generally considered to be essential to the economy of their respective nations. There is a tendency towards inversion that is produced by monetary globalisation in industrialised nations. From one point of view, some companies are moving their research and development efforts overseas in order to move closer to significant markets or to make use of capabilities that are more readily available in remote locations. In contrast, there are some companies who are moving their manufacturing activities offshore in order to concentrate on research and development within their own country. Existing business conditions are a component of the problem that exists between monetary areas in growing to aspects that are becoming increasingly apparent, such as the cost of labour. For instance, lowering security requirements, increasing working hours, or imposing limits on worker guilds all be tempting to multinational corporations, which could lead to a rise in offshore employment. In the same way that we have lately observed in a number of EU countries with regard to working hours, this may have an effect on the business circumstances that exist in the nations that are the source of offshore. Changes in patterns of specialisation brought about by monetary globalisation or technological innovation, such as an economy that is increasingly focused on providing services, may also have unanticipated implications for the conditions of the organisation.

Globalization has assumed a pivotal job in the formation of business openings in India. A cautious investigation of the effect of globalization on work in India guides us toward certain imperative elements. Monetary progression arrangements of the 1990s totally re-imagined the work situation of India. The open market strategies brought about a considerable improvement in the total national output of the nation. There was likewise a fundamentally increment in the fares. This development sought after brought about the requirement for expanded stockpile which thus required more creation; at long last bringing about expanded work. Globalization has likewise determined the development of customer markets. There has been a rambling development of different modern portions obliging shifted needs of purchasers. The appearance of a wide scope of mechanical divisions brought about the financial development of India on one hand and in the popularity for quality workforce on the other. This has led business open doors for the Indian youth at different enterprises and furthermore the presentation of specialized and the board preparing that would get ready possibility to be appropriate to take up different jobs in the corporate world. The development of business openings isn't just found in the sorted out segments but on the other hand is seen in the sloppy and unified parts also. Keeping pace with the positive effect of globalization on Indian work, the Government of India has taken various activities to make manageable business chances to all segments of the general public, the urban just as the country youth.

The impact of globalisation was felt in India in the mid-1990s, when then-Finance Minister Manmohan Singh implemented free market rules. This resulted in a significant increase in the nation's GDP, as well as substantial fare increases. There was a significant increase in the client base, which progressively led to the purchaser showcase, where market adjustments were dependent on interest supply chains. Indeed, the desired development resulted in a positive transformation, and inventories began to expand rapidly. Because supply is directly related to work, having more stock led to more production, which led to more work over time.

Because of globalisation and the expansion of the buyer market, many fragments in various parts of the business have emerged throughout the years. This has led a major increase in the rate of interest and supply. In recent years, various industry segments, such as data innovation, agricultural products, individual and exceptional care, medical services, and others, have entered the market. According to specialists, the presentation of a diverse range of divisions has led the nation's economy to grow positively. With the emergence of additional industrial fragmentation, there has been an increased demand for a competent labour. As a result, a large number of young people are pursuing jobs in each of these sectors in order to launch a successful career. In the messy part, there has also been an expansion in several fields, which has increased the nation's work tempo. According to continuing evaluations, there has been a significant increase in the number of people working in waste and related areas. The pay package in all of these turbulent portions has expanded as well. Because of globalisation and the expansion of the buyer market, many fragments in various

https://doi.org/10.52202/074664-0013

parts of the business have emerged throughout the years. This has led a major increase in the rate of interest and supply. In recent years, various industry segments, such as data innovation, agricultural products, individual and exceptional care, medical services, and others, have entered the market.

3. Conclusion:

The beneficial growth of the economy in the country can be attributed to the presentation of a diverse range of divisions. Considering the growing number of industrial segments that are emerging, there has been a demand for a workforce that is of high quality. As a consequence of this, a large number of young people are seeking employment in each of these sectors in order to launch a successful career. As well as in the sloppy portion, there has been an expansion in a variety of fields, which has resulted in an improvement in the rate of work throughout the nation. The ongoing studies indicate that there has been a significant increase in the number of people working in the sloppy and associated portions of the organisation. As a matter of fact, the compensation package in each of these chaotic portions has also grown in size.When it comes to the first decade of the 21st century, integration with the global economy has been linked to high levels of investment and savings, quick per-capita income growth, and dynamic export outcomes. However, it has also been related with some less advantageous aspects. In the first place, it has not resulted in a significant improvement in the employment situation of the Indian economy.

The second issue, which is connected to the first, is that the manufacturing sector, which is the most interconnected sector in the economy, has only seen a slight growth in its proportion of output and has actually experienced a slight decrease in its percentage of employment. In other words, global integration has been the engine that drives growth, but it has also warped structural change. Third, it is true that globalisation has exacerbated the mismatch between employment structure and output structure. Reason being, it has sparked a multi-faceted agricultural catastrophe. The productivities of agriculture and other sectors have not converged but have instead grown apart due to this crisis. Additionally, land-hunger is caused by the slow expansion of employment prospects both within and outside of agriculture. This, in turn, leads to widespread pushback from below to the purchase of land and other natural resources.This widespread resistance is an unintended consequence of globalisation.

References

1. Akram, M, Faheem, A, Khyzer.: Globalization and its impact on the World Economic

Development.Journal of Buisness and Social Sciences, (2011).Vol 2 No.23.

2. Alam, M., &Siddiqui, M. I.: Effective framework to tackle urban unemployment by e-government: an IoT solution for smart/metro cities in developing nation. *Journal of Science and Technology Policy Management*, *14*(1), 213-238 (2023).
3. Aslam, H., Nazir, A., & Zia, U.: Pakistan's Way Forward towards a Green Economy: Perspectives for a Clean Energy Transition. *JSTOR Sustainability Collection,* (2021).
4. Bue, M. C. L., Le, T. T. N., Silva, M. S., &Sen, K.: Gender and vulnerable employment in the developing world: Evidence from global microdata. *World Development*, *159* (2022).
5. Globalization and its Impact on Employment in India.Retrieved from Ukessays.com/essays/economics/globalization-and-its-impact-on-employment-in-India-economics-essays.php?ref=1.
6. Goldar, B.N.: Globalisation, Growth and Employment in the Organised Sector of Indian Economy, (2014). Working paper no. wp 06/2014.
7. Handbook of Statistics on Indian Economy,2015-2016.
8. Kokkodis, M.: Dynamic, multidimensional, and skillset-specific reputation systems for online work. *Information Systems Research*, *32*(3), 688-712 (2021).
9. Mariotti, S.: A warning from the Russian–Ukrainian war: avoiding a future that rhymes with the past. *Journal of Industrial and Business Economics*, *49*(4), 761-782 (2022).
10. Shahzad, Alisha.: What is Globalization-Historical Background. Jadavpur Journal of International relations, Vol 10, Issue 1, Pp. 204-212 (2006)..
11. Singh, Jasbir.: Employment generation Under Globalization in India: Adverse Implication of Capitalism. World review of political economy.Vol 8, No. 2 Pp.190-202 (2017).
12. Torosyan, K., Wang, S., Mack, E. A., Van Fossen, J. A., & Baker, N.: Assessing the impact of technological change on similar occupations: Implications for employment alternatives. *Plos one*, *18*(9), e0291428, (2023).
13. Xia, W., Apergis, N., Bashir, M. F., Ghosh, S., Doğan, B., &Shahzad, U.: Investigating the role of globalization, and energy consumption for environmental externalities: empirical evidence from developed and developing economies. *Renewable Energy*, *183*, 219-228(2022).
14. Yang, B., Jahanger, A., Usman, M., & Khan, M. A.: The dynamic linkage between globalization, financial development, energy utilization, and environmental sustainability in GCC countries. *Environmental Science and Pollution Research*, *28*, (2021).

 https://doi.org/10.52202/074664-0013

Changing Paradigm of Travelling in Digital Age

Aanchal Arora[1], Dr Manju[2], Dr Parvanshi[3]

[1] PhD Research Scholar, Chandigarh University, Mohali
[1]Assistant Professor, Chandigarh University, Mohali, India
[2]Assistant Professor, Chandigarh University, Mohali, India
[3]Assistant Professor, Chandigarh University, Mohali, India

Abstract:

The advent of the digital era has fundamentally changed how people plan, experience, and reflect on their travels, bringing about a paradigm shift in travelling modes and so the travelling industry. The present paper seeks to explore the transitions in the modes of travelling from traditional to the digital ones based on the qualitative analysis of the selected Insta posts to suit the purpose. By gathering data in the form of feed on Instagram using the hashtag (#) travel life, this study investigates how the ways of traveling is changing in the digital age leading to a new phase in the tourism industry.The study aims to illustrate the ways in which social media has affected the economy of uncharted territories with particular emphasis on the alteration in honeymoon options. The results of this study shed light on the complex interactions between technology and the travel industry and also brings into account the advancements in the thinking of post-modern digitally literate travellers.The paperprovides policymakers, researchers, and stakeholders in the travel industry with important insights into how traveling is taking a drastic leap from industrial perspectives to solo-travelling lens in digital era.

Keywords:digital; hashtag; Instagram; travel; and tourism.

Introduction

Travelling is an integral part of human lives. The whole point of traveling is to see the inner beauty of this cosmos and to encounter new things. Traveling gives one the opportunity to gain a profound understanding of various cultures and landscapes; it's not only about seeing new locations. "Travel-for-leisure or tourism around the globe could thus be seen as a prime example of cosmopolitanism-as-practice, allowing people to accumulate knowledge about "Other" worlds" [2]. Traveling is the finest method to escape a busy schedule, it is an extremely important element of life. It also refers to having diverse life experiences. Actually, tension, anxiety, and depression can all be effectively treated by traveling. It also enhances both physical and emotional well-being. "Leisure tourism, including destination choice, can be viewed as an investment in mental health maintenance" [1].Life is all we have, we should be grateful that it has allowed us to become the most evolved species on Earth. We get to see not just the splendour of nature, but also many topographies, geographies, and human populations. Discovering new locations, cultures, cuisines, customs, and lifestyles is the essence of travel. We travel because, in addition to being invisible from the comforts of home, distance and diversity are the hidden poisons that stifle creativity and development.

Origin

The idea of travel has long history, going back to the prehistoric era when early humans moved around in quest of resources, food, and shelter. "Tourism has been in existence from the beginning of the human civilization yet in different forms, phases, modes and practices" [15]. Travel increased in frequency for trade, exploration, diplomacy, and pilgrimage as civilizations advanced. Long trade and conquest voyages were undertaken by the ancient Greeks, Romans, and Egyptians, while the Silk Road promoted trade and cultural interchange between Asia and Europe. "The initial reconstruction in the first half of the twentieth century of how Indian Ocean trade and Indian culture spread east across the Bay of Bengal to secure not only the Silk, but especially the spices to fulfil Roman demand has been revised in the second half of the twentieth century" [16].

Within the realm of travel and tourism, the segment of honeymoon travelling has taken a significant place since the emergence of the concept itself irrespective of traditional or modern ways of travelling. There are several guesses as to the origin of the honeymoon. One tradition, coming from ancient Babylonian culture, about 4,000 years ago, suggests that "the bride's father would supply his new son-in-law with all the mead (honey beer/wine) that the young man could drink. Their calendar was lunar-based, and, as it turned out, this tradition, called the 'honey month,' was just about the time it took the groom to consume his gift. Ultimately, this period of time just after the wedding became known as the honeymoon." [17]. According to a different belief, the honeymoon originated in the Norse and Teutonic cultures of old, when newlyweds would go into hiding or isolation for a month to avoid trouble from relatives or bad spirits. In order to ensure a happy and successful marriage, the couple would eat and drink foods and beverages containing honey throughout this time.The honeymoon has changed over time to become a romantic retreat or journey that couples take soon after their wedding. "A honeymoon typically begins right after or even during the wedding ceremony when a newly-married couple sets off for holidays to a distant destination, leaving behind—both symbolically and literally—their old lives" [18]. This custom offers a chance for intimacy, bonding, and leisure away from the stresses of daily life, symbolizing the start of their life together as a married couple. "These intentions, either reflexive or embodied, lead to the consumption of various romanticised practices and products, including photos, celebratory rituals and vlog-making" [19]. Consequently, the practice has contributed to the travel and tourism industry in a remarkable way. The newly wed couples have often looked for the honeymoon packages to their desired destinations or countries contributing to the rise of travel industry: agencies, tourist agents, hotels, resorts, and magazines. Later, with the advent of Web 1.0, the industry got a boost by developing websites exhibiting the various features of their packages with alluring hypermedia. However, Web 2.0 has widened up the prospects of travel and tourism industry by shifting the focus from meta traditional narratives of periodicals, magazines, and websites to mini-narratives of individual travellers, solo-travellers, and travel agents. Especially, in contemporary era the outbreak of #travel life on Instagram has made travelling a source of individual income. The entire dynamics of models of Travelling- its modes, benefits, its industry involving economy- are changed making it a potential area to research. The article will

https://doi.org/10.52202/074664-0014

examine how social media, particularly Instagram, has contributed to revolutionize the ways of travelling and contributed in the monetary rise of the travel industry globally through mini-travel narratives. The study also aims to speculate the reasons of alteredchoice in honeymoon destinations from famous to less-travelled spots by analysing the travel Insta narratives about visiting less-travelled or undiscovered locations. By gathering data in the form of feed on Instagram using the hashtag (#) travel life, this study investigates how the ways of traveling is changing in the digital age leading to a new phase in the tourism industry. The study aims to illustrate the ways in which social media has affected the economy of uncharted territories with particular emphasis on the alteration in honeymoon options. The results of this study shed light on the complex interactions between technology and the travel industry and also brings into account the advancements in the thinking of post-modern digitally literate travellers.

Traditional Modes and Traditional Travellers

Traditional travel methods and approaches for geographical exploration frequently arouse feelings of nostalgia and a more straightforward perspective on the world. Traditional travellers frequently adopted the idea of leisurely travel, devoting their time to thoroughly experiencing the people, customs, and natural surroundings of the destinations they visited. They travelled by land or sea more frequently and valued the trip just as much as the destination. Travelers used paper maps and guidebooks to find their way around foreign areas prior to the invention of GPS and internet mapping services. "The two most quintessential tourist publications are the guidebook and the map. These are often used in combination when tourists navigate and find out about what to do in different places and how to get between them" [7]. These sources included insightful details regarding local culture, lodging options, and attractions. "One reason why guidebooks are so useful to tourists is that they catalogue, in a structured and relatively standardised form, relevant aspects of the places that tourists visit. They list accommodation (with phone numbers), attractions (with opening times), recommended bars and restaurants, and so on. This standardisation can make strange places feel considerably safer to tourists by reducing their uncertainty" [7].

Travelers had little access to information about locations, lodging options, and activities prior to the invention of the internet. Their heavy reliance on travel agents, guidebooks, and word-of-mouth recommendations occasionally resulted in fewer options and lost opportunities. It is necessary to supplement what the guidebook provides with other information, especially map information or suggestions from locals. It was common to see travellers using both a guidebook and a map simultaneously. Specifically, travel guides were interactive objects; during a conversation, visitors would point at the guidebook and then point at a map or in a certain direction to connect the enterprises under discussion with their locations.

A few decades back in modern India, easy and pleasurable travel experiences were frequently guaranteed by prepared arrangements and entertainment sources. Travel agencies were essential in helping travellers arrange their lodging, activities, and flights. They offered insightful knowledge and tailored suggestions according to the traveller's tastes and financial constraints. "According to Dubé and Renaghan (2000) traditional travel agencies plays an important role as information brokers, transferring money, booking rooms, give advices to customers and also integrate with their customers and by that, satisfy their needs and requirements which for customers leads to value-added service" [3]. There is a plethora of travel agencies worldwide, each with a distinct range of services and areas of expertise. Booking services for flights, hotels, rental cars, cruises, and holiday packages are offered by Expedia, an online travel agency. Expedia wants to provide travelers with an easy way to look up, evaluate, and reserve travel services at affordable costs. "Expedia.com is a popular booking website for tourists to book flight tickets, accommodation reservations, cruise tickets and also excursions, owned by Expedia Group" [20]. Another well-known internet travel company that focuses on hotel reservations is Booking.com. Users of the portal can make room reservations, read reviews, and search for lodging all over the world. Booking.com wants to offer a variety of lodging choices and a smooth booking process to tourists. "Booking.com is one of the world's largest online travel companies where millions of customers find their accommodation through an extensive and diverse list of properties including hotels, apartments, guest houses, and more" [21]. Advice, reviews, and opinions from travelers worldwide are available on TripAdvisor, a travel website and online review platform. TripAdvisor offers bookings for hotels as well as details on dining establishments, tourist destinations, and travel guides. The website wants to assist tourists in making knowledgeable choices and organizing amazing journeys. "TripAdvisor, which runs on user-generated content, provides a platform for travelers to socialize their opinions on several aspects of hotels" [22]. Travelers can save time and effort by researching, planning, and booking their travels all in one place with the help of online travel agencies. Travelers can compare costs and choose the best options based on their interests and budget by using travel agencies, which offer a wide range of options for flights, lodging, and activities. To help customers save money on their bookings, a lot of travel firms provide unique discounts, promotions, and discounted pricing. Travel companies ensure that reservations are safe, dependable, and supported by customer service in the event of problems or crises, giving travelers piece of mind. Travel firms provide insightful data, evaluations, and suggestions to assist tourists in making wise choices and organizing lifelong memories.

Travelers also frequently turned to travel periodicals as a source of knowledge and inspiration. Since the 1980s, a number of travel magazines have gained popularity, offering readers advice, suggestions, and inspiration for their journeys. Since the 1980s, a number of travel magazines have gained popularity, offering readers advice, suggestions, and inspiration for their journeys. Conde Nast Traveler is a luxury travel magazine that was founded in 1987 and is renowned for its excellent photography, in-

 https://doi.org/10.52202/074664-0014

depth location recommendations, and insider knowledge of the best luxury travel experiences available worldwide. Since its 1984 launch, National Geographic Traveler has been praised for its breathtaking photography, in-depth stories, and compelling narratives about adventure, culture, and the environment. Launched in the 1980s as a result of the popularity of the Lonely Planet guidebooks, Lonely Planet Magazine offered readers travel information, inspiration for destinations, and cost-effective travel recommendations for independent and budget-conscious travelers. These magazines included travel advice, hotel and restaurant reviews, destination stories, and insider information on local customs and landmarks. They gave readers insightful advice and suggestions to assist them properly organize their travels. "Travel and tourist magazines are major sources of information about destinations, events, accommodations, and transportation. The majority of this information is conveyed visually through colourful, stimulating, and seductive photographs of familiar landscapes and destination" [4].

Conventional tourists frequently interacted face-to-face with locals in order to get advice on genuine experiences and insights into the customs and culture of the destinations they visited. They were more likely to stay in bed & breakfasts or guesthouses run by locals, which promoted ties with the neighbourhood. Traditional travellers were less regimented and more flexible with their schedules, which allowed them to happen across unanticipated meetings and hidden treasures along the route. They relished the chance to adjust to novel situations and enjoyed the unpredictable nature of travel. Conventional tourists looked for meaningful encounters that would help them appreciate the destinations they were visiting on a deeper level and valued cultural immersion. They wanted to see the local markets, go to cultural events, and take part in customs and ceremonies. According to Mandala Research "Tourists, who have participated in cultural activities on their most recent trip or within the past three years, represent almost 54% of the US adult population" [8].

Expenses associated with traditional traveling were frequently greater, particularly when using travel agencies or buying printed guidebooks. Furthermore, there were few options for inexpensive lodging and transportation, which could add to the expense of travel, particularly on lengthy itineraries. Travelers frequently had restricted access to communication tools and resources when on the road prior to the widespread availability of the internet and mobile connectivity. It could be challenging to contact loved ones, get emergency help, or make last-minute plans as a result of this.

In the past, couples choose their honeymoon locations for a number of reasons, frequently influenced by practical considerations, personal preferences, and cultural traditions. It was common for couples to ask friends, relatives, or acquaintances who had visited romantic locations or had amazing honeymoon experiences for ideas. They valued the advice from people they knew and trusted their firsthand accounts and insights. "It is likely that individuals interact with others during the decision-making process and those interactions affect or influence the final destination choice, particularly concerning family travel" [9].There were several places that everyone agreed were ideal for a romantic honeymoon, like Paris, Venice, Hawaii, the Caribbean islands, and picturesque European cities. Couples were frequently lured to these well-known locations by their reputations for romance, beauty, and charm.

Conventional travel has its drawbacks despite being nostalgic and enlightening. Language hurdles, fewer options, and increased expenses resulted from reliance on agents and printed materials. Connectivity and flexibility were constrained. Notwithstanding its shortcomings, it promoted authentic cultural immersion and life-long memories, influencing how we travel the world today.

However, with the advent of digital technology and the internet, the travel industry has changed and so has changed the dynamics of travelling. Travellers can now easily plan and book their travels on popular platforms including online travel agencies (OTAs) and travel booking websites. These platforms make it simple for tourists to compare costs and make bookings by providing a large selection of options for travel, lodging, vehicle rentals, and activities. "Online tourism portals are significantly important channel for tourism sales these days" [5]. E-magazines and travel blogs have also grown to be significant information and inspiration sources for tourists. Nomadic Matt is a well-known travel site that helps and inspires independent travelers with affordable travel advice, destination recommendations, and personal travel narratives. "Nomadic Matt has a clear logo and like the blond abroad the names gives tells right away what this SMI is all about: travelling like a nomad" [23]. A thorough resource for optimizing travel rewards and points to reduce costs on airfare, lodging, and other travel-related expenses is explain in The Points Guy. "The points guys reviews credit cards, hotels, and flights and has a links to many great deals" [23]. **Lonely Planet's Thorn Tree Forum**is an online forum where visitors may interact with like-minded people, exchange stories, and ask questions. "And lonelyplanet.com"s Thorn Tree travel forum is „the place where the Lonely Planet travel community get [sic] together to exchange travel information, advice, hints and tips" [24]. AFAR is an online travel magazine that offers location recommendations, insider advice, and immersive storytelling to encourage genuine and meaningful travel experiences. "AFAR's tagline is "the leader in experiential travel" [25]. Travelocity's Travel Blog provides destination guides, guidance from experts, and travel ideas to make travel planning and booking easier.

Change in Mode in Contemporary Times: From Public to Private

There is a marvellous increase in the self-reliant travel ventures with the intervention of blogging on social media platforms like Facebook, Instagram, ands Twitter now X. Nowadays, a lot of travellers use the internet to look up places, read reviews, and get advice from other travellers. "Visitor opinions posted on leading travel blog sites were analyzed to gain an understanding of the destination experience being manifested" [6].

 https://doi.org/10.52202/074664-0014

The technological advancement in tourism industry does not only ease the modes and patterns of organizing the travels but also opens a segment of "travel influencers" in the global travelling community. By utilizing influencers to promote tourism, a destination's visibility, engagement, and visitor count can all rise, which in turn spurs economic development and growth.

Mini-narratives on social media sites like Instagram Reels provide a range of viewpoints and encourage creative travel options in the current digital era. Travelers share their experiences, from luxurious resort getaways to solitary hiking expeditions, highlighting lesser-known locations called "uncanny" and unusual activities. One of the instapages titled, "Littleknowngems" posts for such little-known gems under #travellife:

A swimming lane rockpool in South Australia
Shaped uncannily like a lane in a swimming pool but set in the vast, open embrace of the ocean, CarrickalingaRockpool is a natural wonder that combines the structured fun of lap swimming with the wild beauty of the sea.
The crystal-clear blue waters are simply mouth-watering
Want the exact location? Comment below and we'll shoot you a link with a map. [26]

1. Cosyacademia

Another user,"Thetravelsaaint1," posts about his experience of unexplored Rovinj:

Each corner of Rovinj shares a vibe and a peaceful atmosphere. Was sipping on the wine at this relaxed environment with the soothing music and sounds of waves hitting the shore, feel the life, feel the happiness. [28]

Furthermore, travellers share their insights on lesser-known sites and hidden jewels on platforms like Instagram Reels, which make it easier to discover undiscovered spots. Those who follow the right hashtags and accounts can find hidden locations, beautiful scenery, and cultural gems that are not easily found. Reels are an easy and motivating method to learn about new places through engrossing images and educational subtitles that spark wanderlust and an interest in experiencing new things. A traveller with user identity, "Ellistravelling" posts under #travellife:

Istanbul is definitely must-visit place and one of my favourite
With its delicious food, hospitable people, breathtaking landmarks, warm and welcoming vibe I believe you will love it
Of course, prices vary depending on personal preferences but, it fits to all budgets.
It is solo travel friendly because you will find a lot to do. Whether you're savouring local delicacies or immersing yourself in its rich history, there's something for everyone. But remember, while touristy areas embrace diversity, it's important to respect the dominant religion and culture of the country.
Have you been to Istanbul before? What are your thoughts about the city? [30]

Such variety of travellers posting on Instais a reflection of the changing tastes and goals of contemporary explorers. "The use of smart tourism technologies such as travel-related websites, social media, and smartphones in travel planning has been pervasive and growing" [12].The hashtag #TravelLife has revolutionized the way people view honeymoon stories and discover new places. Travelers share their experiences on social media sites like Instagram, revealing hidden treasures and encouraging others to take on adventurous excursions. "People value visual aspects in travel planning and especially women use Instagram as a source for travel inspiration and information. 82% of the respondents indicated Instagram having had at least some influence in their travel planning" [10]. The standard honeymoon story has also been altered by this online community, which inspires couples to look for unusual events and create priceless memories. The simple crux recommendations by individual travellers to their community and followers hit the crux with readings like:

Palawan is one of my favourite destinations in the Philippines
I would def recommend to visit:
Coron
El Nido
San Vicente
Port Barton
Balabac
Puerto Princesa. [32]

Besides the individual travellers, the travel agencies, guides, and firms have also made their way into the digital tourism world by developing their pages and profiles on Instagram to achieve desired targets. For instance, "Egypttourguide.ahmed.waheeb" posts:

One of my Favourite Shots I have ever taken… King Chephren great Pyramid.. Giza Pyramids Egypt [33].

The popular tourist places gain re-assurance with posts like:

Witness the jinamen pass at Sichuan!
The pass is famous for its steep cliffs and narrow passage, which made it a natural defensive barrier in ancient times. It has been an important military stronghold throughout Chinese history, especially during the Three Kingdom period (220-280 AD) and the subsequent dynastic eras.
Today, Jianmen pass is a populr tourist destination known for its scenic beauty and historical significance. Vistors can explore the ancient pathways, fortified walls, and temples that dot the area while enjoying breath-taking views of the surrounding

 https://doi.org/10.52202/074664-0014

mountains and valleys.[31]

The analysis of the domain of travel influencers on Instagram using #travellife is one of the potential areas to comprehend the marketing strategies of travel agencies and stakeholders having presence on digital platforms transforming the travel-psyche of travellers. A two-way monetary traffic can be seen when an influencer earns by collaborating with online travel firms and further, the travel firms earn with an increase in the number of travellers reaching the promoted destinations. A few samples mentioned below shed enough light on marketing approach using travellers'-oriented linguistics like "cosy," "natural," "friendly and trusting" in their posts. One of the promoters, "Cosyacademia" posts:

Forever seeking adventures and all things cozy
Qotd: what would you spend your life doing if there were no limits?
Some clips from cozy days exploring rainy, gloomy Edinburgh. I'm somehow always yearning for travel while also very much enjoying being perfectly cozy at home. I suppose I can enjoy them both! If I could- I would certainly travel even more- I have so many places I want to see. [27]

Another influencer profiled "Helensuisse", highlights the "friendly and trusting" features like:

In a few months Lucerne will look like this. The swans in Lucerne are very friendly and trusting. They would swim very close to people And look how clear the water is! [29].
Yet another explorer, "The_.explorer_" promotes WadiSheesmountain in Sharjah:

Wadishees natural trail, Sharjah a
Welcome to breathtakingWadiSheesmountain in Sharjah
WadiShees is a picturesque valley located in the emirate of Sharjah in the UAE. It is known for its stunning natural beauty, rugged terrain, and opportunities for outdoor activities.
WadiShees is a popular natural trail located in Sharjah, UAE.
The trail offers stunning views of the surrounding mountains and desert landscape
Visitors can enjoy hiking, bird watching, and photography opportunities along the trail
The trail is relatively east to moderate in difficulty, suitable for all skill levels
It is recommended to bring plenty of water, sunscreen, and sturdy footwear when visiting WadiShees
The trail is open year-round, but it is best to visit during cooler months to avoid extreme [34].

Influencers posting online on sites like YouTube, Instagram, and TikTok to monetize their content and make their love of travel a career in today's digital world. Figure 1 shows the travel content creators.

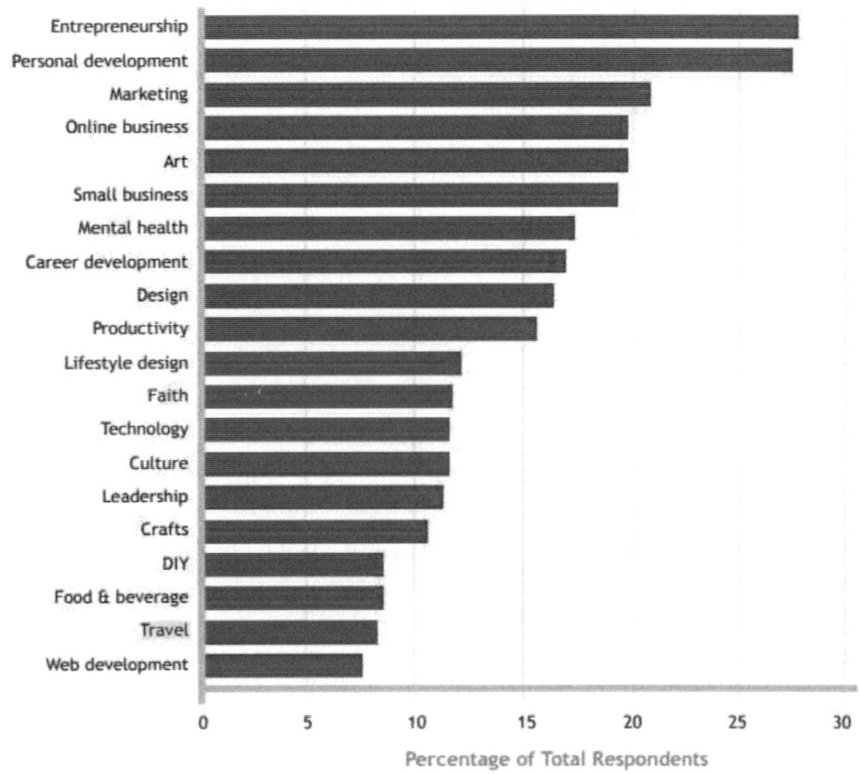

Influencers make money by endorsing goods, services, locations, and experiences to their followers through sponsored posts, brand partnerships, affiliate marketing, and sponsored travel opportunities. In the internet age, the reel-tradition approach has transformed the way that people think about storytelling and travel, providing a new platform for artistic expression for travel-influencers as well as individual travellers.

Moreover, the travel agents on social media gain more prominence when these are given credits using @agent's handlers on social media by the satisfied customers. The 'credit giving' concept replaces the traditional word–of-mouth making it digital-word-of-mouth. One of such satisfied customers, Aaronkbl promotes the travel agent like:

Have you ever been to Moreton Island?

This is honestly the perfect day trip idea if you're visiting Brisbane or simply looking for something fun to do over the weekend! Moreton has so many fun activities do on island, this includes:

1. *Snorkelling the wrecks*

https://doi.org/10.52202/074664-0014

2. *See though kayak tours*
3. *Sand Boarding (Personal Favourite)*

Big thanks to @travelloapp @visitbrisbane for getting me over there! If you're interested or wanting to find out more information about this day trip head over to travello.com for more! [35].

Thus, the conventional travel agency model has changed into a more decentralized strategy in the current travel environment, where independent agents play a significant role. Travelers now have access to independent travel agents that provide individualized services and knowledge catered to their unique requirements and preferences, rather of depending exclusively on brick-and-mortar companies. "In contemporary tourism, technology enables travellers to make travel arrangements in various ways, for example they can reserve tickets directly from airlines company's website or purchase products from online travel agencies" [11].These lone agents use social media and internet channels to establish connections with customers, customize travel itineraries, and offer continuous assistance during the journey. This change is in line with the evolving dynamics of the travel business, which places a premium on flexibility and individualized service to cater to the wide range of demands of contemporary tourists. "This change is in line with the evolving dynamics of the travel business, which places a premium on flexibility and individualized service to cater to the wide range of demands of contemporary tourists" [13,14].

Conclusion

In a nutshell, InstagramReels have completely transformed the travel and tourism sector by offering an engaging medium for inspiration, discovery, and narrative. In contrast to traditional travel, which mainly depended on tour operators and written materials, Instagram Reels provide a wide range of mini-narratives that highlight distinctive locations, experiences, and viewpoints. Influencers utilise their digital platforms to generate revenue from their content and reinterpret the conventional travel narrative, thereby motivating tourists to pursue uncharted territories and genuine encounters. Digital platforms are being used by individual agents to provide customized services and meet the changing demands of contemporary tourists. In the current digital era, Instagram Reels have revolutionized travel by increasing accessibility, diversity, and immersiveness, thereby influencing how people see the world.

References

1. Cooper MA, Buckley R.:Tourist Mental Health Drives Destination Choice, Marketing, and Matching. *Journal of Travel Research*. Published online May 6, 2021:004728752110115, (2021).
2. Urry J.: *Consuming Places*. Routledge, (1995).
3. Dubé L, Renaghan LM.: Marketing Your Hotel to and through Intermediaries. *Cornell Hotel and Restaurant Administration Quarterly*. 2000;41(1):73-83(2000).
4. Kingsbury PT, Brunn SD.:Freud, Tourism, and Terror. *Journal of Travel & Tourism Marketing*. 15(2-3):39-61(2004).
5. Srivastava S, Kaushal V, Ali MdN.:Role of Online Travel Agents in increasing the frequency of traveling habit in Gen Y. *Journal for Studies in Management and Planning*. 1(7):100-108 (2015).
6. Pan B, MacLaurin T, Crotts JC.:Travel Blogs and the Implications for Destination Marketing. *Journal of Travel Research*.46(1):35-45(2007).
7. Brown B, Chalmers M.:Tourism and mobile technology. In: *ECSCW* Springer;335-354.
8. Mandala Research.(2015).Mandalaresearch.com. Published 2015,(2003). Accessed March 12, 2024. https://mandalaresearch.com/
9. Jang H, Lee S, Lee SW, Hong S.:Expanding the individual choice-sets model to couples' honeymoon destination selection process. *Tourism Management*. 28(5):1299-1314 (2007).
10. Terttunen A. *The Influence of Instagram on Consumers' Travel PlanNing and Destination Choice*.
11. Liao R. *The Value of Traditional Travel Agency in the Digital Age*.
12. Huang CD, Goo J, Nam K, Yoo CW.:Smart tourism technologies in travel planning: The role of exploration and exploitation. *Information & Management*. 54(6):757-770(2017).
13. Buhalis D.:Strategic use of information technologies in the tourism industry. *Tourism Management*.19(5):409-421(2018).
14. Egger R, DimitriosBuhalis.:*ETourism Case Studies : Management and Marketing Issues*. London ; New York Routledge, Taylor Et Francis Group, (2015).
15. Ranasinghe R, Gangananda N, Bandara A, Perera P.: Role of Tourism in the Global Economy: The Past, Present and Future. *Journal of Management and Tourism Research* . 4(1):6-12().
16. Guan K.: The Maritime Silk Road: History of an Idea.
17. Enduring Wedding Tradition.:Customs and their Origins, (2010).
18. Isański J. HONEYMOON.:AS A RITE OF PASSAGE: SOCIOLOGICAL ANALYSIS OF CHANGES IN THE PHENOMENON. *Folia Turstica*.28(2):109-129(2013).
19. Yan W, Liu J, Zhu Y.: Honeymoon Tourism as Romantic Consumption in China, (2023).

https://doi.org/10.52202/074664-0014

20. Shen Y.: *How to Improve Customer Loyalty to Online Travel Agencies -A Research on Expedia, an Online Travel Booking Platform,* (2018).
21. Mavridis T, Hausl S, Mende A, Pagano R.:Beyond algorithms: Ranking at scale at Booking.com.
22. Nilashi M, Ibrahim O, Yadegaridehkordi E, Samad S, Akbari E, Alizadeh A.:Travelers decision making using online review in social network sites: A case on TripAdvisor. *Journal of Computational Science*.28:168-179(2018).
23. Laczkovich I.:*Social Media Influencers in the Travel Industry: An Analysi,* .(2018).
24. Young T, Hanley J.:Virtual mobilities: backpackers, new media and online travel communities(2010).

25. About AFAR.: Retrieved from http://about.afar.com/about/advertise/afar-mediakit/,(2015).

26. Littleknowngems.:A swimming lane rockpool in South Australia,(2024).
27.Cosyacademia.:Forever seeking adventures and all things cozy, .(2024).
28. Thetravelsaaint1.:Each corner of Rovinj shares a vibe and a peaceful atmosphere, .(2024).

29.Helensuisse.:In a few months Lucerne will look like this,(2024).

30. Ellistravelling.:Istanbul is definitely must-visit place and one of my favourite,(2024).

31. Worldbook.:Witness the jinamen pass at Sichuan,(2023).

32.Leyanebula.:Palawan is one of my favourite destinations in the Philippines,(2024).

33.Egypttourguide.ahmed.waheeb.: One of my Favourite Shots I have ever taken… King Chephren great Pyramid.. Giza Pyramids Egypt,(2024).

34. The_.explorer_.Wadishees natural trail, Sharjah.: Welcome to breathtakingWadiShhesmountain in Sharjah,(2024).

35. Aaronkbl.:Have you ever been to Moretob Island?,(2024).

Investigating Regulatory Challenges in India's Wine Industry: APath to Policy Reform for Sustainable Growth

Balraj Benedict[1], Dr. Kiran Shashwat [2]
[1]Assistant Professor, UITHM-Chandigarh University, balraj.benedict@gmail.com
[2]Associate Professor, UITHM-Chandigarh University, Kiren.e12368@cumail.in

Abstract

The investigation of regulatory issues in India's wine business, as summarized in the research "Investigating Regulatory Challenges in India's Wine Industry: A Path to Policy Reform for Sustainable Growth," explores the complexities and obstacles that the industry must overcome. The goal of the research is to find opportunities for policy reform that can lead to sustainable growth through a thorough analysis. This study explores the complex regulatory environment of the Indian wine sector, highlighting obstacles from the perspectives of law, economy, and culture. The report puts up a thorough reform agenda that calls for simplified licensing procedures, more sensible taxation, and a transformation in society's perception of wine drinking. To provide a consistent business environment, it recommends looking into standardized rules among states. Using expert insights and foreign models, the research seeks to provide policymakers with well-informed suggestions to support a flourishing and sustainable wine market in India. The study presents a comprehensive reform plan that includes a range of actions to address the impediments that have been identified.

Keywords: Investigating, obstacles,transformation,sustainable,Policymaker.

Introduction

Growth of Indian wine industry

Over 90 officially recognized wineries are currently producing more than 17 million liters of wine annually throughout the nation. Over 450.00 crores have been invested in this area thus far. Wine from Maharashtra is exported to the USA, UK, Singapore, France, Italy, and Belgium. Mumbai (30%), Delhi (20%), Goa (20%), Bangalore (15%), Punjab (5%) and the remaining 10% of India make up the country's wine consumption trend. But given that the OIV data for 2017 states that the world produced 25 million t of wine, with France, Italy, and Spain accounting for over 40% of that total, we are lagging far behind. Compared to 30-45, the average person's annual wine intake in India is less than 10 mL. At present more than 90 registered wineries are operating in the country crossing the 17 million L of wine production. Investment made in the sector has now reached to more than 450.00 crores. Maharashtra is exporting wine to France, Italy Germany, USA, UK, Singapore, and Belgium. The wine consumption pattern in India stands at – Mumbai (30%), Delhi (20%), Goa (20%), Bangalore (15%), Panjab (5%) and the rest of India accounts for 10%. However, considering the global wine production of 25 million t as per OIV report for the year 2017 and the contribution of France, Italy, and Spain to the extent of around 40%, we are far behind. The wine consumption in India per person per year is less than 10 mL as against 30-45 L in France and Italy, 30 L in Australia, 10 L in the USA and less than 2 L in China (2015). The Indian Grape Processing Board (IGPB) (Ministry of Food Processing Industry) was also playing a vital role in deciding market policy of Indian wines in the past until 2013. APEDA is presently monitoring export of Indian wines to European and other countries with the help of NRC for Grapes and other nominated laboratories and certifying agencies. After becoming member of OIV, the Indian wine industry has a chance to occupy its footprint on the global arena. Leading wineries are also engaged in import of quality plant material in terms of grapevine and their compatible rootstocks and arranging the production of grafted plant material of high quantity in their own

https://doi.org/10.52202/074664-0015

nursery units to supply to their contract farmers to produce quality wine grapes and ultimately high-quality wine[1].

Background and significance of the Indian wine industry

Worldwide, grapevine (Vitis vinifera L.) is mostly planted for wine production; nevertheless, commercial wine grape production has only recently started in India. India has had remarkable success producing table grapes, and global grape yield levels have risen to somewhat higher levels. Roughly 78% of grapes cultivated in India are eaten raw. 1.60 million tonnes of grapes are produced annually on an area of 60,000 hectares (FAO, 2005). In the states of Maharashtra, Karnataka, and Andhra Pradesh, there are about 1200 hectares planted with wine grapes. Owing to low domestic wine consumption and the lack of standard wine varietals to make high-quality wines meeting international standards, research to boost wine production was not given significant priority. [2] In ancient times, wine was commonly called samarasa; the Vedic scriptures also mention soma. Additionally, Drakshsava is mentioned in Ayurvedic texts. Drakshsava was essentially a digestive preparation made with natural alcohol, cinnamon, vidanga, tejpatra, pippali, and black pepper, as well as matured grapes (Rathish, 2009). The oldest recorded reference of grape-based wine dates to Chanakya's writings from the late BC, when he was Emperor Chandragupta Maurya's chief minister. In his writings, Chanakya denounces alcohol consumption while documenting the emperor's and his court's regular consumption of Madhu, a type of grape wine (Robinson, et al., 2006).Over the ensuing years, wine evolved into the preferred beverage of the kshatriya, or noble class, while alcohol derived from wheat, barley, and millet was usually consumed by the lower caste. As per Islamic dietary law, alcohol was forbidden during the Muslim Mughal empire's reign. In the 16th and 17th centuries, early European visitors to the palaces of the great Mughal emperors Akbar, Jehangir, and Shah Jehan recounted sampling wines from their royal vines. Large amounts of wine were also made for Hindu holy festivals. Aside from this, certain regions have only produced wine in small amounts due to colonial influence, most notably the Portuguese in Goa in the sixteenth century. The arkesham vine was used to make red wine, and the arkawati and bhokry grapes were used to make white wine. During the 19th century, Kashmir and Baramati in Maharashtra saw the establishment of vines due to British influence. Many Indian wines were showcased during the Great Calcutta Exhibition of 1884, when they were well-received by attendees [4]..India has been producing wine for up to 5000 years [2]. However, grapevines are thought to have been brought to India from Persia during the Indus Valley Civilization, when viticulture was first established. It was thought that around the fourth millennium BC, Persian traders brought viticulture to India. According to historians, table grape juice was produced primarily from the early plantings rather than alcohol [5]

Limitations imposed by regulatory challenges

There are disparate regulations governing the sale and distribution of alcoholic beverages all over India. State-level regulatory agencies are established by the states to monitor the sale of alcohol within their borders. These organizations have the authority to control the distribution and manufacturing of alcoholic beverages inside their borders. They are also in charge of upholding the licensing laws that are in effect in respective states. These state organizations are generally in charge of granting licenses to state-based alcohol manufacturers and wholesalers. They have the authority to put limitations on the distribution and sale of alcohol inside the state. For instance, selling alcohol during specific hours of the day or week is strictly prohibited in several states. [6]. Apart from the regulatory mandates enforced by state authorities, there are several legal constraints that are applicable across the nation. The purpose of these limitations is to safeguard the rights of minors who drink alcohol and those who aren't allowed to by law. If these limitations are broken, there may be harsh consequences. If found guilty of giving alcohol to a minor, a person faces severe fines and/or up to five years in jail. In a similar vein, giving alcohol to someone without a legal license carries a huge financial penalty and/or a maximum three-year prison sentence.

The Indian alcoholic beverage business has regulatory constraints in the form of liquor taxes, licensing regulations, trade barriers, and other market-related impediments. In contrast to other established markets, India experiences less of these problems. The main obstacle that India's beverage business must overcome is the extensive level of government interference. To guarantee that people can eat food and beverages safely, the Indian government also continuously enacting new regulations and standards. Despite these obstacles, several prosperous businesses have succeeded in breaking into the market and becoming well-known. These businesses are creative and have modified their business plans to address the difficulties associated with doing business in India. Getting over the obstacles posed by regulations.The National Wine Board encourages contract farming, supports the grape, apple, plum, apricot, and wine industries, and supports farmers who grow grapes, apples, plums, apricots, and makers of raisins, juice, and wine. In addition to other areas, the Agricultural and Processed Food Products Export Development Authority promotes "Wines of India" throughout the world, with a focus on France, Italy, Germany, the United States, the United Kingdom, and Singapore. Numerous wine clubs, wine tourism, and wine magazines have emerged in different Indian towns.

Chapter's objective: To identify obstacles and propose policy reforms for sustainable growth.

Regulatory Landscape of the Indian Wine Industry

Regulatory structure governing the wine industry in India- national, state, local.

License to Manufacture Wine in India

The Constitution of India has provided the power to State Government to levy or collect duties of excise and make the rules on import, purchase, sale, or manufacturing alcoholic liquor including wine in their respective state. The purchase, sale, import, export, bottling or manufacturing of wine prohibited in the state of Gujarat, Bihar, Nagaland, Mizoram, and Union Territory Lakshadweep.

License or Registration under Food Safety and Standard Act 2006 (FSSAI) regulation

The company involved in importing, bottling, and marketing of Wine in India require to obtain the license under the Food Safety and Standard Act 2006. The license issued under the regulation is valid for a period of 1 to 5 years as selected by food business operator, from the date of issuance of license after paying the applicable fees for the period.

Bureau of Indian Standards

The applicant is required to submit the BIS Certification application with required documents and requisite fee. After applying, a preliminary factory evaluation is carried out by a BIS officer. Then samples are tested in the factory and drawn for independent testing. BIS certification is provided if the sample is acceptable. BIS Certification is expected to be granted within 4 months of submission of application.BIS Standards – IS 7058: 2005 for Wine- This standard prescribes the requirements and the methods of sampling and test for table wines. The specified standards contain provisions which through reference in this text, constitute provision of this standard. At the time of publication, the editions indicated were valid. All standards are subject to revision and parties to agreements based on this standard are encouraged to investigate the possibility of applying the most recent editions of the standards.

Sample & Quality Check Regulations

The Food Safety and Standards Authority issued the Food Safety and Standards (Laboratory and Sample Analysis) Regulations, 2011 for sampling and quality check of the specified goods. The sample of any imported article will be sent by the Authorized Officer for analysis. The Food Analyst of any of the notified laboratories or any other laboratories notified by the Food Authority from time to time having jurisdiction over the area in which the sample was taken.

Packaging and Labelling Regulations

The Packaged Commodities Rules are applicable to alcoholic beverages which covers as below

https://doi.org/10.52202/074664-0015

Any alcoholic beverage when labelled as "matured", shall be matured for a period of not less than one year in oak or other suitable wood vats or barrels or with wooden chips.Where an age claim is made in conjunction with the word "aged", the age must refer to the youngest spirit in the blend. The bulk containers shall have no upper limit for alcohol content and shall meet the safety parameters of the product standards. Such products shall also carry a label declaring, "For Manufacturer of Alcoholic Beverages only. Declaration of alcohol content: Alcohol content shall be expressed as per cent. abv or as proof.

Labelling of standard drink: One standard drink is the amount of beverage that contains in it 12.7 ml of abv as measured at 200C. The label may include a statement of the approximate number of standard drinks in the package. Geographical indicators or names may be used on the label solely for the products originating from that geographical region. There shall be a statutory warning "CONSUMPTION OF ALCOHOL IS INJURIOUS TO HEALTH" and "BE SAFE-DONT DRINK AND DRIVE".

Foreign Direct Investment Policy

Foreign Direct Investment (FDI) up to 100% is allowed under the automatic route for brewing and distillation of liquor subject to the license by the appropriate authority.

Taxes & Duties

The Central government levies following Custom Duties and Social Welfare Surcharge on import of Wine into India and Central Sales Tax on inter-state sale of Wines. The State government levies the duties such as Excise Duty, Value Added Tax, Transport Fee, Import Fee, Export Fee, Toll Tax, Label Fee etc on import, purchase, sale, or manufacturing in relation to alcoholic liquor including wine in their respective state.

Regulatory Areas- licensing procedures, taxation structures, inter-state movement restrictions

The National Wine Board promotes farmers of grapes, apples, plums, apricots, manufacturers of raisins, juice and wine and encourage contract farming and promote grape, apple, plum, apricot processing, and wine industry. The Agricultural and Processed Food Products Export Development Authority promote "Wines of India" across the globe, especially France, Italy, Germany, USA, UK, and Singapore, apart from other regions. Several wine clubs, wine tourism and magazine have been formed in various cities of India.

The Maharashtra government unveiled a groundbreaking Grape Policy that includes export promotion, quality control regulations, certification, and wine excise tax reductions, among other incentives. The Wine Institute was established to supply skilled labour and guarantee international excellence. For the grapes in Nashik, Pune, Sangli, and Solapur, the government established Agri Export Zones. The Karnataka government determined which regions are ideal for cultivating the grape varietals needed to make wines of superior quality. Bengaluru, Kolar, Ramanagar, Chikkaballapur, and Tumkur districts have been designated as "Nandi Valley" wine grape cultivation regions; likewise, the districts of Bijapur, Bagalkot, Belgaum, Gulbarga, Koppal, Raichur, Gadag, Bellary, and Bidar have been designated as "Krishna Valley" wine grape cultivation regions; and Mysore,

The Government of Karnataka covers the scheme as below:

Declaring Wineries as Horticulture and Food Processing Industries

Simplification of rules while issuing License and permit to the Wineries.

Allowing the sale of wines in the premises of wineries to promote "wine – tourism".

Capital Investment Subsidy.

Subsidy for Grape Processing

Training will be rendered to the wine grape growers through seminars, workshops, study tours, export – consultations, demonstrations, etc.

Incentive for marketing and distribution of wines

Concession in Excise Duty

The Label registration Fee is waived off for all the brands of wines, and this will not be reviewed for the next 5 years.

Impact of Government Regulations on Wine stakeholders - wineries, distributors, consumers analysis based on the real data and facts provided by the Excise departments of Government of Delhi and Karnataka. In Karnataka, the stakeholders of the wine Industry as such farmers, wine makers, Distributors and retailers expected many benefits and very hopeful of getting the benefit from this policy and hence invested large amount of finances in expanding the cultivation of wine grapes and in setting up of the wineries, most of the farmers and wine makers have taken large amount of loans from the banks at high interest rates. Finally, after five years of implementation, instead of making the profits, most of the farmers and winemakers incurred losses as the farmers could not get the good returns by selling the wine grapes. In the analysis of Karnataka Grape Processing and wine Policy 2007. there is not much help in taking care of reducing the cost of cultivation of grapes and most of the subsidies and concessions are kept for setting-up of the Wineries and for the marketing and sale of wine. Only one subsidy is accorded to the cultivation of Wine Grapes from National Horticulture board, Ministry of Food processing Govt. of India. Moreover, the cultivation of wine grapes requires the knowledge of modern technology, and this crop is always gamble with the climate and pests. Even after passing through all these difficulties, the farmers are not getting the better remunerative prices for their produce as the consumption of wine is not picking- up as examined in my research. I have made sincere efforts to suggest the measures to enhance the growth of Wine Industry in India which in turn, help to improve the condition of Grape- farmers and promotion of Healthy Alcoholic Drinks i.e. Wine. The Government should give policy help to formulate and develop a Policy Platform for Wine and transform state policies, delink/ differentiate the Wine from other beverages and spirits, focus on Lowering duties and taxes on wine, rationalizing Inter-State Regulatory and tariff barriers, promotion of single window clearances facility, strategize to reduce costs, develop export markets, innovative packaging, develop branding and marketing of Indian Wines and other different suggestions are being made in my research for the growth of wine industry in India.

Regulatory Challenges

Legal

licensing procedures - Complexities for wineries

establishing a wine shop in India involves a meticulous approval process encompassing several crucial steps. Commencing with the acquisition of an alcohol license from the local government is not just a prerequisite but a fundamental step. This licensing procedure is not swift; it unfolds over weeks or even months. Therefore, it demands diligent monitoring to avoid nearing deadlines and potential complications that might impede the progress.

Following the procurement of the alcohol license, another critical step is securing a wine shop permit from the local police department. This permit is pivotal for the official opening and subsequent sale of alcohol from the establishment. It adds a layer of regulatory compliance to ensure that the operation aligns with the legal framework and maintains public safety.

Application and Approval Process for Wine Shop License in India

Securing a Wine Shop Act license in India involves a structured process comprising application processing and certification. To initiate this procedure, prospective applicants must undergo an initial application processing period. While the standard duration for this phase is six months, there is an option for expedited processing by furnishing evidence of commendable business practices. After completing the application review, the next step involves applying for a Wine Shop Certificate through the Regional Wine Regulatory Authority (RWRA). Upon successful application, the RWRA will then issue a certificate affirming the licensing and operational status of the wine shop. This comprehensive process ensures the legal and regulatory compliance necessary for establishing and operating a wine shop in India.

Approval and Issuance Procedures under Varied State Regulations in India

https://doi.org/10.52202/074664-0015

Navigating the wine retail landscape in India involves a stringent approval process governed by diverse state regulations. Acquiring a wine shop license is mandatory, with inspections scrutinizing premises and inventory. Successful establishments receive a two-year license, subject to annual renewal. Varied state laws dictate the regulatory framework, emphasizing the need for compliance with specific licensing prerequisites and additional business permits.

Approval Process: Obtaining a wine shop license is obligatory for any wine retailer conducting business in India. The approval procedure entails a thorough examination of both the premises and the wine inventory. The license may face cancellation or suspension if any issues are identified during the inspection.

Issuance of License: The license will be granted once the establishment successfully passes the inspection and fulfils all necessary documentation requirements. This license remains valid for two years and necessitates annual renewal.

Diverse State Regulations: Various state laws, each with its own set of regulations and requirements for licensing, regulate the nation's wine business. In India, obtaining a license from the appropriate state government is typically necessary to operate a wine shop. It's probably also necessary to get a trade permission and business license.

State-Specific Regulations Governing the Wine Industry in India: Licensing Authorities and Requirements Across Different Regions

Various Indian states have distinct laws governing the wine industry, each administered by its regulatory authority:

Gujarat:

Gujarat Wine Regulatory Authority is the regulatory body in charge (GWRA)

Getting a GWRA license is necessary to operate a wine shop in Gujarat.

Extra requirements include the need for a trade authorization and a business license.

Tamil Nadu:

Tamil Nadu Wine Regulatory Authority is the regulatory body in charge. (TWRA)

Obtaining a TWRA license is necessary to own a wine shop in Tamil Nadu.

Extra Requirements: A business license and a trade permit are required in addition to the TWRA license.

Karnataka:

The regulatory body in charge is the Karnataka Wine Regulatory Board (KBRB)

Obtaining licenses: Obtain a KBRB license to run a wine shop in Karnataka.

Extra prerequisites: Besides the KBRB license, a trade permit and a business license are necessary.

Andhra Pradesh:

The Andhra Pradesh State Road Transport Corporation is the regulatory body in charge. (APSRTC)

Operating a wine shop in Andhra Pradesh necessitates possessing an APSRTC license.

In addition to the APSRTC license, it is necessary to obtain a trade permission and a business license.

Maharashtra:

The Maharashtra State Waqf Board (MSWB) is the regulatory authority.

Overview of Licensing: In Maharashtra, the MSWB regulates the wine business, with district branches caring for things like fruit permits.

Additional Requirements: Permits or compliance may be required to comply with certain district regulations.

Goa:

Regulatory Authority: Goa Wine Regulatory Board (GWRA)

Licensing Requirement: Obtain a GWRA license to operate a wine shop in Goa.

Additional Requirements: Alongside the GWRA license, a business license and a trade permit are obligatory.

Adherence to these state-specific regulations, including obtaining the requisite licenses and permits, is crucial for legal and operational compliance within the diverse landscape of India's wine industry.

Labelling and quality control regulations

Countries have very different laws governing wine. Some nations have very no wine taxes, while others have rates that are higher than 100%. While the EU regulates wine extensively in terms of both quantity and quality, other major producers, like the US and Australia, have far less stringent laws. Trade and trade policy have been impacted by some of these distinctions. Specifically, in the context of international trade talks, the regulatory approach to geographical indications has been and still is the subject of contentious discussions. For instance, in the ongoing EU-Australia Free Trade Agreement discussions, GIs are probably going to be the most disputed topic. The conflict between special interests and the general interest is also evident in wine laws. EU laws limiting vineyard planting appear to be intended to Laws prohibiting the planting of vineyards appear to be intended to limit supply and drive-up costs for customers. The stress is more subdued in other situations. While quality requirements (such as those found in laws governing geographical indications, for example) can increase economic efficiency, they can also affect how rent is distributed. Power battles between "insider" and "outsider" producers characterized the delineation of the production zones, as the history of Europe's major wine regions demonstrates. Similarly, it is noteworthy that big wine-producing nations usually have substantially lower wine consumption taxes, even if it is impossible to prove causation. These taxes can aid in reducing the negative externalities associated with alcohol consumption.[11]

Legal restrictions on inter-state movement of wine

The following states forbid alcohol use: Gujarat, Mizoram, Bihar, and Nagaland. The selling of alcohol is legal in every other Indian state and union territory.[12].Article 47 of the Indian Constitution states under the directive principles of state policy (DPSP) that "the State shall endeavour to bring about prohibition of the consumption of intoxicating drinks and drugs which are injurious to health, except for medicinal purposes."[2] The Directive Principles are essential to the nation's government but are not citizens' rights that can be challenged in court. In accordance with Article 47, the State shall use these guidelines when drafting policy laws. According to Article 38, it is the responsibility of the state and federal governments to create further comprehensive laws and policies for execution that take DPSPs as a basic policy.A growing urban middle class, rising disposable incomes, and changing customer preferences have all contributed to India's wine industry's impressive expansion over the past few decades. But along with this expansion have come a plethora of regulatory obstacles that have prevented the business from reaching its full potential. In order to identify major roadblocks and suggest legislative changes that will promote sustainable growth, we will be examining the regulatory environment that oversees the wine industry in India in this study. There have been challenges as well as opportunities along the way for the Indian wine business. India is home to some of the world's oldest wineries, dating back thousands of years, but the modern wine business didn't really take off until the late 20th century. Since then, it has seen substantial change, with domestic. The wine industry in India has had remarkable growth in the last few decades, attributed to factors such as shifting consumer preferences, growing disposable incomes, and an expanding urban middle class. However, a number of regulatory barriers have also come with this expansion, keeping the company from realizing its full potential. In this study, we'll be looking at the regulatory framework that governs the wine sector in India in order to pinpoint significant obstacles and recommend legislative changes that will encourage sustainable growth. India has faced both possibilities and challenges in the wine sector. The contemporary wine industry in India didn't really take off until the late 20th century, despite the country being home to some of the oldest wineries in the world, with roots stretching back thousands of years. Considering.

- Economic

https://doi.org/10.52202/074664-0015

o Excise duties and taxes on affordability.

The maximum allowable rate of 150% is now the customs duty on wine, up from the previous rate of 100%.· India removed extra customs duties on beer, wine, and liquor imports after the US and the EU filed a protest with the World Trade Organization. For wine, the excise duty has been waived. · In 2007, the education cess that was applied to wine imports was eliminated.

Local Levies:

· A maximum state excise of 25% on wine has been mandated by the Central Government for certain states. · The import charge for wine is two rupees every bulk litre in Kerala. · Wine excise duty has been eliminated in Maharashtra. The excise duty on wine in Haryana has been lowered from Rs. 32.25 per proof litre to Rs. 20.50 per proof litre. Furthermore, it has lowered the export tax to 50% for domestic businesses.

Tax on Value Added (VAT):

Maharashtra State: 20%

Delhi State: 20%

Haryana State: 20%

Chandigarh State: 4%

Karnataka State: No VAT

Tamil Nadu State: 53%

Kerala State: 12.5%

Foreign Direct Investment Policy: Press Note 4 (2006) introduced the policy for alcohol distillation, which allows FDI up to 100% on the automatic route for alcohol distillation and brewing, subject to licensing by the relevant government.[13]

Land acquisition and grape sourcing.

Nonetheless, grapes are grown in India for their superior quality in tropical climates as well. A total of 64.3 thousand hectares are used to cultivate grapes, with a yield of 25.4 tons/ha and a total production of 1,630.7 thousand tons. The output is highest among the world's grape-growing nations in India due to the country's unique arbour training procedures. Maharashtra is the state that produces the most grapes in the entire nation. The districts of Nasik and Sangli have the highest concentration of agricultural area used for grape cultivation and grape production in the state. In addition to these, the districts of Ahmednagar, Pune, Satara, Solapur, and Osmanabad also grow grapes. These days, the Marathwada district of Latur also produces grapes. But the districts of Nasik and Sangli are forward in the scientific cultivation of grapes. Maharashtra has 45.1 thousand hectares of planted grapes, with an annual production of about 1,275 thousand tons. In 2006–07, India exported 5.5 thousand tons of grapes valued at Rs. 300.58 crores; Maharashtra accounted for roughly 80% of this total.[13].

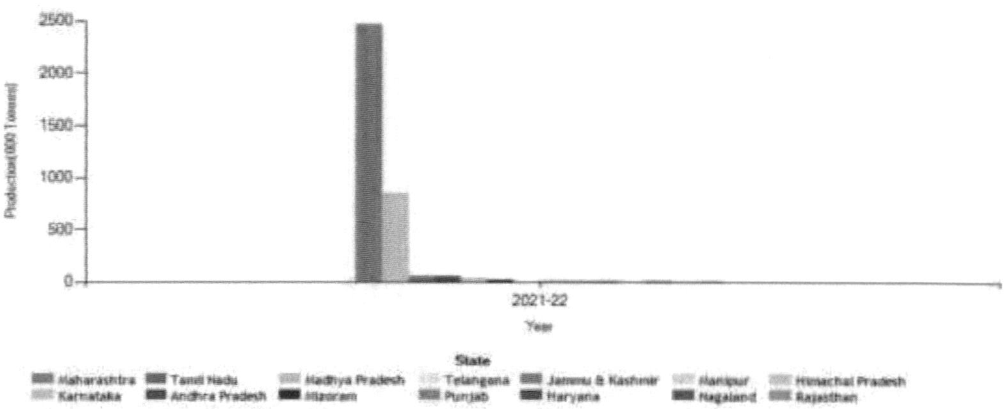

Source: National Horticulture Board (NHB)

- Credit and financial resources.

The grape wine production business requires between Rs 4 crore and Rs 7.5 crore in financing for every 100,000 litres produced (100 KL). And while this covers land, buildings, equipment, and operating capital, it does not include expenses for marketing or for purchasing land and establishing vineyards. NABARD can provide financial support. Additionally, APEDA offers aid with exports. However, an investment of between Rs 1.25 crore and Rs 2.5 crore per 100 KL is needed for fortified wine of lower grade. In addition to NABARD, there are several banks where you can obtain financing for the plant.The cost components of such a model along with the basis for costing are exhibited in Annexure I. A summary is given in the figure below. The project cost works out to Rs. 3.20 lakhs.

Sl. No.	Component	Proposed Expenditure
1.	**Cultivation Expenses**	
	(i) Cost of planting material	10,000
	(ii) Manures & fertilizers	9,000
	(iii) Insecticides & pesticides	4,000
	(iv) Cost of Labour	8,800
	(v) Others, if any, (Power)	3600
		35,400
2.	**Irrigation**	
	(i) Tube-well/submersible pump	56,000
	(ii) Cost of Pipeline	-
	(iii) Others, if any, please specify	-
		56,000
3.	**Cost of Drip/Sprinkler**	**35,000**
4.	**Infrastructure**	
	(i) Store & pump house	30,000
	(ii) Labour room	-
	(iii) Agriculture Equipments	10,000
	(iv) Others, if any. Please specify (Bower system)	1,20,000
		1,60,000
5.	**Land Development**	
	(i) Soil Leveling	4000
	(ii) Digging	-
	(iii) Fencing & gates	29600
	(iv) Others, if any, please specify	-
		33,600
6.	**Land, if newly purchased** (Please indicate the year)	@
	Grand Total	**3,20,000**

Source: National Horticulture Board (NHB)

Societal perceptions of wine consumption in India

Between 2022 and2027, the size of the Indian wine market is projected to increase at a CAGR of 30.92%. It is anticipated that the market will grow by USD 688.16 million. The market was estimated to be worth USD 68.80 million in 2017. The segmentation of the Indian wine market by Type (domestic and imported) and Product (red, fortified, white, and sparkling) is covered in detail in this study. It also features a thorough examination of the obstacles, trends, and drivers. Patterns The primary trend in the market is the growing number of millennials across the globe. Millennials have enormous purchasing power, and this trend is only expected to continue. In 2022, millennials accounted for more than 40% of working-age people in India and earned some of the highest salaries of any generation. Because millennials are more willing to explore new flavors and varieties than other consumer segments, alcoholic beverages, particularly wine, are growing more and more popular among this generation of consumers. In 2022, millennials made up one-third of

https://doi.org/10.52202/074664-0015

India's overall population. This presents a significant opportunity for local players to grow their clientele and geographic reach. Throughout the projected time, the millennial population is anticipated to be crucial to the expansion of the Indian wine industry.
Campaigns for Challenges Against alcohol consumption is the main obstacle preventing the market from expanding. In an effort to stop the rising number of alcohol-related mishaps and misuse, particularly among younger people, organizations are starting campaigns against alcohol use and players. The strong regulations on alcohol-related advertising combined with the growing anti-alcohol efforts in India may hinder the expansion of the wine industry there over the course of the projection period.
Extended intake of alcoholic beverages, such as wine, can result in a number of health problems. Alcohol abuse can have an impact on a person's physical and emotional health. Consumption of healthy drinks and beverages has increased due to rising health consciousness and people's propensity for leading healthier lifestyles.
Evolution and current state of Regulation
The great wine-producing countries of the world have different wine traditions to match their varietals. What was formerly a collection of autonomous nations producing wine according to their own policies has gradually evolved into a comprehensive network of global accords intended to fortify the wine sector overall. The integration of global markets has forced the rest of the globe to the negotiation table to preserve cultural variety, even though the EU remains the strongest coalition. Standardized laws have not, however, resulted from the worldwide integration of the wine business. For example, France's stringent producer regulations, in contrast to the relatively lenient positions adopted by the United States and Australia, have started to have major consequences.

Impact of Regulations on wine Industry Growth

A great industry to research government involvement in the food and agriculture markets is wine. Government rules governing the wine sector vary greatly between nations. There is also a lengthy history of wine restrictions. It is particularly evident how long-standing wine restrictions have been. In Europe, laws passed between the eighteenth and the early twentieth century are largely responsible for many of the rules in place today.There are some qualities of wine that distinguish it from other agricultural products. Unlike traditional agricultural commodities, wines are frequently highly distinctive goods where quality criteria are a determining factor. The main position. Since wine is an alcoholic beverage, it is also directly associated with certain negative health effects. Furthermore, wine markets are subject to numerous particular regulatory frameworks, as this article documents. These and other factors have contributed to the current surge in interest in the discipline of "wine economics," which has resulted in specialized conferences and journals (Storchmann 2012). Perspectives on Applied Economics and Policy 642. The analysis of wine laws reveals commonalities with "normal" agricultural policy analysis notwithstanding these distinctions. For example, wine laws are frequently distinguished by their strong historical tenacity, significant national variation. (Giulia Meloni et al.,2019)

Regulatory Framework in India's Wine Industry

This theme focuses on examining the laws that now control India's wine business. It talks on the difficulties and limitations that the current regulatory frameworks present to wine producers, distributors, and retailers. The impact of regulations on industry pricing, distribution, market entry, and marketing techniques is examined. (Chakravorty, U., & Lall, S. (2007). Fragmented power: Political institutions and the challenges of reform in the Indian electricity sector. World Development.This paper examines the challenges of reform in the Indian electricity sector, focusing on the role of political institutions in shaping regulatory frameworks. The authors argue that fragmented power structures in India hinder effective policy reform and regulatory implementation. The main conclusion of the paper is that addressing the fragmented power dynamics within political institutions is crucial for successful policy reform in the Indian electricity sector. (Saxena, S., & Khandelwal, A. (2018). Regulatory challenges in the Indian wine industry. International Journal of Wine Business Research, Paper specifically explores the regulatory challenges faced by the Indian wine industry, highlighting issues such as licensing, taxation, and distribution regulations. The authors analyze the impact of these challenges on the growth and sustainability of the industry. The main conclusion of the paper is that addressing regulatory challenges, such as simplifying licensing procedures and reducing tax burdens, is essential for promoting sustainable growth in India's wine industry. (Bhattacharya, S., & Chakraborty, I. (2015). Regulatory challenges in the Indian pharmaceutical industry. Journal of Health Management,this paper examines regulatory challenges in the Indian pharmaceutical sector, offering insights into the broader regulatory landscape in India. The authors discuss issues such as pricing regulations, intellectual property rights, and quality control. The main conclusion of the paper is that regulatory frameworks in India need to be more transparent, consistent, and efficient to support industry growth and innovation. (Kumar, A., & Singh, S. (2019). Policy reforms and sustainable growth in the Indian agriculture sector. Journal of Agricultural Economics. While focusing on the agriculture sector, this paper discusses the importance of policy reforms for achieving sustainable growth in Indian industries. The authors highlight the need for regulatory changes to address challenges such as land acquisition, subsidies, and market access. The main conclusion of the paper is that policy reforms, including regulatory changes, are essential for promoting sustainable growth and competitiveness in Indian industries, including the wine sector.

Sustainability and Environmental Regulations

In the context of India's wine business, this theme emphasizes the significance of environmental rules and sustainability. It looks at the steps taken by industry players to follow environmental regulations, embrace sustainable practices, and lessen their carbon footprint. The literature examines the difficulties the industry faces in striking a balance between environmental preservation and economic expansion. (Joshi, V., & Rahman, Z. 2017). Sustainability challenges in the Indian wine industry. International Journal of Wine Business Research,the sustainability challenges faced by the Indian wine industry, focusing on environmental regulations and practices. The study highlights the need for policy reforms to promote sustainable growth in the industry. Addressing sustainability challenges through policy reforms is crucial for the long-term growth and success of the Indian wine industry. (Sharma, S., & Singh, S. 2018). Regulatory challenges and opportunities for the Indian wine industry. Journal of Wine Economics,Paper explores the regulatory challenges and opportunities in the Indian wine industry, analyzing the impact of government policies on the industry's growth and sustainability. The study emphasizes the importance of policy reforms to address regulatory hurdles. Policy reforms are essential to overcome regulatory challenges and create a conducive environment for sustainable growth in the Indian wine industry. (Gupta, A., & Kumar, R. (2019). Environmental regulations and sustainable practices in the Indian wine industry. Journal of Sustainable Wine Business Research, investigates the relationship between environmental regulations and sustainable practices in the Indian wine industry, examining the impact of regulatory compliance on environmental sustainability. The study suggests that policy reforms are necessary to promote sustainable growth. Aligning environmental regulations with sustainable practices is essential for the long-term viability of the Indian wine industry. (Chopra, N., & Verma, A. 2020). Policy implications for sustainable growth in the Indian wine industry. International Journal of Business and Wine Research.This paper discusses the policy implications for sustainable growth in the Indian wine industry, analyzing the regulatory challenges and opportunities faced by industry stakeholders. The study highlights the need for policy reforms to drive sustainable development. Policy implications play a crucial role in shaping the future trajectory of the Indian wine industry towards sustainable growth.

Numerous problems with distribution, marketing, licensing, and taxes are among the regulatory obstacles to the wine industry's expansion that are highlighted in the literature on examining

 https://doi.org/10.52202/074664-0015

regulatory hurdles in India. Everyone agrees that the industry's potential for sustainable development is being hampered by the antiquated and cumbersome regulatory structure that currently exists. The need for policy reform to simplify rules and create an atmosphere that fosters the expansion of the wine business has been recognized by researchers. To further harness the wine industry's economic potential and generate employment opportunities in the production and hospitality sectors, the literature highlights the significance of overcoming regulatory difficulties in the sector. Research indicates that an environment with favourable regulations can promote innovation, draw in capital, and boost global market competitiveness.

Suggestions & Recommendations

The Indian government has recognized the wine industry as a thriving sector and has offered subsidies for processing facilities along with lower VAT rates. However, further support in terms of funding research and development, particularly in wine biotechnology, and encouraging academic institutions to undertake related studies is suggested. This support would aid in genetic enhancements of grape varieties, wine yeasts, and wine bacteria, thereby contributing to the industry's growth and sustainability.

Moreover, areas such as storage and transportation infrastructure, wine exports, media campaigns, wine festivals, genetic resources and improvement, production technology, plant health management, and the cascading effect of climate change on grape cultivation require focused attention from both the industry and research institutes. By addressing these areas comprehensively and aligning with sustainable practices, the Indian wine industry can continue its exciting journey of growth and development.

In summary, while the Indian wine industry faces regulatory challenges, there's a consensus among researchers about the potential for sustainable growth. Through policy reforms, investment in research and development, and a focus on sustainable practices, the industry can overcome obstacles and realize its economic potential while contributing to environmental preservation and competitiveness in the global market.

Conclusion

The storage and transportation infrastructure, wine exports, media campaigns, wine festivals, genetic resources and improvement, production technology, plant health management, pre- and post-harvest technology, cascading effect of climate change on grape cultivation, composition of grape juice, yeast ecology of wine production, fermentation kinetics, and wine character are the areas that the Indian market and research institutes should focus on. Wine has been recognized by the Indian government as a thriving industry, and as such, subsidies for processing facilities and lower VAT have been offered. For the purpose of genetically enhancing grape wines, wine yeasts, and wine bacteria, the government ought to fund research and development as well as encourage academic institutions to undertake wine biotechnology. Indian wine originated in Persia and has made a very exciting journey through time.

https://doi.org/10.52202/074664-0015

References

1. Adsule, P. G.:Grape wine industry in India: development, status and prospect. In II International Symposium on Beverage Crops 1274, 57-64(2018).
2. Karibasappa, G.S. and Adsule, P.G.:Evaluation of wine grape genotypes by National Research Centre for Grapes at their farm at Pune, Maharashtra, India. In International Symposium on Grape Production and Processing 785, 497-504(2006).
3. Gade, A.D.: WINE INDUSTRY IN INDIA A REGIONAL SURVEY. Ashok Yakkaldevi, (2022).
4. Kaddy, A.K. and Suklabaidya, P., Review of Maharashtra's Wine Policy and Regulations and its impact on Wine Tourism. Introspecting Problems and Prospects,254.
5. Robinson, J., Harding, J. and Vouillamoz, J.:Wine grapes: a complete guide to 1,368 vine varieties, including their origins and flavours. Penguin UK, (2013).
6. Regulatory challenges for alcoholic beverage industry. Thursday, 01 December, 2022, 12 : 00 PM [IST]Ashwin Bhadri
7. Talwade, Amarnath R.:The Digital Institutional Repository of IIM Bangalore.December 11, 2023Categories: Licenses & Government RegistrationsTags: Get a Wine Shop License in India, legal licensing, Wine Shop License, Wine Shop License in India, (2014).
8. Meloni, G., Anderson, K., Deconinck, K. and Swinnen, J.:Wine regulations. Applied Economic Perspectives and Policy, 41(4), 620-649(2019).
9. "States with total and phase-wise prohibition of alcohol in India". The Indian Express. 6 April 2016.
10. https://www.indialawoffices.com/ilo_pdf/industry-reports/wineindustry.pdf.
11. https://agriexchange.apeda.gov.in/prodgallery/prdprofile_moa.aspx?hscode=08061000
12. Mathur, A.K.: Alcoholic beverages industry in India: An exploratory study. Eduved Global Management Research, 1(1), (2014).
13. Yadav, M.K. and Dixit, S.K.:The growth and evolution of global wine tourism. In Routledge Handbook of Wine Tourism, pp. 15-28(2022). Routledge.
14. Sheth, J.N.:Impact of emerging markets on marketing: Rethinking existing perspectives and practices. Journal of marketing, 75(4),166-182(2011).
15. McHugh, J.:Grape wine in ancient and early Medieval India: The view from the centre. The Indian Economic & Social History Review, 58(1), 113-144(2021).
16. Jha, J.:Wine in India: Changing Consumer Perceptions. ICFAI Journal of Consumer Behavior, 2(2), (2007).
17. Adsule, P.G.:Grape wine industry in India: development, status and prospect. In II International Symposium on Beverage Crops 1274, 57-64(2018).
18. Bheemathati, S.:An Overview: Recent Research and Market Trends of Indian Wine Industry. J. Food Process. Beverages, 3(1), pp.1-5(2015).
19. Bargain, O., Cardebat, J.M., Chiappini, R. and Laffitte, C.:Long-term prospects of the international wine trade. Journal of Wine Economics, 13(4), pp.442-450(2018).

 https://doi.org/10.52202/074664-0015

20. Hussain, M., Cholette, S. and Castaldi, R.M.:An analysis of globalization forces in the wine industry: implications and recommendations for wineries. Journal of Global Marketing, 21(1), 33-47(2008).
21. Joshi, V.K. and Attri, D.:Panorma of research and development of wines in India, (2005).
22. Srivastava, R.K.:Perception Study of Wine Market Culture of an Emerging Market-India-Will it Emerge from Infant to Growth Stage. Journal of Hotel & Business Management, 3(01), 1-6(2013).
23. Gupta, V. and Duggal, S.:Impact of perceived risks and benefits: a case study of wine consumption patterns in the Indian context. Journal of wine research, 31(1), 67-83(2020).
24. Gupta, V. and Sajnani, M.:Risk and benefit perceptions related to wine consumption and how it influences consumers' attitude and behavioural intentions in India. British Food Journal, 122(8), 2569-2585(2020).
25. Khot, V.:Breaking The Monopoly of Wine Imports in India. Spawton, L.L.T., International Journal of Wine Marketing, (2007).
26. Morrison, A. and Rabellotti, R.:Gradual catch up and enduring leadership in the global wine industry. Research Policy, 46(2), 417-430(2017).
27. Costopoulou, C., Ntaliani, M. and Ntalianis, F.:An analysis of social media usage in winery businesses. Adv. Sci. Technol. Eng. Syst, 4, 380-387(2019).
28. Shinde, G.S., Singh, D.K. and Jaju, D.R.:Wine Consumption Patterns of Indians with Special Reference to The Nashik District: A Review. International Journal of Management (IJM), 14(1), (2023).
29. Mason, M.C. and Paggiaro, A.:Investigating the role of festivalscape in culinary tourism: The case of food and wine events. Tourism management, 33(6), 1329-1336(2012).
30. Joshi, V.K. and Kumar, V.:Importance, nutritive value, role, present status and future strategies in fruit wines in India. Bioprocessing of Foods. (Eds. PS Panesar et al.), 39-62(2011).
31. Sharma, P., Singh, S. and Misra, R.: Wine-related lifestyle segmentation in the context of urban Indian consumers. International Journal of Wine Business Research, 32(4), 503-522(2020).
32. Sheth, H., Pereira, K.A., Thomas, T.S., Kumar, V.K., Kelapanda, M.C. and Dant, M.G.:Wine festival and wine consumerism-A perspective study on influence and awareness in the growth and acceptance of wine in Indian market. Int J Eng Technol Manag Appl Sci, 2, 20-36.(2014).
33. Kansara, S.:Exploring the wine sector in the Nashik district of India. International Journal of Wine Business Research, 32(2), 203-217(2019).
34. India Wine Market by Type and Product - Forecast and Analysis 2023-2027
35. Shinde, G.S., Singh, D.K. and Jaju, D.R.:Wine Consumption Patterns of Indians with Special Reference to The Nashik District: A Review. International Journal of Management (IJM), 14(1), (2023).
36. Roy, D.: GRAPES TO GLASSES: UNVEILING INDIA'S WINE ODYSSEY-CHALLENGES AND OPPORTUNITIES IN GRAPE CULTIVATION AND WINE PRODUCTION. International journal of social sciences, 3(08), 01-05(2023).
37. Insel, B.:The evolving global wine market. Business Economics, 49(1), 46-58(2014).
38. Campbell, G. and Guibert, N.:Introduction: Old World strategies against New World competition in a globalising wine industry. British Food Journal, 108(4), 233-242(2006).
39. Gade, A.D.:WINE INDUSTRY IN INDIA A REGIONAL SURVEY. Ashok Yakkaldevi, (2022).
40. Payini, V., Mallya, J. and Piramanayagam, S.:Indian women consumers' wine choice: A study based on conjoint analysis. International Journal of Wine Business Research, 34(4),469-494(2022).
41. Sharma, P., Singh, S. and Misra, R.:Wine-related lifestyle segmentation in the context of urban Indian consumers. International Journal of Wine Business Research, 32(4), 503-522(2020).

42. Ray, P. and Chowdhury, S.:Popularizing grape cultivation and wine production in India–challenges and opportunities. International Journal of Social Sciences, 4(1), 9-28(2015).
43. Castaldi, R., Cholette, S. and Hussain, MA.: country-level analysis of competitive advantage in the wine industry, (2006).
44. Sharma, P., Singh, S. and Misra, R.:Wine-related lifestyle segmentation in the context of urban Indian consumers. International Journal of Wine Business Research, 32(4),503-522(2020)..

 https://doi.org/10.52202/074664-0015

Exploring the Cultural Capital Dimension in Tourism: Study of Punjab State

[1]Rajshree Bhargava, [2]Sandeep Guleria, [3]Arvinder Kaur
UITHM Chandigarh University, Mohali Punjab, India

Abstract.
The growing recognition of cultural resources as vital economic assets has driven this study, which seeks to understand and leverage cultural capital for tourism development. Zanzibar, renowned for its rich cultural heritage, inspires this research, which aims to replicate and expand upon its established framework to explore the cultural dynamics in Punjab. The study conducted in Zanzibar highlights seven dimensions known as tourism cultural capital dimensions (TCCDs), which significantly attract tourists to the region for their vacations (Salim, 2015). Using the same framework, Exploratory Factor Analysis (EFA) was employed to identify the dimension of cultural capital that impact tourists visiting Punjab. The researcher has adopted mixed-method approaches, combining qualitative and quantitative research techniques, with data collected through a comprehensive questionnaire covering almost 42 variables contributing to tourism cultural capital. SPSS was used to conduct factor analysis in order to reduce the number of variables to significant factors. Valuable insights are provided by the study regarding the correlation between cultural heritage and the development of sustainable tourism in varying geographical and cultural backgrounds. It emphasizes the importance of cultural custodians, tourism developers, and policymakers working together to develop effective strategies that leverage cultural capital to drive economic growth.

Keywords – Cultural Capital, Tourism Cultural Capital Dimension (TCCD), Cultural Heritage, Sustainable Tourism Development.

1. Introduction

India's cultural richness showcases an array of historic monuments, pious religious sites, and vibrant traditional forms of arts and crafts. These cultural aspects make India a magnet for global travellers drawn to its unique blend of history, spirituality, and artistic expression. Tourism contributes to the global economy, generating significant revenue for many countries and destinations worldwide. In 2022, India experienced a remarkable recovery in its tourism sector, with Foreign Tourist Arrivals (FTAs) soaring to 6.44 million, marking a staggering 321.54% increase from 1.52 million in 2021. Business professionals, leisure holidays and recreation, the Indian Diaspora, and medical professionals were the primary purposes of visiting India. This surge was also reflected in the Non-Resident Indians (NRIs) and International Tourist Arrivals (ITAs), with NRIs arrivals up by 43.9% and ITAs by 104.4% from the previous year. India's global Travel and Tourism Development Index (TTDI) position stood at 54 in 2021, highlighting its significant role and potential in global tourism. (India Tourism Statistics 2024, MoT)

As of September 2023, India recorded 6.43 million foreign tourists arriving (FTAs) and a remarkable 305.4% annual growth rate in FTAs in 2022. Additionally, the country achieved $16.92 billion in foreign exchange earnings from tourism in 2022, holding a 1.64% share in global International Tourism Receipts. Punjab emerged as a notable destination, securing the 7th position nationally with 5% of total foreign tourist visits. At the same time, Gujarat and Maharashtra attracted the most foreign

https://doi.org/10.52202/074664-0016

tourists, signalling a robust post-COVID-19 recovery and an upward trend in hotel room rates towards pre-pandemic levels. (Invest India 2024)

Culture is an economic asset that can be commercialised and presented as an appealing option for tourists. Its ability to generate income positions culture as a form of capital, encompassing certain intangible elements. Recognising culture's capacity to yield substantial revenue has contributed to a noticeable surge in global tourism, making it a key driver for travellers seeking to fulfil their cultural interests and expectations (Alinejad & Razaghi, 2012). Understanding the dynamics of this economic asset, referred to in this study as 'tourism cultural capital,' is crucial for devising strategies to maximise its benefits. The effectiveness of these strategies ultimately determines the level of advantages derived from cultural tourism, which constitutes an essential component of India's tourism industry, attracting over 70% of its visitors. India's allure lies in its deep cultural heritage, boasting a wealth of historical sites, sacred locations, and a vibrant arts and crafts tradition, making it an enticing destination for global travellers. In 2023, popular domestic tourist destinations included the Taj Mahal, Sun Temple, Red Fort, Qutub Minar, and Agra Fort, while international tourists favoured the Taj Mahal, Agra Fort, Qutub Minar, Humayun's Tomb, and Fatehpur Sikri, underscoring India's widespread cultural appeal. This diverse cultural landscape enriches the visitor experience and significantly contributes to the national economy, solidifying India's position as a premier cultural tourism destination. Punjab, renowned for its vibrant culture and historical significance, ranks seventh in India for foreign tourist arrivals, constituting 5% of the national total. 2019, the state attracted approximately 1.1 million foreign tourists, showcasing its growing popularity among international visitors. Despite the challenges posed by the global COVID-19 pandemic, Punjab has witnessed a robust recovery in its tourism sector.

Cultural attractions in most destinations encompass various dimensions, some rooted in tangible objects and others in intangible characteristics. For instance, Jallianwala Bagh stands out as a significant attraction due to its historical association with the infamous shooting incident by General Dyer during India's freedom struggle. Meanwhile, annual festivities like 'Holla Mohalla' and 'Kila Raipur' offer a distinct experience to visitors, particularly those who find immense satisfaction in cultural phenomena. To optimise returns as cultural products in tourism, managing and sustaining tangible and intangible cultural resources effectively involves promoting and preserving monuments while enhancing their historical narratives. Conversely, inadequate management can lead to the gradual decline or even disappearance of the potential inherent in these resources, impacting their ability to generate tourist revenue. Consider the example of the Kila Raipur and Holla Mohla festivals. Suppose the current organisers and performers, who may belong to an older generation, pass on their skills and knowledge to younger generations. In that case, the festival can maintain its value and thrive.

Cultural resources are often overlooked as economic assets which play a crucial role in tourism development. These tangible and intangible resources require expert management to maximise their potential returns. One significant advantage lies in aligning specific characteristics of this cultural capital with productive activities. As tourism revenue is derived from visitors, it is essential to recognise that tourist markets in source countries exhibit varying segment characteristics.

For Punjab, a region with a substantial Non-Resident Indian (NRI) population, focusing on cultural resources that resonate with ancestral legacies becomes logical. These resources include ideologies, performance arts, sports, architectural heritage, artistic enterprises, cultural artefacts, and artworks.

https://doi.org/10.52202/074664-0016

Destination managers can observe interrelated attributes and efficiently group them for targeted exploitation in specific markets.

Consider Punjab's prime attractions: the mud and stone architecture of Taxila and Harappan civilisations, the rich historical narrative, aromatic spices, ancient sites, customs, vibrant festivals, exquisite antique carpentry, soulful local music, musical instruments, and folk dances. While festivals, music, and dances represent one dimension, the stone and mud architecture, intricate wood carvings, and Jharokhas constitute another facet of this diverse cultural capital.

Given its complexity, cultural capital cannot be considered one-dimensional. Instead, it comprises various facets that lend themselves to distinct marketing strategies. Identifying these facets accurately can lay the groundwork for tourism marketing professionals' strategic planning, encompassing resource allocation, sustainability, and enhancement considerations. Therefore, the primary aim of this research is to conceptualise cultural resources as multifaceted capital for tourism, with each facet exhibiting unique features. These features largely stem from tourists' perceptions, who seek out these cultural offerings to fulfil specific desires. This study focuses on Punjab, a renowned tourist destination in India celebrated for its rich cultural history, which has fascinated global audiences. The research commenced with an exploratory phase, conducting in-depth interviews with various tourism stakeholders in Punjab. This phase was succeeded by a descriptive study surveying tourist. Exploratory factor analysis was then applied to identify the dimensions of cultural capital.

2. Review of Literature

The Concept of capital emphasises both its economic and non-economic value. As defined by Macbeth, Carson, and Northcote (2004), capital encompasses resources that empower communities, ranging from natural resources to knowledge, skills, and political power. This broad understanding of capital facilitates examining how non-economic aspects, like a community's cultural and social attributes, can be leveraged for economic gain, particularly in tourism.

Bourdieu's (1960) cultural capital theory further expands on this by examining how culture influences social class disparities, particularly in educational outcomes. His later work (Bourdieu, 1986) breaks down cultural capital into institutionalised, embodied, and objectified forms, which have been applied to various fields, including tourism. Researchers argue that cultural assets can be commodified and marketed as tourism products, transforming culture into an economic resource (Brubaker, 1985; Macbeth et al., 2004; Addo, 2011).

Cultural tourism is highlighted as an area where tourists seek unique experiences that differ from their everyday lives (Alinejad and Razaghi, 2012). It involves engaging with local arts, history, and traditions, underlining the inseparability of culture and tourism (OECD, 2009). Despite the complexity and varied definitions of culture, it is universally recognised as encompassing a specific group's behaviours, beliefs, and products (Hofstede, 1997; Hall, 1976).

The review also illustrates how cultural experiences, such as visiting historical sites or engaging with local traditions, contribute significantly to the economies of tourist destinations. Examples include the draw of cultural figures, sports, and religious pilgrimages in regions like Zanzibar, Europe, and the Middle East, respectively, underscoring culture's potential as a lucrative economic resource (Lareau and Weininger, 2003; Addo, 2011).

The interplay between cultural capital and tourism activity underscores a pivotal aspect of destination competitiveness, as elucidated through various scholarly contributions. Akama (2000), Vengesayi (2003) collectively assert that the absence of cultural capital precipitates a diminished need for tourist facilities and services, thereby suggesting that its effective management is instrumental in augmenting a destination's competitive advantage. This assertion is further expounded by exploring cultural capital's dimensions, which vary across countries due to differing cultural resources that attract tourists. Mwaipopo and Salim (2012) propose a conceptual framework for tourism cultural capital, as illustrated by Zanzibar, encompassing cultural behaviour, historical objects, and an artistic dimension, enriching the tourist experience together. Issa Seif, Lufumbi Mwaipopo (2015) discuss the economic value of culture as a tourist product, its impact on tourism revenue, the role of cultural capital in promoting destinations and creating economic opportunities and acknowledge the study's limitations. This literature review accentuates the critical role of cultural capital in the tourism industry, advocating for enhanced understanding and strategic utilisation to bolster the allure and competitiveness of tourism destinations.

The prosperity of tourism in Punjab is intricately linked to preserving and promoting its rich cultural heritage, religious beliefs, and community values. The key to establishing the state as a premier destination lies in building cultural awareness and fostering community involvement to enhance its competitive edge. The Department of Tourism and Cultural Affairs in Punjab has identified a multitude of cultural resources that are highly attractive to visitors. These resources include religious shrines like Sri Harminder Sahib (Golden temple), historical sites, museums, cultural landmarks, nature trails, ecological hotspots, border tourism, and the Indus Civilization - the birthplace of two ancient Harappa sites. There are also forts, palaces, wetlands, fairs, festivals, customs, language and dressing styles. Artistic works such as antique wood Inlay Craft of Hoshiarpur and local music like folk dances such as Bhangra Giddha are also popular. The goal of this study is to collect data that will help identify the different dimensions of cultural capital in tourism based on specific attributes.**3.**

3. Research Methodology

During the initial investigation, an exploratory approach was taken, involving interviews with 24 individuals representing various stakeholder groups in Punjab. This process identified 42 distinct cultural attributes contributing to the region's appeal to tourists. These attributes formed the foundation of a survey instrument developed using a 7-point Likert Scale. The survey was subsequently administered to 164 individuals who had visited Punjab.

After confirming all the necessary preconditions, including checking reliability and validity by ensuring that Cronbach's Alpha Coefficients ($\alpha \geq 0.6$), as recommended by Tabachnick and Field (2001) for exploratory research, were met. Before proceeding to distribute the survey. Other preconditions were checked: Kaiser-Meyer-Olkin's measure of sampling adequacy (KMO > 0.5) and Bartlett's test of sphericity scores ($p < 0.001$) for both surveys were significantly met. The expectation-maximization algorithms address missing values and any outliers not found.

Exploratory Factor Analysis (EFA) was conducted on the data to identify the hidden cultural capital dimensions of tourism in Punjab's tourism industry. The reason behind choosing EFA was the

 https://doi.org/10.52202/074664-0016

uncertainty regarding the number of dimensions and their corresponding elements (Voon & Lee, 2009).

Table I: Respondent's Demographic Profile

Demographic Profile	Categories	Percentage
Gender	Male	41.5
	Female	58.5
Age	18-29 Years	28.3
	30-49 Years	47.2
	50-59 Years	24.5
	60 Years & Above	Nil
Region of World	Asia	92.8
	Other	7.2
Employment Status	Full Time	46.2
	Part-Time	7.7
	Self Employed	17.3
	Unemployed	Nil
	Student	4.7
	Retired	7.7
	Homemaker	13.5

The respondents filled out a total of 164 questionnaires, out of which 32 turned out to be unusable due to not visiting Punjab or inconsistent responses. Table I provides the Demographic profile of respondents, mainly from Asia; other regions include Australia, New Zealand, North America and Europe.

The Finding of the Exploratory Analysis (EFA)

An EFA was performed using principal component analysis and varimax rotation. The minimum factor loading criteria was set to 0.50. The communality of the scale, which indicates the amount of variance in each dimension, was also assessed to ensure acceptable levels of explanation. The results show that all communalities were over 0.50.

An important step involved weighing the overall significance of the correlation matrix through Bartlett's Test of Sphericity, which measures the statistical probability that the correlation matrix has significant correlations among some of its components. The results were significant, x2 = 2693.632 ($p < 0.001$), which indicates its suitability for factor analysis. The Kaiser–Meyer–Olkin measure of sampling adequacy (MSA), which indicates the appropriateness of the data for factor analysis, was 0.707. In this regard, data with MSA values above 0.700 are considered appropriate for factor analysis. Finally, the factor solution derived from this analysis yielded Seven factors for the scale, which accounted for 54.377 per cent of the variation in the data.

In the initial EFA, five items ("HO5: Touring the historical places was a hassle-free experience with adequate information." "CE7: Participating in local sports and games was fun filled activity." "TMSP1: The traditional medicine (Ayurveda) and spiritual practices were accessible for observation or participation." "HC4: The process of creating handicrafts was demonstrated or explained engagingly." and "HC5: A wide variety of souvenirs were available in memory of the destination" failed to load on any dimension significantly. Hence, the five items were removed from further analysis. Total 38 latent variable finalised out of 42 variables.

The authors repeated the EFA without including these items. The results of this new analysis confirmed the seven-dimensional structure theoretically defined in the research. The Kaiser–Meyer–Olkin MSA was 0.806. The Seven dimensions explained a total of 53.321 per cent of the variance among the items in the study. Bartlett's Test of sphericity proved to be significant, and all communalities were over the required value of 0.500.

Table II: Tourism Cultural Capital Dimensions (Extracted) and Factor Loading

The Seven Factors structure retained a meaningful ground for conceptualising the dimensions of tourism cultural capital for Punjab. Factor 1 consists of items related to visiting historic buildings, native architecture, monuments, places, iconic structures of residents, museums and galleries, narrow lanes, archaeological ruins and curved doors. Because each of these items represents an element of "Historical Objects", it was named as such. Factor one was the strongest component based on the percentage of variance explained (53.321%), the highest eigenvalue (19.729), the number of factor loadings (9) and Cronbach's alpha coefficient (α=.899). Factor two had seven variables with a 9.652% percentage of variance explained, an Eigenvalue of 3.571 and Cronbach's alpha coefficient of α=.941. This factor was labelled "cultural events" as the variables explain related events that local communities engage in and reflect their traditional activities. Such events are dance, drama, music, and social and cultural activities. Factor three was cultural behaviour; factor four was cultural medicine and spiritual practices. Factor five was traditional sports and adventure activities, factor six was food and culinary experiences, and factor seven was labelled handicrafts, identified as part of this EFA and aligned with the theoretical proposition in this research. All seven factors exceeded the minimum alpha value (α= 0.6) Tabachnick and Field (2001) recommended for an exploratory study. The specific characteristics of all the retained dimensions are summarised in Table III.

https://doi.org/10.52202/074664-0016

Table III: Summary of Key Aspects of Cultural Capital in the Tourism Sector (Extracted)

Therefore, through exploratory factor analysis, verified by reliability tests of the dimensions' constructs, it was revealed that seven dimensions likely represent the dimensions of tourism cultural capital that Punjab is endowed with.

4. Discussion and Findings

These findings suggest that the cultural capital of tourism is complex and made up of multiple factors that can be analysed and manipulated from a theoretical and strategic perspective. What this means is that the reasons why tourists are drawn to a particular destination are not one-dimensional. Instead, there are several distinct hidden variables, each with its own unique attraction and level of visitor satisfaction based on their expectations and experiences. For instance, when it comes to tourists visiting Punjab, the historical objects dimension appears to have the most significant impact, which can be directly attributed to its rich history of bravery and freedom struggle, embodied by landmarks such as Jallianwala Bagh, the Golden Temple, Wagh Border and the Partition Museum. These sites help to reinforce visitors' perceptions and expectations of Punjab's unique cultural heritage.

From a capital resources perspective, the dimension in question is considered to be of utmost significance in contributing to the revenue of Punjab's tourism industry. This factor holds the most influential role in attracting visitors to the region, as it serves as a key driver of tourism. In its absence, the number of tourists attracted to Punjab would be notably less. Therefore, it is important for destination managers to take stock of the items that constitute this dimension and develop comprehensive and sustainable plans to ensure optimal exploitation. These plans should be strategically targeted with a focus on core messages that promote the destination's historical object dimensions.

The dimension of cultural events is highly influential, with festivals, celebrations, Beating the Retreat Ceremony at Borders, theatres, drama performances, indoor social activities, cultural events, folk music and dance all falling under its purview. This dimension is centred around cultural institutions, in contrast to the historical objects dimension. To effectively cater to tourists' interests along this dimension, it is essential to understand precisely what draws them to these events and create appropriate positioning messages. Destination managers must also coordinate programs for all these cultural events to ensure they meet tourists' expectations. Further research is needed to determine how various events relate to visitor encounters, whether chronologically or hierarchically.

It is imperative to determine the significance of tourist segments on various dimensions. Without this knowledge, we cannot fully comprehend the factors that contribute to a successful tourism industry.

The cultural behavior, known as the third dimension, is entirely attributed to the state's native people. Throughout history, the Punjabi and Sikh communities have created a unique cultural melting pot that is unmatched anywhere else in the world. This undoubtedly adds to the allure of the destination, emphasizing the need to preserve and maintain these defining characteristics.

Tourists may find themselves drawn to traditional medicine and spiritual practices, whether out of curiosity or a genuine interest in alternative medicine. This unique characteristic of the state is a major draw for visitors, and destination managers should conduct further research to fully leverage its potential.

Tourists are increasingly attracted to the dynamic cultural facets of traditional sports and adventure activities, notably Marshal Art-Gatka and Rural Olympics, along with eco-biodiversity explorations.

These have become pivotal in defining Punjab's tourism cultural capital, offering thrilling, immersive experiences. Moreover, the region's wetlands present invaluable historical and ecological wealth for exploration and appreciation.
The study has revealed that two additional factors, namely culinary experiences and handicrafts, have emerged as distinct dimensions. This comes as no surprise since Punjab is renowned for its delectable cuisine, and numerous visitors are keen on exploring the diverse food options that feature locally sourced handmade products and spices. Moreover, locally crafted handicrafts offer a unique dimension that highlights the dynamic cultural capital resources of the tourism industry, which in turn, contributes to the revenue generation of the destination.

5. Conclusion and Implication for Further Research

This research has systematically unveiled the intricate layers of Punjab's cultural capital, revealing its substantial impact on the region's tourism sector. In this study, the cultural capital of tourist destinations was explored, focusing on the cultural resources that can be used as capital for economic gain. Punjab served as the case study, where seven underlying dimensions were identified through exploratory factor analysis. By dissecting dimensions such as historical objects, cultural events, cultural behavior traditional medicine & Spritual practices, traditional sports & adventure activities, culinary arts, and handicrafts, the study not only amplifies the understanding of Punjab's rich cultural heritage but also underscores its potential to significantly boost tourism. These findings highlight the necessity for strategic emphasis on preserving and promoting these cultural assets, ensuring they serve as sustainable drivers of economic growth and cultural preservation within Punjab's evolving tourism landscape.

The findings have theoretical implications for conceptualizing cultural resources as capital and can provide destination managers with valuable insights for tourism strategies.

However, the limitations of the study should be considered, such as the random sample selection process and the majority of tourists being from India. These results are specific to Punjab and other destinations may have different dimensions of cultural capital depending on their endowments. Further research is recommended to determine factor structures of other destinations and validate these dimensions through confirmatory factor analysis. Separate studies on these latent variables can also provide a better understanding of their role as pull factors for visitors.

https://doi.org/10.52202/074664-0016

References

1. Addo, E.: Diversification of the tourist landscape on Ghana's, (2011).
2. Akama, J.: Cultural Tourism in Africa: Strategies for the New Millennium: Proceeding of the Atlas Africa International Conference. Mombasa. Kenya, (2000).
3. Alinejad, M. E. and Razaghi, Z.: Culture and its role in tourism development. Life Science Journal, vol.9, no.3, 1593-1597 (2012).
4. Atlantic coast: Forts, castles and beach hotel/resort operations in the tourism industry. Journal of Tourism Consumption and Practice, vol.3, no.1, pp. 1-26.
5. Bandarin, F., Hosagrahar, J. anqad Albernaz, F, S.: Why development needs culture. Journal of Cultural Heritage, (2011).
6. Bourdieu, P.: The form of capital, in J. Richardson (Ed), (1986).
7. Brubaker, R.: Rethinking Classical Social Theory: The Sociological Vision of Pierre Bourdieu. Theory and Society, vol. 14, pp. 745-75 (1985)..
8. Hall, E.: Beyond culture. Doubleday, Anchor Press, New York, (1976).
9. Handbook of theory and research for sociology of education (New York, Greenwood), pp.241-258
10. Hofstede, G.: Culture Consequences: International Difference in Work-Related Values. Sage Publications, Inc., California, (1980).
11. K. Palwinder & Rekhi.S.: Factors influence tourist decision-making on destination selection phase: an Empirical Analysis on Punjab Tourism, *International Journal of Research in Social Sciences* Vol. 8, Issue 10(2), (2018).
12. Kant, S.: A SWOT Analysis on Farm Based Tourism-A Case of Punjab Regional State, India. *India* (2022).
13. Kim, M., & Kim, M.: Factors Influencing the Continued Intent to Use Virtual Interactive Platforms in Korean Small- and Medium-Sized Enterprises for Remote and Hybrid Work. Sustainability, 15(13), 9972, (2023).
14. Krippendorf, J.: The holidaymakers: Understanding the impact of leisure and travel. *Heinemann Professional Publishing,* (1987).
15. Kumar, N., Singh, R., & Aggarwal, A.: Spatial Analysis of Rural Tourism Sites in Punjab. International Journal of Applied Business and Economic Research, (2017). http://www.serialsjournals.com
16. Lakha, R & Avikshit. : A Study On Tourism And Hospitality Industry In Punjab-An Overview, (2022).
17. Lareau, A. and Weininger, E. B.: Cultural Capital in Educational, (2003).
18. Macbeth, J., Carson, D. and Northcote, J.: Social capital, tourism and regional development: SPCC as a basis for innovation and sustainability.: Current Issues in Tourism, vol. 7, no.6, pp.502- 522. Management and Sustainable Development, vol. 1, no.1, pp.15-25 (2004)..
19. Ministry of Tourism, Government of India.: India tourism statistics at a glance 2023 - English version (2023). Retrieved from https://tourism.gov.in/sites/default/files/2023-07/India%20Tourism%20Statistics%20at%20a%20glance%202023%20-%20English%20version_0.pdf
20. Mwaipopo, L. J. and Salim, I. S.: Exploring destination branding through utilizing the capital inherent in a culture: The case of Zanzibar". Challenges in tourism development and sustainability. International Conference Zanzibar. Tanzania, (2012).
21. OECD.: The Tourism of Culture on Tourism. OECD publishing, Paris, (2009).
22. National Centre for Biological Sciences. (n.d.). Overview of Indian healing traditions. Retrieved from https://www.ncbs.res.in/HistoryScienceSociety/content/overview-indian-healing-traditions

23. Punjab Bureau of Investment Promotion. (n.d.). [Tourism report]. Retrieved from http://investpunjab.gov.in/assets/docs/Tourism.pdf
24. Research: A Critical Assessment. Theory and Society, pp. 567-606.
25. Salim, I. S., & Mwaipopo, L.: Exploring Cultural Capital Dimemsions in Tourism: The Case of Zanzibar. Global Advanced Research Journal of Management and Business Studies, 4(10), 390–397 (2015).
26. Singh, N., & Kaur, N.: Punjab tourism: A SWOT analysis. *International Journal of Multidisciplinary Research and Development*, *3*(1), 131-133. society, vol. 10, no. 1, pp. 65-82 (2016).
27. Tabachnick, B.G, Fidel, L.S.: Using Multivariate Statistics, 4th Ed. New York: Allan and Bacon, (2001).
28. The Times of India.: Punjab launches adventure, water tourism policies, (2023, March 19). Retrieved from https://timesofindia.indiatimes.com/city/chandigarh/punjab-launches-adventure-water-tourism-policies/articleshow/89764494.cms
29. Vengesayi, S.: A Conceptual Model of Tourism Destination Competitiveness and Attractiveness. ANZM 2003 Conferences Proceedings Adelaide 1-3 December 2003, (2003).
30. Voon, B.H, and Lee, M.: Identifying dimensions of tourist satisfaction for a cultural destination: The case of long house on Sarawak (Borneo). International journal of business and and society, vol. 10, no. 1, pp. 65-82, (2009).

 https://doi.org/10.52202/074664-0016

Performance and Information Technology: An Empirical Analysis Affecting the Hotel Sector in Mures County

Koontru Rajeev[1], Allamraju S Narayana Teja[2],Dr Praveen Kumar[3][0000-0003-0294-3830]

[1,2]Research Scholar, Vivekananda Global University, Rajasthan,
[3]Dr Praveen Kumar, Assistant Professor, UITHM, Chandigarh University,

Abstract.

The hotel sector needs to create, deploy, and improve information technology due to the constantly changing competitive environment. Since client contentment is crucial, hotels must adopt new technologies that help them comprehend what their patrons need to consume and that help them meet the demands of the present service delivery framework. In light of the situation, we believe that information technology—a claim backed by worldwide study in the field—should be a vital tool used by hotels in their efforts to improve performance. With the use of parametric and non-parametric statistical techniques, the current study thus demonstrates the significance of information technology implementation for managers as well as the relationship between the elements that characterize information technology in hotels in MureʀŸ county and economic performance.
The findings partially validate the research hypotheses, thus the hotel sector in Mureʀث county ought to think about advancing information technology to support the sector. The study's conclusions highlight how crucial it is for hotel management to satisfy guests and produce results that are implicitly beneficial to the business.

Keywords: hotel industry, information technology, performance

1. Introduction

For hotels to perform better financially, information technology ought to be a core resource. Research conducted in the field has supported this claim by emphasizing the positive relationship between information technology implementation and performance (Rai et al., 1997; Byrd and Turner, 2001; Kim et al., 2008; Abu Kasim and Badriyah Minai, 2009; Wu and Lu, 2012). The competitive climate that exists at the national level for travel service providers, including hotels, is strongly correlated with policy and strategy that are established through the service delivery process, which necessitates the adoption of new information technologies. Because the county of MureʀŸ offers a diverse variety of travel reasons (medical, dental, spa, business, conference, entertainment, etc.), it contributes significantly to the nation's tourism supply. This has increased competition in the hotel market. Despite the fact that economic factors affected every aspect of the economy, including the travel and tourism sector, the National Institute of Statistics reports that between 2009 and 2011, the number of visitors staying in hotels in Mure퀄 county increased at an average annual rate of 13.71%. Due to the advantages of information technology—gaining devoted clients through the use of an application called Customer Relationship Management, increasing profitability, and streamlining procedures related to service delivery—this scenario led to a rise in competition and the requirement to employ it.
As a result, the focus that academics are currently giving to the hotel business on a national and worldwide scale, along with the statistical information that foreign organizations working in the field supply, highlight the significance of this research, which is directed in two directions: Performance in the hotel sector and information technology

In light of the situation, we would like to note that this study is a component of a larger quantitative research project that aims to enhance management of hotel services in MureŀŸ county. The research includes the following elements: hotel staff services, food services, entertainment services, extra services, and customized services, as well as managerial techniques through service quality, human resources, and information technology

1.1. Information technology in the hotel industry

In order to boost production, maintain management decisions, and promote job efficiency, service providers have made significant investments in the development of information technology in recent years (Kim et al., 2008, p. 500). In order to increase staff productivity and customer happiness, hotels have implemented information technology (Ham et al., 2005; Lam et al., 2007).

Since technology has advanced steadily over the past few years, customer focus is becoming more and more important in Internet, software, and mobile technologies (Aldebert et al., 2011, p. 9).

At present, information technologies in Romania that serve hotel services are the following:

1. Front-office information systems type Fidelio. The system facilitates registering of tourists, inventory and management of rooms, marketing of tourist products or record of income (http://www.gtd20.ro/2010/01/sisteme-informatice-ce-deservesc-serviciile-turistice/);
2. 2. Systems that combine sales and booking services with information authorization, allowing data to be transmitted and received from the global distribution systems (GDS) World span and Amadeus (http://www.gtd20.ro/2010/01/sisteme-informatice-ce-deservescserviciile-turistice/). These two systems are connected to the Internet, which is utilized by both hotels and customers. Additionally, the rapid growth of the Internet has had a significant impact on the hotel industry (Abrate et al., 2012, p. 160), leading to a rise in competition;
3. Hotel management information systems:
   a. The Windows version of the Medallion Property Management System (Medalion PMS) includes all the modules needed to manage a hotel's operations, regardless of the number of rooms or the structure of the establishment (http://www.tcnet.ro/solutii-hoteliere/solutii-pentruhoteluri/expressoft-interface-manager.html). The system increases revenue, boosts profitability, and enhances the reputation of the hotel;
   b. Another system that hotels utilize is Expressoft Interface Manager, which comes in the following versions: eXpresSoft Wireless Check (http://www.tcnet.ro/solutii-restaurant/solutii-pentru-restaurante/expressoft-event.html), which enables waiters to use wirelessly connected PC pocket devices in place of notebook computers to accept orders directly from the customer table. eXpresSoft Event, a system that executes menus for parties, festal meals, catering, etc., and eXpresSoft Master, a system designed to provide precise inventory control and improve planning, management, and administration of resources (http://www.tcnet.ro/solutiihoteliere/solutii-pentru-hoteluri/medallion-pms.html);
4. Applications for customer relationship management (CRM), which stem from client orientation, are utilized by travel service providers, particularly in hotels, to strengthen the bond between the business and its patrons (Wu and Lu, 2012, p. 276). Some authors (Buttle, 2004; Abu Kasim and Badriyah Minai, 2009) contend that CRM is a strategy and is necessary to draw in and retain clients. CRM, according to Buttle (2004), is a fundamental business strategy that combines internal and external networks with procedures to generate and deliver value to customers and boost profits. towards the conclusion (Wu and Lu, 2012, p. 277). According to Minghetti (2003), the creation of a CRM application necessitates a thorough understanding of consumer behavior,

https://doi.org/10.52202/074664-0017

wants, and preferences. Additionally, the development of new technologies is the primary driver of change in the hotel sector (Minghetti, 2003, p. 141).

In light of the situation, we believe that new information technology—such as the Internet, Fidelio, CRM, and Global Distribution System—has revolutionized the hotel sector and produced both favorable customer relations and financial outcomes.

1.2. Performance in the hotel industry

According to Laitinen (2002), the term "performance" literally refers to the assessment of metrics like profit, costs, and market share. Three dimensions are used by Philips (1996) in an effort to adopt a multidimensional approach in the evaluation of performances in hotels (Reichel and Haber, 2005, p. 683):

- Effectiveness: occupancy, average daily rate per room, etc.;
- Efficiency: profit; profitability of investments; and
- Adaptability: income from the successful introduction of new services and products; respectively.

Effectiveness: Sink and Tuttle (1989) point out that non-financial measures (such as employee and customer happiness) should be taken into account when evaluating performance, as opposed to solely using financial indicators. (Page 148, Avci et al., 2011). According to Reichel and Haber (2005), hotel managers have specific goals and proprietary techniques for measuring performance, which they use to assess the effectiveness of their work. Performance can be measured both objectively and subjectively (Reichel and Haber, 2005, p. 683).

Given the situation, we believe it is crucial to note that hotel performance is still a relatively new idea in specialty studies conducted around the country, and managers create their own systems for assessing employee performance. Therefore, we employed within the research the general indicators used in evaluating performance, indicators that focus on the efficiency of work, respectively, such as: turnover for each category of activity, market share, occupancy of accommodations, value of income and value of spending.

2. Objectives, assumptions and methodology of research

The study objectives serve as the foundation for defining the research assumptions, which are then tested using statistical parametric and non-parametric techniques (table 1). The following software has been used to process the data: Microsoft Office Excel and SPSS 17.0 (Statistical Packages for the Social Sciences).

Table 1. Definition of objectives and research assumptions:

Objectives	Objective related assumptions	Research method used
O1: An examination of the significance of information technology implementation in hotels	H1: Information technology adoption has been deemed important or very important by at least 50% of the hotels included in the survey.	Test of binomials
O2: Examining relationships between performance and the characteristics that characterize information technology in hotels	H2: Performance metrics and information technology implementation in hotels are strongly connected	parametric correlations of Pearson

The following statistical techniques are applied when analyzing and interpreting data:

- To determine the strength of a linear relationship between two quantitative variables, use the parametric statistics tool Pearson correlation (Novak, 2004, p. 127); and

- Using study data, the binomial test (a non-parametric technique for evaluating statistical hypotheses) verifies research hypotheses related to the binomial distribution, p and q, for a collectivity (Gravetter and Wallnau, 2009, p. 658).
-

3.Description of studied population and research variables

The 42 hotels in the county of MureRK comprise the analyzed population. The small size of the studied population (42 hotels) necessitated extensive research for this study, which used data from the Ministry of Regional Development and Tourism in Romania's "List of tourist reception structures with classified accommodation functions" report. Data were gathered from June to September of 2012.
A noteworthy phase of the study involved pre-testing questionnaires for each classification category of hotels in Târgu-MurelŸ, which was conducted in May 2012 with participation from a management representative. This led to the modification of certain variables, and this time, the time allotted for filling out the questionnaires was also tested.

The elements studied within the research are the following

- Aspect of information technology implementation in hotels using the subsequent factors: CRM.
- software programs, common software (Word, Excel, databases), Fidelio type front-office information systems
- systems of record for reservations (Worldspan, Amadeus, Galileo, etc.), Medallion PMS hotel management systems, platforms such as eXpresSoft, Utilizing some specialist websites that offer us online promotion and consultation;
- The study's hotels' performance as measured by the following indicators: Using the space in lodgings
- Market share, Turnover for food services, Turnover for lodging services, and Turnover for other services

Volume of income and Volume of expenditure.

Respondents (hotel managers) have been asked to describe the importance rate of factors on a Likert scale ranging from 1 (not important) to 5 (extremely important) in order to better understand the dimension of information technology.

Respondents were asked to list the efficiency rate of indicators using a range scale (from very small – below 20%, to very big – over 80%) and to indicate the value of income and expenditure for 2011 using the same range scale (from below 0.5 million, to over Lei 15 million) in order to assess the performance of the hotels included in the study

4.Results of research

We believe that it is critical to present the findings of the descriptive analysis for the variables that make up the information technology dimension of the research in order to test the research hypotheses.

As a result, the following characteristics of the analysis done over factors characterizing information technology inside the hotels included in the study are present (figure 2):

- Variables have produced the highest average results in the study. Hotels believe these variables to be very significant for their work; we use ordinary software (Word, Excel, databases), and we use services of some specialized sites that provide us online consultancy and advertising, 4.64 and 4.55, respectively.

https://doi.org/10.52202/074664-0017

- • Changeables We employ the Medallion PMS hotel management system and the eXpresSoft system, which both had the lowest average survey scores—1.5 and 1.6, respectively—indicating that very few hotels use these kinds of IT for hotel administration.
- • Changeables We have a built application called Customer Relationship Management, and we use front-office information systems like Fidelio and booking-specific information systems like Worldspan, Amadeus, and Galileo. We also received poor average survey scores, between two and three.

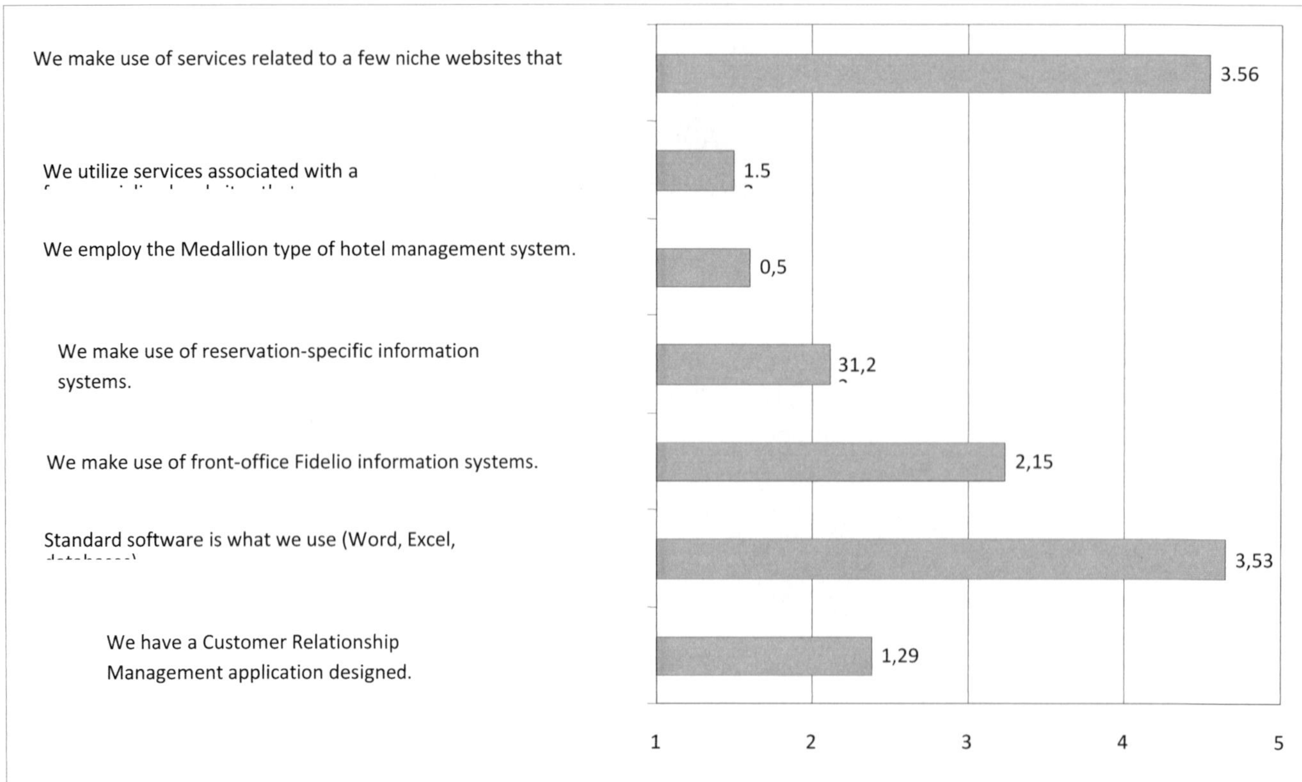

Fig.1. Graphical representation of average score achieved by each variable in category *information technology* within the entire study

The binomial test was used to evaluate the H1 assumption, determining if 50% of the hotels included in the study had information technology-focused applications deployed (table 2).

Table 2. Results of binomial test for variables that define information technology in surveyed hotels

Variables	Group	Criterium	Absolute frequency	Noticed weight	Tested percentage	Level of statistical significance
TesWe have a Customer Relationship Management application designed.	1 2	<= 5 > 4	6 36	.86 .14	.50	.00

https://doi.org/10.52202/074664-0017

	Total		42	1.00		
Word, Excel, and databases are the common applications that we utilize.	1	<= 4	27	.64	.50	.045
	2	> 4	15	.36		
	Total		42	1.00		
We employ Fidelio front-office information systems.	1	<= 4	14	.33	.50	.045
	2	> 4	28	.67		
	Total		42	1.00		
We make use of booking-specific information systems, such as Galileo, Amadeus, and Worldspan.	1	<= 4	3	.07	.50	.000
	2	> 4	39	.93		
	Total		42	1.00		
We employ the Medallion PMS kind of hotel management system.	1	<= 4	4	.10	.50	.000
	2	> 4	38	.90		
	Total		42	1.00		

https://doi.org/10.52202/074664-0017

We employ eXpresSoft-type systems.	1	<= 4	42	1.00	.50	.000
	Total		42	1.00		
We employ the services of a few niche websites that offer us online advertising and consulting.	1	<= 4	26	.62	.50	.164
	2	> 4	16	.38		
	Total		42	1.00		

Based solely on variable findings from the binomial test, the test-related assumption has been validated. We use databases, Word, Excel, and other common software. We also use services from a few specialist websites that offer us online advertising and consulting, the latter of which has a statistical significance of a=0.16. As a result, the assumption is partially verified.

Table 3 displays the findings of our usage of the Pearson parametric correlation coefficient to assess the degree and direction of the relationship between the information technology category variables and the performance indicators in hotels in order to evaluate the H2 assumption.

Table 3. Results of Pearson correlation analysis between information technology category and efficiency indicators in hotels

VARIABLES			Occ of accommodation			Market	Income	Expenditure
We have a developed	Correlation		.326*	.348*	,258	.328*	.364*	.448**
Customer Relationship Management	Level of significance	,035	,024	,099	,034	,018	,003	,001
	N	42	42	42	42	42	42	42
	Correlation	-,136	-,136	**-.333***	-,108	-,154	-,237	-,222
software (Word, Excel, data bases)	Level of significance	,391	,391	,031	,495	,330	,131	,159
	N	42	42	42	42	42	42	42
front-office Software Fidelio	Correlation	,079	,138	-,019	,043	-,020	,200	,197
	Level of significance	,617	,382	,906	,786	,898	,205	,212
	N	42	42	42	42	42	42	42
information Systems used for bookings (Worldspan, Amadeus, Galileo etc.)	Correlation	-,091	-,068	-,206	-,184	-,192	,159	,212
	Level of significance	,565	,667	,190	,244	,223	,313	,177
	N	42	42	42	42	42	42	42
hotel management system PMS	Correlation	,156	,257	,206	,224	,252	**.390***	**.452****
	Level of significance	,325	,100	,190	,154	,107	,011	,003
	N	42	42	42	42	42	42	42
We use systems type eXpresSoft	Correlation	-,003	,058	,119	,143	,088	,146	,223
	Level of significance	,985	,714	,452	,366	,581	,355	,155
	N	42	42	42	42	42	42	42
We use services of some specialized sites that provide us online	Correlation	,166	,016	-,085	,138	,052	,027	-,011
	Level of significance	,292	,922	,593	,382	,741	,866	,943
advertising	N	42	42	42	42	42	42	42

.484*

https://doi.org/10.52202/074664-0017

consultancy and

*** Significant correlation at level 0.01*

**Significant correlation at level 0.05 N = number of employees*

The findings highlight the fact that values that are statistically significant for the study have only been attained for the following variables:

• Positive correlations of medium strength between the variables: we use the hotel management system Medallion PMS and the expenditure indicator (0.452); we have a developed application for customer relationship management and indicators for income (0.448) and expenditure (0.484);

• We use the hotel management system Medallion PMS and Income (0.390) indicator; we have a developed application Customer Relationship Management and Occupancy (0,326), Turnover for accommodation services (0.348), Turnover for other services (0.328), and Market share (0.367) indicators.

• Low intensity negative correlation between the variables: Turnover for food services (-0.333) and Word, Excel, and databases are used. We note that this assumption is somewhat supported by the results.

1. Conclusions

Both the national decision-making authorities and the hotel managers in Mureş County value individual empirical contributions.

The analysis's findings highlighted the fact that just two factors—using specialized websites for online advertising and consulting and standard software like Word, Excel, and databases—that characterize information technology in the Mureş County hotel industry have been rated as important or extremely important by more than 50% of managers. Furthermore, the study has highlighted low intensity positive correlations for the following variables: we use the hotel management system Medallion PMS, we use standard software (Word, Excel, databases), and we have a developed application called customer relationship management. For all other variables, the results have

Individual empirical contributions are valued by both the Mureş County hotel managers and the national decision-making authorities.

The analysis's conclusions brought to light the fact that just two characteristics of information technology in the Mureş County hotel industry—using specialized websites for online advertising and consulting and standard software like Word, Excel, and databases—have been rated as important or extremely important by more than 50% of managers. Low intensity positive correlations have also been found in the study for the following variables: we use databases, Word, Excel, and the hotel management system Medallion PMS; we also have a developed application called customer relationship management. Regarding every other variable, the outcomes have

We believe that the research boundaries are where future research opportunities originate. Due to the limited size of the surveyed population, we will be able to include a greater number of hotels from a wider range of geographic locations in our future research. Furthermore, we believe that future research on hotel ownership is necessary to determine whether information technology deployment differs amongst hotels that belong to an international hotel group.

References:

1. Choi, D.-H.: Impact of organizational culture on the organizational commitment in hotel employees. Indian Journal of Public Health Research and Development. 9, 841 (2018). https://doi.org/10.5958/0976-5506.2018.00838.0.
2. Barreto, M., Ryan, M.K., Schmitt, M.: The glass ceiling in the 21st century: Understanding barriers to gender equality. In: American Psychological Association eBooks (2009). https://doi.org/10.1037/11863-000.
3. Bertrand, M., Goldin, C., Katz, L.F.: Dynamics of the gender gap for young professionals in the financial and corporate sectors. American Economic Journal. Applied Economics. 2, 228–255 (2010). https://doi.org/10.1257/app.2.3.228.
4. Calinaud, V., Kokkranikal, J., Gebbels, M.: Career advancement for women in the British hospitality industry: the enabling factors. Work, Employment and Society. 35, 677–695 (2020). https://doi.org/10.1177/0950017020967208.
5. Crites, S.N., Dickson, K.E., Lorenz, A.: Nurturing Gender Stereotypes in the face of Experience: A study of leader gender, leadership style, and satisfaction. Nurturing Gender Stereotypes in the Face of Experience: 19, 1 (2015).
6. Crompton, R., Lyonette, C.: Women's career success and work-life adaptations in the accountancy and medical professions in Britain. Gender, Work and Organization. 18, 231–254 (2011). https://doi.org/10.1111/j.1468-0432.2009.00511.x.
7. Da Silva, H.B., Yukongdi, V., Kusakabe, K., Zimmermann, W.: Career advancement in the hotel industry in Timor-Leste: Study of individual, organizational, and gender differences. SAGE Open. 11, 215824402110326 (2021). https://doi.org/10.1177/21582440211032683.
8. Dashper, K.: Mentoring for gender equality: Supporting female leaders in the hospitality industry. International Journal of Hospitality Management. 88, 102397 (2020). https://doi.org/10.1016/j.ijhm.2019.102397.
9. Eagly, A.H., Karau, S.J.: Role congruity theory of prejudice toward female leaders. Psychological Review. 109, 573–598 (2002). https://doi.org/10.1037/0033-295x.109.3.573.
10. Elhoushy, S., El-Said, O.A., Smith, M.F., Dar, H.: Gender equality: caught between policy reforms and manager beliefs. Journal of Sustainable Tourism. 1–24 (2023). https://doi.org/10.1080/09669582.2023.2218067.

 https://doi.org/10.52202/074664-0017

11. Elomami, A.: Job satisfaction and employees' turnover in Libyan oil companies: the application of the job characteristics model in a different cultural setting, https://ethos.bl.uk/OrderDetails.do?uin=uk.bl.ethos.678804, (2015).
12. Gill, A., Fitzgerald, S.P., Bhutani, S., Mand, H.S., Sharma, S.: The relationship between transformational leadership and employee desire for empowerment. International Journal of Contemporary Hospitality Management. 22, 263–273 (2010). https://doi.org/10.1108/09596111011018223.
13. Gupta, V.: Talent management dimensions and their relationship with retention of Generation-Y employees in the hospitality industry. International Journal of Contemporary Hospitality Management. 31, 4150–4169 (2019). https://doi.org/10.1108/ijchm-10-2018-0859.
14. Li, L., Leung, R.W.: Female managers in Asian hotels: profile and career challenges. International Journal of Contemporary Hospitality Management. 13, 189–196 (2001). https://doi.org/10.1108/09596110110389511.
15. Sigler, T.H., Pearson, C.M.: Creating an empowering culture: examining the relationship between organizational culture and perceptions of empowerment. Journal of Quality Management. 5, 27–52 (2000). https://doi.org/10.1016/s1084-8568(00)00011-0.
16. Yan, Z., Couch, S., Blum, S.C.: Factors affecting women's career advancement in the hospitality industry: perceptions of students, educators and industry recruiters. Journal of Hospitality & Tourism Education. 23, 5–13 (2011). https://doi.org/10.1080/10963758.2011.10697015.

Pondicherry Cuisine as a Culinary Identity: An Exploration of Community Endorsement and its Alignment with Sustainable Development Goals in Puducherry

Velvizhi. R[1] and Dr. Gujjeti Thirupathi[2]
[1]Ph.D. Research Scholar, Department of Tourism Studies, School of Business Studies, Central University of Kerla.Trjaswini Hills, Periye, Kasaragod District, Kerla
[2]Assistant Professor, Department of Tourism Studies, Department of Tourism Studies, School of Business Studies, Central University of Kerla.Trjaswini Hills, Periye, Kasaragod District, Kerla.

Abstract.
Culinary tourism is the only form of tourism that has a unique feature of providing direct access to a culture and delivers its core experience in no time. While this is the fact, the regional food of a particular destination is considered to best reflect the culture of the destination.

Pondicherry which is referred as "the window of French culture" by former Prime Minister, Pandit Jawaharlal Nehru, is a well- known heritage and cultural destination. The city has been reaping its benefits in terms of tourism by showcasing Indo- French architecture in the form of buildings and urban planning. Further the destination has been diversifying its attraction towards, beaches, water sports and so on, since last 10 years. Nevertheless, the destination is yet to tap its potential through culinary tourism.

This study aims explore the stance of host community of the destination in making the "Pondicherry cuisine" that carries a unique feature, as the culinary identity for the destination. An exploratory study is conducted among the practicing community and their level of acceptance and the benefits they value in making Pondicherry Cuisine is studied and evaluated. The outcome of the study endorses the social innovation that aims to utilize untapped potential of "Pondicherry Cuisine" practiced by the disintegrating Franco-Tamil community, which will in turn lead to regeneration and revival of Franco-Tamil culture back into the society thereby supporting the city to achieve two SDG goals viz. 8 and 11.

Keywords – Culinary Tourism, Host community, Sustainable Development.

1. Introduction

Tourism today is a multi-dimensional, multi-faceted activity which touches many lives and various economic activities. It has developed into a mass activity turning it into an immense enterprise. It is also being recognized as a source of employment of opportunities. The by- products of tourism were the creation of jobs in the form of guides, escorts, travel agents, tour operators, transport, hotels, motels, resorts etc. In fact the aim of tourism was to promote physical invigoration, mental rejuvenation, cultural enrichment and spiritual elevation. In addition in helping in the regional economic development, tourism also makes a tremendous contribution to the improvement of socio-political understanding and cultural exchanges.

According to WTTC's annual research, the percentage share of Travel and Tourism sector on global GDP is 7.6% and accounts for 1 in 11 of global employment during 2022. While on the national scenario, Invest India has released a report stating Travel and Tourism sector as the 6th major contributor to the country's GDP for the year 2021 and accounts for about 15.34% of total employment share. Compared to other industries, tourism industry has a peculiar economic feature which is the income earned at the place of residence of a person is spent in a totally different place where the person is visiting, which is away from his residence (Holloway & Robinson, 1995). It means in order to make the maximum utilization of the tourism economic benefits; the host community needs to be agile and proactive to create attractive, authentic and unique products and services that ensure quality experience to the tourists in addition to regular the services.

Engaging the host community directly or indirectly in the production of various niche products and services for the tourists like producing handicrafts, acting as a guide, providing culinary experience etc. are few of the ways to optimize the economic benefits. Of these, culinary tourism is the most effective and fastest way as it does not require any extra training or new skill to produce it, and it involves sharing and showcasing their regular food and way of living to the tourists. While this is the basic advantage of culinary tourism, its momentum and fame increases manifold when it is

https://doi.org/10.52202/074664-0018

attached with culinary identity, which refers to the rare, unique, non-imitable and non-substitutable culinary practices of the destination.

2. Background of the study

Pondicherry also known as Puducherry is a destination at the East Coast of India, an enclave within Tamil Nadu, famous for its Indo-French Architecture, lined with different beaches along with its international recognition for spiritual retreat. While the Indo-French architecture as unique identity of the city was recognized and restored by the tireless effort of INTACH (Indian National Trust for Art and Cultural Heritage) in support with organizations like PPA (Puducherry Planning Authority) and VMF (Vieilles Maisons Françaises), the culinary identity of the city is yet to be recognized. Though the residents of the Union Territory follow different culinary practices and offers different cuisines to the tourists, till date there is no one cuisine identified as culinary identity of the destination also no such scholarly studies have been conducted in this ground. Also, when it comes to the uniqueness and non-imitable features, there exists a culinary practice called "Pondicherry cuisine" among the Franco-Tamil community of the city.

Hence the aim of this study is to analyze the practicing community perspectives on making this cuisine as culinary identity of the city and the benefits they value more. The results of the study will further aim to form a base for achieving the SDG goals 8 & 11 which are sustainable economic growth and sustainable cities and communities respectively, which benefits the host community practicing the cuisine and in-turn the Union Territory of Puducherry.

2.1 Literature Review:

Folklorist Lucy M Long [2013] 1 was credited for coinage of the term 'culinary tourism' during 1996 and defined it in humanities perspective which meant eating out of curiosity and later refined the definition in both humanities and tourism perspectives which meant culinary tourism is a negotiation of exotic and familiar with otherness. She also stated that exploring the role of culinary tourism in disciplines like identity construction can be utilized for sustainable tourism growth. Further Hairi Jalis [2014] 2 opine that, a destination must identify and build up a clear and distinct cuisine image to differentiate themselves from the rest of the neighboring destinations especially from those sharing similar culture and cuisine.

To make the culinary identity more meaningful, Linman Li [2017] 3 proposes, planning and development of tourism must be based on aesthetic orientation of the destination by proper utilization of resources in order to achieve higher satisfaction of the tourists. While this is the scenario from tourist side, Ritchie [1988] 4 opines that, a significant degree of community participation is indispensable for a long-term success of tourist destination while still deeper analysis of Garrod [2003] 5 implies that achieving consensus of community by involving them during tourism planning would support experts community which done otherwise might cause resistance and opposition. Finally, to achieve the sustainable goals proposed by UN by 2030, the national strategy for sustainable tourism formulated by Niche tourism Division of Ministry of Tourism, of Government of India [2022] 6, has proposed three basic guiding principles that are to be focused while practicing sustainable tourism which are environmental, socio-cultural and economic sustainability.

3. Research Gap

Based on the above literature, the growing importance of culinary tourism owing its economic benefits along with its ability to support sustainability of a destination can be understood. Also the importance of local community participation from the development stage of a tourism product is also emphasized. But there is no study that has focused on importance and methods of achieving consensus of practicing community of a particular cuisine in order to promote their cuisine as a culinary product of a particular destination.

4. Statement of Problem:

Puducherry is a thriving tourism destination with Indo-French culture as their unique attraction compared to the nearby tourism destinations. Due to its growing importance and increased tourist arrivals manifolds in the recent past, many local, domestic and international food service providers have started mushrooming in the city. Therefore the need for identifying and promoting a culinary identity matching with the aesthetic beauty of the destination has become indispensable. In connection with it, achieving consensus of the practicing community to share their culture and cuisine with the tourist becomes the first step in the development of the product.

a. Aim of the study:

Referring to the above stated problem, this study aims at evaluating the opinion and the degree of acceptance of the Franco- Tamil community of Puducherry to share their cuisine culture with the tourist, as their cuisine culture matches with the major aesthetic beauty of the destination which is Indo-French culture.

b. Objectives:

i. To evaluate the practicing community's acceptance level to make Pondicherry Cuisine as the culinary identity of Puducherry

ii. To study the opinion on tentative benefits and their priority by the practicing community from making Pondicherry Cuisine as culinary identity.

5. Methodology

5.1 Research Approach:

This study adopts exploratory research for the following reasons

- no previous studies are orienting to the research problem which is identification and recognition of culinary identity of the destination.
- To understand the perspective of the practicing community towards providing such identification for "Pondicherry Cuisine' for the reason them being one of the major stakeholder and the future supplier.

5.2. Research Design:

The study uses qualitative research design in the form of experience survey, with the exponents and members of the practicing community.

Population: About 250 households approximately comprising of 1200 individuals practice this cuisine as their way of life. Among them and average of 500 individuals only reside in the town during normal days.

Sample: Totally 37 respondents were interviewed. Purposive sampling technique was adopted by dividing them into clusters based on geography and from the clusters respondents were randomly selected.

https://doi.org/10.52202/074664-0018

5.3 Research Methods:

Data collection:

Among them 3 of them were interviewed extensively with structured questionnaire with both close ended and open ended questions, as they are core practitioners and influencers, while the other 34 respondents were interviewed with structured close ended questionnaires. The close- ended structured questionnaires were in the form of Google forms.

Data were collected my meeting the respondent's personally either at their residence or at their workplace, especially the exponents. The time period was spread over a period of two weeks. The concept of the research was explained first and then the data were collected once they understood the aim of the study.

Data analysis: The questionnaire consisted of 3 sections, the first section for demographic data, second section on the awareness of destination and the concept of culinary identity and the third section was on their opinion of Pondicherry Cuisine. A self-introduction and the purpose of the study was provided along with informed consent at the 1st section, short description of the concept of Culinary Identity was given at the 2nd section, while a description of Pondicherry Cuisine was given at the 3rd section. The question types include 1 open-ended, 1 dichotomy, 7 single select multiple- choice and 8 scaling. Therefore total of 17 questions were posed to the respondents.

The data collected were of categorical and was analyzed using Microsoft-Excel to calculate Frequency distribution, contingency tables and their respective bar charts.

The population of this study is confined to a particular geographical location called the "Tamil

6. Results

Precinct" as categorized by INTACH, concentrated at the center of the city of Pondicherry, which is the residing area of almost 95% the practicing community. First the exponents who have in- depth knowledge on the Franco-Tamil culture and the ones who practice the "Pondicherry Cuisine" dedicatedly were contacted and discussed about the potential of the study. The structured questionnaire was simplified by the suggestion of the exponents to make the respondents feel at ease.

The data collection commenced by contacting the members of practicing community at different places. Most of the houses remained closed, few of them were initially skeptical about such a survey; few were reluctant to spare time. Majority of respondents at the age group 19-35 were interviewed either directly or over zoom meeting at one place on two Saturdays at the AIFPR (Association of Indian French Professionals and Researchers) office and hence provided with the limited time period total of 37 responses were secured. The respondents were provided with sufficient background information and questions were carefully placed to bring out their true opinion. The categorical data collected were both nominal and ordinal from dichotomy, multiple choice and scaling responses. The results of the analyzed data are furnished below.

https://doi.org/10.52202/074664-0018

	Community Opinion on culinary identity			
View Point		Need for culinary ID	Matching cuisine	Pondicherry Cuisine as culinary ID
	Strongly agree	17	13	22
	Agree	13	22	15
	Neutral	6	2	0
	Disagree	1	0	0
	Strongly Disagree	0	0	0
	Total	**37**	**37**	**37**

Table: 1. Community Opinion on culinary identity

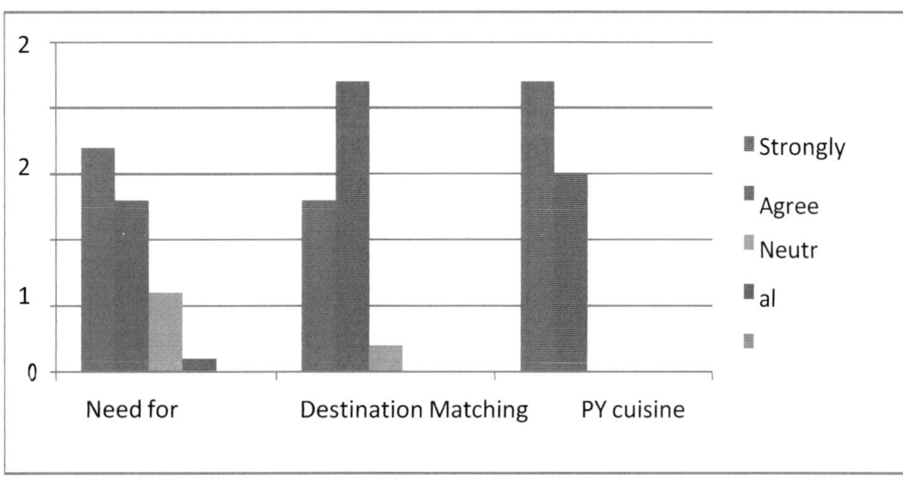

Column Chart: 1. Community Opinion on culinary identity

A contingency table, which summarizes the relationship between several categorical variables, is used to analyze the data related to level of acceptance of practicing community to make Pondicherry Cuisine as culinary identity of Puducherry. The above contingency table is derived on the basis of 3 different questions asked at different sections of the questionnaire. The 2nd section which aims to assess the general awareness and community's perception on the concept of culinary identity 1 warm-up question was posed whileother 2 questions were posed as transition and core questions at the 3rd section. The respondents

 https://doi.org/10.52202/074664-0018

were provided with sufficient background information and questions were carefully placed to bring out their true opinion.

7 Output 1:

The table displays that among 37 respondents

i. 30 respondents are agreeing that Pondicherry must have a its own culinary identity, while 17 strongly agree and 13 agree.

ii. All the 37 respondents agree that the culinary identity must match the attraction of the destination, while 13 strongly agree and 22 agree.

iii. Also all the 37 respondents agree to make Pondicherry Cuisine as culinary identity of Pondicherry, while 22 strongly agree and 15 agree.

https://doi.org/10.52202/074664-0018

View Point	Opinion on Community advantage					
		Business opportunities	**Emotional attachment**	**Community pride**	**Preservation of cuisine and culture**	**Curiosity for Franco-Tamil younger generation**
	Strongly agree	19	13	18	20	18
	Agree	18	18	15	15	16
	Neutral	0	6	4	1	2
	Disagree	0	0	0	1	1
	Strongly Disagree	0	0	0	0	0
	Total	**37**	**37**	**37**	**37**	**37**

Table: 2. Practicing community's opinion on the perceived advantages of developing Pondicherry Cuisine asculinary identity of Puducherry

https://doi.org/10.52202/074664-0018

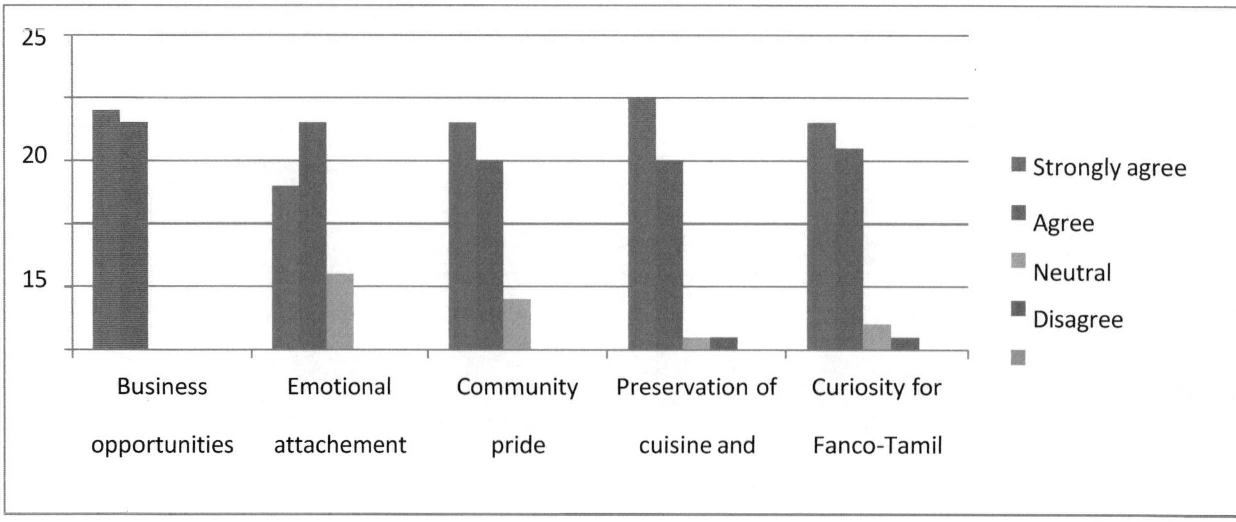

Column Chart 2. Practicing community's opinion on the perceived advantages of developing PondicherryCuisine as culinary identity of Puducherry

The above contingency table is derived on a scaling question with 5 dimensions that is asked at 3rd section of the questionnaire in order to assess the true sentiment of practicing community towards the tentative benefits they would obtain if their cuisine is being identified as culinary identity of Puducherry.

a. Output 2

The following are the opinion on the tentative benefits of making Pondicherry Cuisine asculinary identity. Among the 37 respondents,

i. All the 37 respondents expect creation of business opportunities.

ii. 31 respondents opine that it will encourage emotional attachment.

iii. 33 of them opine that it will boost the pride to the practicing community.

iv. 35 of them opine that it will help in preserving the cuisine culture.

v. 34 of them opine that it will indirectly motivate the young Franco-Tamil generation topay regular visit to their native community.

	Age to PY cuisine as culinary identity					
Age category		19-30	31-45	46-60	61-75	75 and above
	Strongly Agree	7	9	5	1	0
	Agree	4	10	1	0	0
	Neutral	0	0	0	0	0
	Disagree	0	0	0	0	0
	Strongly Disagree	0	0	0	0	0
	Total	11	19	6	1	0

Table: 3 Opinion on Pondicherry Cuisine as culinary identity based on age-category

https://doi.org/10.52202/074664-0018

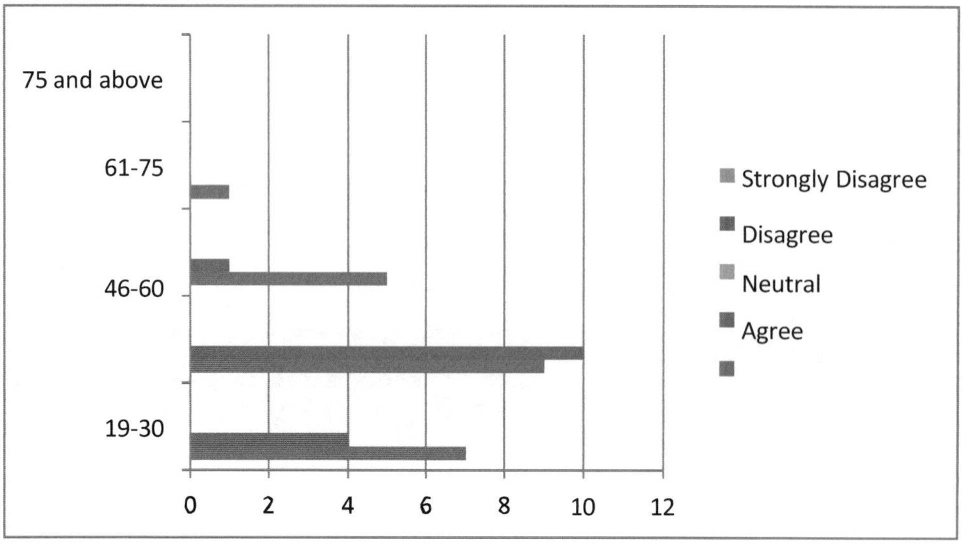

Bar Chart: 1 Opinion on Pondicherry Cuisine as culinary identity based on age-category

The above table is derived to figure out the opinion of f making Pondicherry Cuisine as culinaryidentity based on different age categories.

Output 3

Among the 37 respondents

- All the 11 respondents who belong to the age group 19-30 accept the concept while 7 ofthem strongly agree and 4 of them agree.
- All the 19 respondents who belong to the age group 31-45 accept the concept while 9 ofthem strongly agree and 10 of them agree.
- All the 06 respondents who belong to the age group 31-45 accept the concept while 05 ofthem strongly agree and 01 of them agree.
- 01 respondent who belong to the age group 46-60 strongly agrees the concept.

Education Category	Education to PY cuisine as culinary identity						
		No formal Education	SSLC/ Intermediate	HSC	Graduate	Post Graduate	Doctorate
	Strongly Agree	0	0	1	7	10	4
	Agree	0	0	0	4	10	1
	Neutral	0	0	0	0	0	0
	Disagree	0	0	0	0	0	0
	Strongly Disagree	0	0	0	0	0	0
	Total	0	0	1	11	20	5

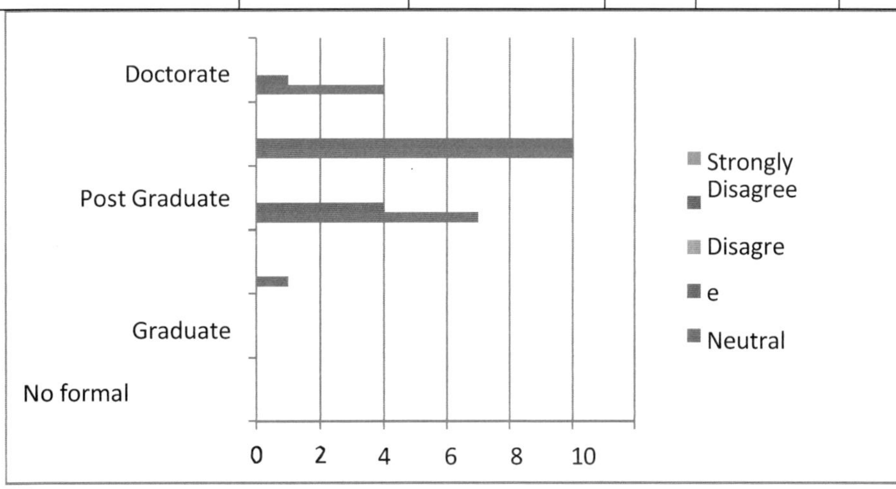

Table: 4 Opinion on Pondicherry Cuisine as culinary identity based on Education-category

Bar Chart 2. Opinion on Pondicherry Cuisine as culinary identity based on Education-category

The above table is derived to figure out the opinion of making Pondicherry Cuisine as identity based on different education categories.

Output 4:

Among the 37 respondents

 https://doi.org/10.52202/074664-0018

- All the 20 respondents who belong to the Post Graduate category accept the conceptwhile 10 of them strongly agree and 10 of them agree.

- All the 11 respondents who belong to the Graduate category accept the concept while 07of them strongly agree and 04 of them agree.

- All the 05 respondents who belong to the Doctorate category accept the concept while 04of them strongly agree and 01 of them agree.

- 01 respondent who belong to the HSC category strongly agrees the concept.

8. Discussions

- The study aims at analyzing the practicing community's perspectives on making Pondicherry Cuisine as culinary identity of Puducherry. Two dimensions of their perspectives are studied.

- First, assessing their view and acceptance level to make the cuisine as the culinary identity which indirectly tries to explain to some extent their readiness and willingness to share their culinary practices and contribute in different ways to promotion of culinary tourism as a whole.
- Second, evaluating their opinion on the tentative benefits expected by the practicing community, which might form as their motivation to get involved in culinary tourism, if the cuisine is made as the culinary identity of Puducherry.

8.1 Interpretation of results:

Output 1: Though according to output 1, all the 37 respondents agree to make Ponidcherry cuisineas culinary identity of Puducherry, it has to be noted in the Table 1 that 1 respondent disagreed and 6 respondents remained neutral on accepting the concept of culinary identity. Since the concept was new to them, they were not able to accept it initially and gradually as they got the clear picture of the concept by answering the consequent questions, at the end they understood the concept clearly and gave their acceptance to make their cuisine as culinary identity of Puducherry.

Output 2: Despite all the tentative benefits are accepted by the practicing community, from the column chart 2 it is clear that creation of business opportunity, preservation of cuisine and culture are assessed as stronger benefits expected by them, which is followed by boosting of community pride and encouragement of revisit by younger generation of Franco-Tamil community from territories of France.

Output 3: This output attempts to identify the age-group that is more inclined and interested in making the Pondicherry Cuisine as culinary identity of Puducherry. Hence it is observed after ignoring category that has no or very minimum number of respondent, the age-group 46-60 (83.33%) is more inclined (strongly agree), followed by age-group 19-30 (63.6%) and age group 31-45 (47.4%).

Output 4: This output attempts to identify the Education category that is more inclined and interested in making the Pondicherry Cuisine as culinary identity of Puducherry. Hence it is observed after ignoring the category that has noor very minimum number of respondent, the Doctorate (80%) is more inclined (strongly agree), followed by Graduate category by (63.6%) and Post Graduate (50%).

8.1 Outcome:

The final outcome of this study is that practicing community is willing to accept their cuisine to be identified as culinary identity of Puducherry.

8.2 Finding:

The major finding derived through this study is that though few of the members were skeptical about the concept, after clear explanation by the researcher orally and mentioning the same in thequestionnaire in the written form, they understood it gradually in the right manner and acceptedit at the end of the interview.

8.3 Limitations of the study

The study supports to find a novel way to valorize Puducherry through a unique product that is Pondicherry Cuisine, as the city is not yet endorsed with any culinary identity till date.

In addition, through this study, a new knowledge is brought in to the academic world which is the existence and significance of Pondicherry Cuisine. The study adopted only qualitative research, provided with limited time limit. Therefore an extensive empirical research is required in order to validate the strength of the results.

Unlike a normal day, the head count of practicing community tends to increase during festivals, like Christmas, New Year and other function seasons, when the friends and relatives flock to the city to celebrate together. But it becomes hardly possible to meet them during those times. Also the population displays a conservative attitude during a direct approach and feels at ease only if any of their community members accompanies or recommends the researcher.

Based on the available categorical data itself various useful combinations can be made to get more meaningful interpretations.

9. Conclusion and Implications

The study has focused on Pondicherry Cuisine that matches with of major attraction of the city which is the Indo- French architecture. Though French cuisine is one of the other competing cuisines that match with the major attraction of the destination, technically, the cuisine is not home to the city and it is also available in other destinations. Therefore, the promotion of Pondicherry Cuisine, not only adds unique flavor to the city but also will help the Union Territory to achieve the sustainable development goal 11 which is to make cities and human settlements inclusive, safe, resilient and sustainable, particularly the Target 11.3- to enhance inclusive and sustainable urbanization and capacity for participatory, integrated and sustainable human settlement planning and management in all countries and 11.4- to strengthen efforts to protect and safeguard the world's cultural and natural heritage, as the community that practice this cuisine are slowly losing their cultural values as the members are getting mixed with other cultures and are preferring to settle in the territories of France for both cultural and economic reasons.

Developing Pondicherry Cuisine as a new tourism product can bring in two majoreconomic benefits. The first is that the revenue earned by it will be immediately realized and circulated in the host environment compared to the French cuisine and second is that of the multiplier effect, as the cuisine uses mostly the home ingredients, compared to French cuisine, that demands import of authentic French ingredients from France or other international suppliers leading to leakage of the earned revenue. In addition, promotion of this cuisine will also help the Union Territory to achieve the sustainable development goal 08- which is to promote sustained, inclusive and sustainable

https://doi.org/10.52202/074664-0018

economic growth, full and productive employment and decent work forall.

To be more precise, it helps in achieving Target 8.2- achieve higher levels of economic productivity through diversification, technological upgrading and innovation, including througha focus on high-value added and labour intensive sectors, Target 8.3- promote development- oriented policies that support productive activities, decent job creation, entrepreneurship, creativity and innovation, and encourage the formalization and growth of MSME, including access to financial services, Target 8.9-By 2030, devise and implement policies to promote sustainable tourism that creates jobs and promotes local culture and products. The qualitative study on the practicing community has revealed that majority of them are willing and accepting to endorse Pondicherry Cuisine as culinary identity of Puducherry. In addition, from their responses towards the tentative benefits, the major benefits they value more, is also revealed through this study.

Owing to its economic benefit the Government of India can develop the cuisine as culinary tourism product. Considering its uniqueness and its aesthetic similarity with the established image of the Union Territory; the Indo- French architecture, the Government can takesteps to recognize the cuisine as culinary identity of Puducherry that will attracts both domestic and inbound niche tourists. On top of all the above implications, a careful consideration on this subject will also support the city to achieve efficiently the sustainable development goals 8 & 11 that are sustainable economic growth and sustainable cities and communities respectively.

This study can serve as a foundation and supportive knowledge for future research in the culinary tourism of Puducherry and Pondicherry Cuisine. An in- depth ethnography study can also be performed to study the attitude of the practicing community, in- order to derive the definite motivation factors to make them involved in culinary tourism extensively.

Disclosure of Interests:

The authors have no competing interests to declare that are relevant to the content of this article.

References:

1. Long, L.M. Culinary Tourism. In: Thompson, P., Kaplan, D. (eds) Encyclopedia of Foodand Agricultural Ethics. Springer, Dordrecht (2013).
2. Hairi Jalis, M. Utilising local cuisine to market Malaysia as a tourist destination. In: 5thAsia Euro Conference 2014 on Procedia - Social and Behavioral Sciences, 144, 102 – 110. Elsevier, Australia (2014)
3. Linman Li. Analysis on Relationship between Tourism Aesthetics and Development and Utilization of Tourism Resources. In: Proceedings of the 2017 9th International Economics, Management and Education Technology Conference (IEMETC 2017), Atlantis Press (2017)
4. Ritchie, J. R. B. Consensus policy formulation in tourism. Tourism Management, 9(3),199–212 (1988).
5. Garrod B. Local participation in the planning and management of ecotourism a revisedmodel approach. Journal of Ecotourism 2(1): 33-53 (2003).
6. National strategy for sustainable tourism. Formulated by Niche Tourism Division ofMinistry of Tourism, Government of India (2022).
7. Anand Ballabh, Towards Sustainable Tourism, Akansha Publishing House, New Delhi,2007.
8. Aseem Bhattacharya, Build Your Career in Travel and Tourism, Abhishek Publications,Chandigarh (India), 2009.
9. Batra. G.S, Tourism in the 21st Century, Annual Publication Pvt. Ltd, New Delhi, 1996.
10. Bezbaruah, M.P, Indian Tourism beyond Millennium, Gyan Publication, New Delhi,1999.
11. Bhatia. A.K, Tourism Development and Practices, Sterling Publishers, New Delhi, 1998.
12. Chib. S.N, Perspectives on Tourism in India, Govt. of India, New Delhi, 1983.
13. Garima Chawla, Planning and Development Management, A.K. Publications, Chandigarh(India), 2008.
14. Gupta. S.K, Travel and Tourism Management, Mohit Publications, New Delhi, 20

https://doi.org/10.52202/074664-0018

The Creative Tourism - A New Approach of Sustainable Practices: A Case Study of Kinnaur, Himachal Pradesh

[1]Dr. Vikesh Kashyap, [2]Dr. Ashutosh Sharma, [3]Dr. Abhimanyu Sharma

[1] Chandigarh College of Hospitality, Chandigarh Group of Colleges, Landran

[2] Himgiri Zee University, School of Hotel & Hospitality Management, Dehradun

[3] Chandigarh College of Hospitality, Chandigarh Group of Colleges, Landran

Abstract:
With the growing popularity of tourism in Himachal Pradesh, it has become imperative to adopt sustainable practices to minimize negative impacts and ensure long-term development. One such approach gaining traction in the region is creative tourism. [1] briefs that the development of experiences that enable tourists to interact with local culture, art, and heritage in a destination utilizing encounters, conversations, and facilitated exchanges represent creative tourism. The power to foster innovation, assets, and services seems to be demonstrated by creative tourism. Creative tourism alludes to the advancement of tourism encounters that permit sightseers to effectively take an interest in the neighborhood culture, expressions, and legacy of a goal, through experiences, discussions, and co-creation encounters. Creative tourism appears to illustrate the control to cultivate the advancement of developments, resources, and services.
Integrating creative tourism into the Himachal Pradesh tourism industry can promote sustainable development, preserve cultural heritage, and empower local communities. This research paper represents potential creative activities of the region with the findings to be developed as a creative place.
The potential outcome shows the various activities that can be developed as a creative approach for sustainable development in the region.
Keywords – Creative tourism, Co-creation, Sustainable Practices, Cultural Heritage, Kinnaur

Introduction:

Creative Tourism is a catalyst that gives the opportunities to develop creative clusters and resources with creative experience in many domains. These domains manifest the characteristics of the destinations to involve tourists in activities such as cookery, agriculture, painting, architecture, photography, crafts, and arts. Creative tourism is knowledge-based tourism with a creative economy and intangible tourism resources. Creative tourism has a twofold association between tourists and the community that links economic, social, relational, and intellectual capital networks. Creative tourism, which enables visitors to immerse themselves in local traditions and culture, is a growing concept of cultural tourism. For the growth and development of communities, this is a way to use. Creative tourism, which enables visitors to immerse themselves in local traditions and culture, is a growing concept of cultural tourism. [2] says this is a sustainable way that can be used for the development of communities.

"Tourism which offers tourists the opportunity of developing their creative potential through active participation in learning experiences that are characteristic of a holiday destination where they take place." A new integration of culture and creativity, which leads to creative tourism, is transforming traditional approaches into modern ones. With the increasing interest, creative tourism is developing at a pace and emerging as a new form of cultural tourism. The UNESCO report 'Our Creative Diversity (1995) points out, that there has been a fundamental shift from exchanging ready-made products to joint exploration, experimentation, and co-production.

a. Creative mode
b. Convergence mode
c. Chains mode
d. Cluster mode
e. Concordance mode

Tourism does not follow the boundaries and traditional paths; it changes with the pace of the period. The integration of the tourism industry allows the development and acceptance the new concepts for growth and development. [3], defined the new concepts of 5C mode for the development and growth of tourism.

The development of creative tourism in past years has shown growth and acceptance. The creative mode hastaken momentum around the globe by attracting and creating a creative experience and involvement. With diversity in creativity, UNESCO has provided a unique international implementing platform to assist countries in developing capacities and policies for developing their cultural and creative sectors.

Table 1 Elaborated on the development phases of Creative TourismSources- [4]

Table 1 Elaborated on the development phases of Creative TourismSources- [4]

Stage Stages	Start Dates	Forms	Emphasis	Projects/ creative destinations
Creativ e tourism 1.0	200 0	Learning activities and workshops	A bridge between supplyand demand	EUROTEX
Creativ e tourism 2.0	200 5	Destination creative experiences	Creative Cities Network (Recognizing creativity as a significant factor in their urban development)	Creative Tourism Barcelona Creative Tourism NewZealand Creative Tourism Austria
		Community-based tourism	Developing Community thinking	Hallyu Tourism, or the'Korean Wave,' Designated Areas for Sustainable Tourism

https://doi.org/10.52202/074664-0019

				Administration (DASTA), Thailand
Creative tourism 3.0	2010	Links to the Creative Economy	Creative economy and profiling of the destinations	The Ministry of Tourism, Indonesia created a separate organization Ministry of Tourism and Creative Economy
Creative tourism 4.0	2015	Relational networks and co-creation of experiences	Shift from consumption to prosumption, servicescape to experience cape, and mass tourism to relational tourism	Thailand, Indonesia, CREATOUR in Portugal

Table 1 describes the development phases of creative tourism with the term –

Creative Tourism 1.0 featured and centered on providing an experience through participation in workshopsand learning activities. The focus was on developing the small-scale creative experience and creating entrepreneurs. These activities take place in small groups that allow visitors to explore the authenticity anduniqueness of the culture. Creative Tourism 1.0 provides the basis for a more sustainable form of tourism with production focused and developing a bridge between supply and demand.

The Creative Tourism 2.0 stage created a destination-based network and linked community-based tourism with creative tourism. Community thinking developed to involve themselves and attract the tourists to the destination with creative activities (reflects a macro consumption) and develop policies related to creative experiences. The vision of developing the economy and social development through tourism was put forward to the attention in 2004; the United Nations Educational, Scientific and Cultural Organization (UNESCO) launched the project "Creative Cities Network" (UCCN) to build the network between the creative destinations. By November 2019, 246 cities in the World were part of the UNESCO Creative Cities Network. Figure 1 Elaborated on the development phases of Creative Tourism Sources- [4]

With the sequence of growth in Creative Tourism, several creative destinations have extended the scope ofcreativity and more indulgence of creative recreation. Creative Tourism

The new creation of cultural tourism products allows production and consumption [5] [6]. In this sense, it can be observed in Figure 3, that tourists need new creative areas: wherein they have the role of 'Prosumer' which represents tourists who involve themself in the activities of learning, experiencing, and developing aconnection with the community and their culture.

Creativity is multidisciplinary and includes three creatives: Technological creativity (invention), Economic creativity (entrepreneurship), and Cultural or artistic creativity (Community). It is interdependent and reinforced by all these dimensions of creativity. The link between technology, art, and business is the drivingforce behind the creativity economy.

The [7] recognizes the inventiveness characteristics in several regions as follows: artistic inventiveness involves the creative ability and production capacity of most thoughts and other ways of deciphering the world communicated in content, sound, and picture. Interest and eagerness to explore and make unused connections to fathom issues are key components of logical inventiveness. Financial imagination incorporates an energetic preparation driving development in innovation, commerce hones, promoting, and picking up a competitive financial advantage.

The new generations such as Gen Y, Z are attracted by creative tourism and play a key role in generating revenue for countries by focusing on creating value for the socio-cultural heritage of the locals and helpingthe creative economy [8].

Creative Tourism is a modern and innovative form of Cultural Tourism. Creative tourism shows, cooperation, and partnerships between tourism and creative industries [9]. This industry can play a significant role in protecting, preserving, and promoting tangible and intangible heritage, and generating socio-economic benefits for all stakeholders. The different cultural aspects, learning perspectives, and futureconsumption of tourism products were defined and showcased the movement of creating many opportunities for destinations, its cultural uniqueness in creative content, and creative performance role in tourism to attract visitors [10]. (Table 3) also gives the notion of the consumption of tourism products and services to learning with active involvement.

Table 2 Consumption to LearningSource- [11]

Type of tourism	Time	Cultural Focus	Consumption	Learning
Cultural tourism	Past and present	High culture, popular culture	Product, Process	Passive
Creative tourism	past, present, future	Creative Process	Experience, co: production between tourist and local	Interactive, Active skill development

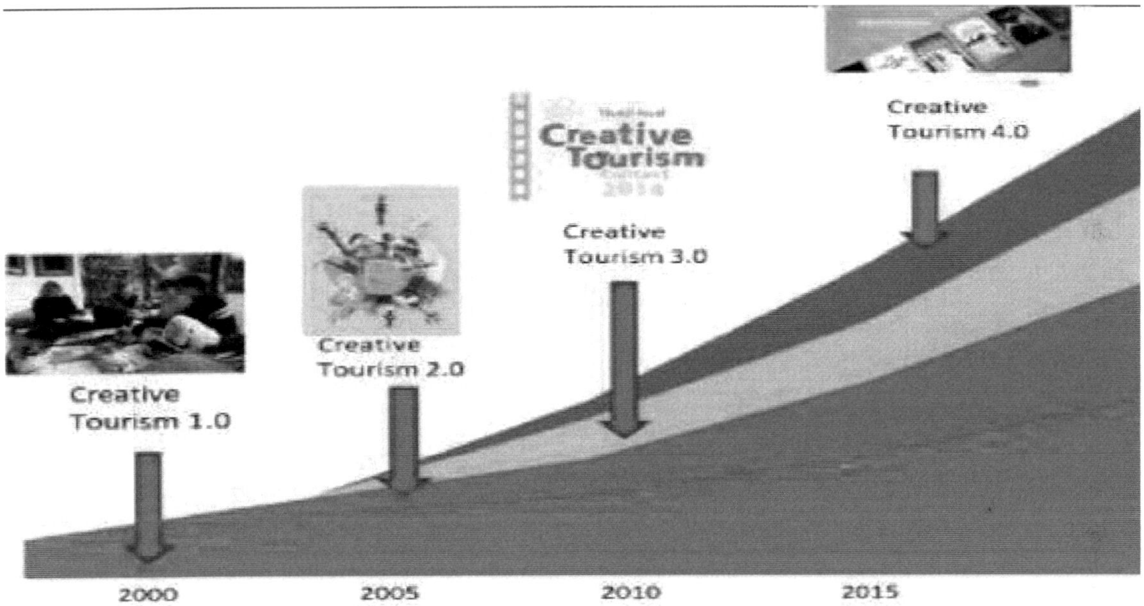

Figure 2 Culture to Creative Tourism Source [6]

Creative tourism is expected to attract tourists through various cultural values, the goal is to develop the creative community and encourage the development of the creative economy, and the income from tourism would significantly improve the local economy. With the shreds of evidence from the literature review, it is observed that creative tourism has a great potential to be developed as a part of tourism. Tourism has demonstrated the art of traditionalism in various communities and it can be observed as a success in various parts of the world. The motivation of the tourists is changing from commercial investment to traditional investment with their involvement in traditional activities. Considering the creative tourism in the state of Himachal Pradesh this paper is an attempt to find –

- The potential of activities that can be seen as a creative activity in the region. Methodology- The conceptual study was done to understand the potential of creative tourism activities in Kinnaur. The study was designed from secondary data and took a certain view from different people (professionals and tourists) through a telephonic interview.

Study Area

Kinnaur district is in the eastern part of Himachal Pradesh. The district is divided into five tehsils: Nichar, Kapla, Sangla, Pooh, and Moorang, and one sub-tehsil, namely Hangrang [12]. [13] [14], described Kinnaur district as in the eastern part of Himachal Pradesh. There are five tehsils in the district: Nichar, Kapla, Sanglarang, Pooh, Moorang, and one sub-tehsil called Hangah. The district of Kinnaur, with its rough terrain and isolated culture, is a far hinterland in the high mountains of Himachal Pradesh, where the lifestyle and ethnic community of the state is located. [15]. The unexplored dimension reveals an organic image of the region, which is entirely rural and constitutes a planned area of Himachal Pradesh. As stated above, there are no urban centers [16] and it is an unexplored dimension that reveals an organic image of the region, which is entirely rural and constitutes a planned area of Himachal Pradesh. As stated above, there are no urban centers.

 https://doi.org/10.52202/074664-0019

Table:3 Creative Tourism Potential in Kinnaur District

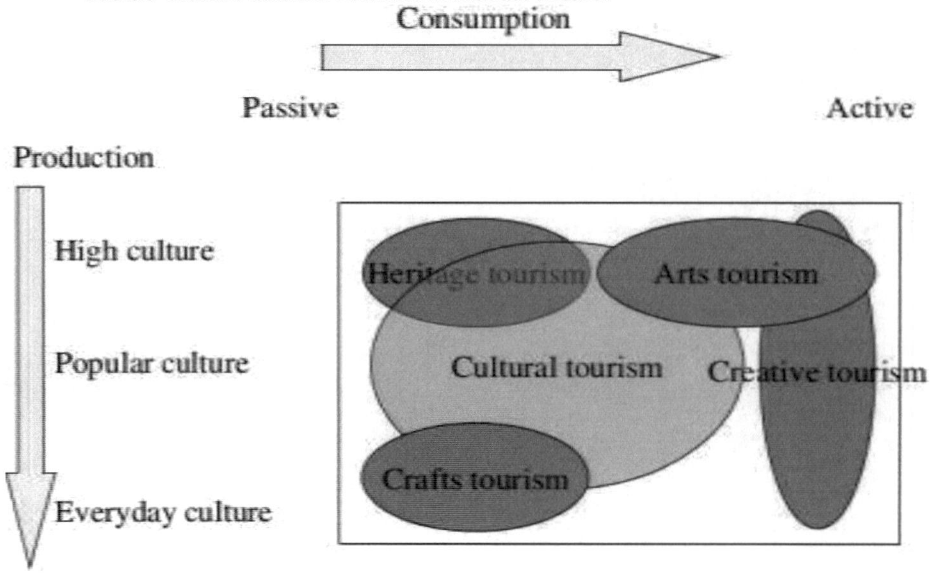

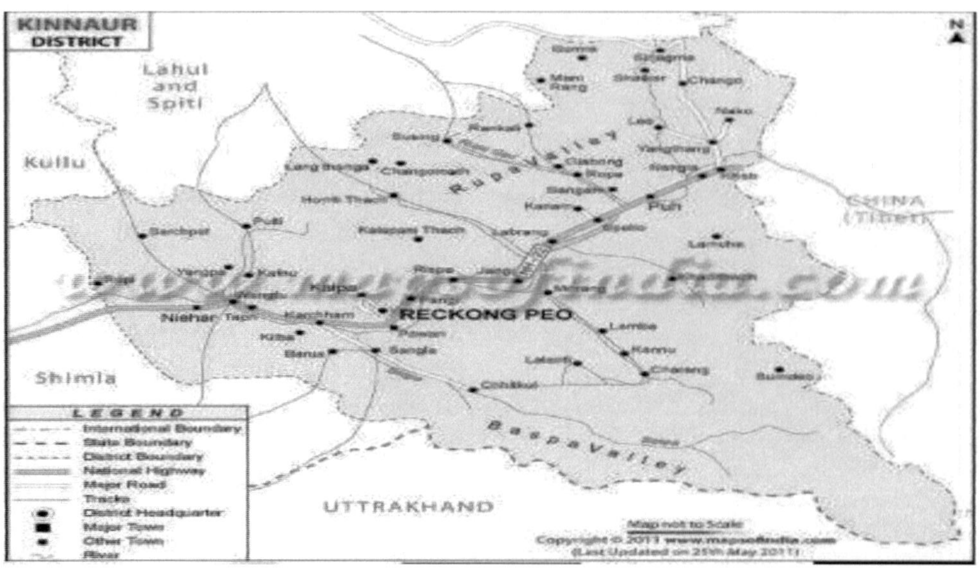

Figure 3 Geographic Map of Kinnaur District

Source: Official website of Maps of India

Table:3 Creative Tourism Potential in Kinnaur District

Creative tourist productsfor Kinnaur	Places/examples/factors	
Apple/ Grape Wine ➢ Home mad e apple/grape wineproduction ➢ Tasting ➢ Excursi onsto workshops	There is a traditional way of making apple wineat their home to allow tourists to learn the manufacturing and tasting. The Purest form extracted from Grapes is known as Moori. (Anguri). The utilization of wine on different occasions- Given to Deity (Namit), Given to Buddist temples (Sankim), and during ring ceremony (Quyang)	
Orchard Farms ➢ Harvesti ng (New & Old Technique) ➢ Cam pingin Farms ➢ Pic kingyour apples	The Kinnaur District produces juicy and qualityapples, which are famous all around the world. Creative tourism has a great potential to make orchard farms used as creative products to attract tourists to the farms and allow them to learn the techniques of harvesting, production, and staying within the farms. The community can also allow the tourists to be involved in picking the apples for them.	 

https://doi.org/10.52202/074664-0019

Gastronomy ➢ TraditionalFoods ➢ Dishes prepared fr omapples	The gastronomy of the Kinnaur is unique, authentic, and traditional. The various cuisine dishes are created with organic farming. The meals of the day are known as Ken or Tshema (Morning), Shod or Chicken (Lunch), and Yangskin or Gongal (Dinner). The various famous cuisines of the Kinnaur region are Juthey- is made from buckwheat flour. It can beprepared sweet or salty as per taste. Chilta- also known as Poli. We can eat poli withhoney, dughti, and with kadi. Dughti- it is a kind of chatni. It is made from walnuts, apricot seeds, and almonds. From ancient times, we made dughti on a stone called paan (a flat stone)and kollon(a round stone). Saachaa/Namakeen Chai- known as the drink of Kinnaur. Also known as butter salt tea. Saa - salt and chaa - tea that's why salt tea. We consume saa chaa mostly in winter to keep our bodies warm. Chull Fanting- also called kulthoo in our dialect. It is made from apricot. We can eat chull fanting with sattu (sampa). We usually make chull fanting in summers. This pole is made from Ogal and Phapra (buckwheat). It is known as a jalebi of Kinnaur. Chulli ka Tel – Apricot oil used for the health benefit. Panthang (fireplace) is the house's fireplace, where all family members sit together and enjoy the heat from the fire. It is commonplace for all family members. Kala Jeera- is one of the unique spices that is found in the Kinnaur region. It has obtained the geographic indicator tag. Damo (Chai maker) is a piece of cylindrical equipment in which chai is mixed with butter andmilk. All these dishes are well known in the Kinnaur region and have a unique way of preparation. The organic ingredient used in the preparation shows the charm of the region. The region shows the creative potential that can involve tourists in understanding the utilization and preparation of the various dishes.	

Festivals & Fairs ➢ Spiritual Devotions	Festivals & fairs of the regions are very spiritual and can be used as a creative product to allow tourists to understand and participate in the devotions. These are some famous Events that can attract tourists and learn about the creation of therituals. ➢ Dakreni or Dakrain- is a famous festival celebrated in July. ➢ Ukhyang or Phulaich - is a festival of flowers celebrated in September. ➢ Sazo – is a festival celebrated in January. In this festival, the deity is brought out and worshiped with wine and halwa. ➢ The loser is celebrated in December to welcome the New Year. ➢ Phagul or Suskar It is celebrated in February/March. In this festival, the spirit of Kanda (Peaks) called Kali is mainly worshiped. The festival lasts about a fortnight and is celebrated all over Kinnaur. ➢ Tribal Festival has beencelebrated since 1994 from 30 October to 2 November every year at District Headquarters Reckong Peo, and this festival has been declared as a State Level Festival.	  
➢ Architectural of Temples	The architecture of temples is unique &traditional. (Devta /DietyTemples)	

https://doi.org/10.52202/074664-0019

Linguistic Affair ➢ Learn theLocal Language ➢ Lea rnLocal Dance	The dominant language spoken by Kinnaur inhabitants is 'Kinnauri' or 'Kanauri.' language is the way of communication. The Kinnauri language can be a creative product by allowing tourists to learn the language through variousworkshops. Kinnaur district is well known for its traditionaldances and known as Kayang. There are two typesof dance forms prevalent in this district. One is dedicated to the Gods and Goddesses, and the second is performed during wedding ceremonies and other festivals. In these dance forms, men and women form a semicircle and dance specific steps. This dance is a vital part of tribal life, and it reflects the culture and tradition of the region. (Devil Dance Chaamb, Kyang or Kiang, Baakyaang, Byangchchhu, Jaatru Kiang, Japro). The tourists get attracted to the folk and seem eager to learn the steps done by the locals. These positive gestures show the interest and potential of creative tourism for the regions.	
Traditional Craft ➢ Gold and Silver Jewel ryCrafting	The Kinnaur district has ornaments that are precious and old traditional. The region uses silver for making jewelry. The women wear this jewelry on various occasions (Marriages, Festivals, and other functions at home). This jewelry crafting is done and known by the local craftsman. The creation of these ornaments shows the old and uniqueness of the region. These Jewelry are- Trimonia is the traditional jewelry of the region and is only created by the jewelers in the Kinnaurdistrict. Sanglong, Brkri, and Shapotwa are silver jewelry used by the ladies to hold the dodhu. It istypical jewelry that is manufactured by the sonarsof Kinnaur. Chak- is made from gold, and it is known as Suhag of the women. It is worn on the hair by theladies. Doroli- is a combination of silver, orange stones (Shlik), and blue stones (Ting). Digra- is a chain with a mix of gold and silver with old silver coins on it.	

	Ghau- is a piece of jewelry made with gold and silver. Known as Son eke Ghau and Chandi ke Ghau. Kanta- is an earring. Sunago are known as kangan Daglo is known as round thick silver ornamentsworn on the hands. Muh Chanogh is the silver jewelry worn on thefront of the head of the bride. Bakuh is worn on the Nose, and it is the most prominent jewelry of the bride. Chhullkshu is a necklace made with orange stones. Koashmal is a necklace made with black and orange stones. These jewels can be shown to the tourists, as itwill attract tourists and encourage them to know how to make and wear the jewelry in various traditions. The crafting has a great potential of showcasing the craft done on silver and gold jewelry. The workshops and seminars can involvethe tourists to understand and apply themselves in the activity.	
Fish (Trout) Farming	Trout is a freshwater fish, and with massive water resources, trout production is developing quickly. The production of Trout fish is done in the Sangla in Kinnaur District. Tourists who are fish lovers would love to learn and understand the production of trout fishing. So, Aquaculture has a great perspective to be developed as a creative tourism destination.	

Conclusion:

With the potential for creative tourism, artists might attempt to engage visitors in activities, either actively or passively. The creator of the itineraries can engage visitors with the locals to showcase the socio-cultural aspects of the areas. Any region's culture is its heart, and visitors need to be able to see and feel its authenticityto learn about the location. Communities must be leveraged to foster creativity and innovation when they have the knowledge and abilities to play to their strengths. Such actions will promote community development and sustainability through tourism.

https://doi.org/10.52202/074664-0019

Reference

1. V. Kashyap, "The Dormant of Creative Tourism in Pabbar Valley (Kothkai-Jubbal), Himachal Pradesh," in FUTURE TRENDS IN HOSPITALITY INDUSTRY: New Age Hotels and Alternate Accommodation, Bhopal, (2020).
2. V. Kashyap, ""The Comprehension of the Nichar Community for the Development and Growth of the Destination As a Creative Tourism Destination?"," ASEAN Journal on Hospitality and Tourism, 109, (2023)
3. W. Hui, "New Concept of Tourism Industry Development: 5C Mode," China Industrial Economy, (2007).
4. N. Duxbury and G. Richards, A Research Agenda for Creative Tourism, UK: Edwards Elgar Publishing, (2019).
5. E. Kakiuchi, "Culturally creative cities in Japan: Reality and prospects. City, Culture and Society,"101-108, (2016).
6. G. Richards, "Creative tourism: Development, trends and opportunities INTO Mallorca October," 2011. [Online]. Available: https://www.slideshare.net/gregbarca/creative-tourism-into. [Accessed 22 01 2022].
7. D. O. E. United Nations, United Nations E-Government Survey 2010: Leveraging e-government at a time of financial and economic crisis, United Nations Publications, (2010).
8. OECD, "Tourism and the creative economy," in OECD Studies on Tourism, http://dx.doi.org/10.1787/9789264207875-en, (2014).
9. UNWTO, "International Congress on World Civilizations and Creative Tourism," Bulgaria, (2016).
10. M. H. Zaal, F. Doosti and . M. R. Lasbuie, "Factors Involved in Assessing Capacity of Creative Tourism: Case Study -Tabriz City in Iran," Journal of Tourism & Hospitality, (2017).
11. G. Richards and J. Wilson, "culture?, Developing creativity in tourist experiences: A solution to the serial reproduction of," Tourism Managemen, 1209–1223, (2006).
12. K. District Administration, "Kinnaur," 21 04 2021. [Online]. Available: https://hpkinnaur.nic.in/dc-kinnaur/.
13. J. S. Parmar, "Tourism Development in Himachal Pradesh: Emerging Dimensions," International Journal of Hospitality & Tourism Systems, 66-76, (2012).
14. C. Mehta, A Comparative Study of Traditional Costumes of the Pangwala tribe residing in Pangi Valley of Himachal Pradesh through three generations, Chandigarh, (1995).
15. Sharma and G. Kalotra, "OUT-MIGRATION IN A MOUNTAINOUS VILLAGE: A CASE STUDY OF VILLAGE LIPPA," International Journal of Research and Analytical Reviews, (2018).
16. O. I. "District Census Handbook, Series-03, Part XII-B, Himachal Pradesh," 2011. [Online]. Available: https://www.censusindia.gov.in/2011census/dchb/0211_PART_B_DCHB_SHIMLA.pdf.
17. Wikipedia, "Kinnaur district," 8 05 2021. [Online]. Available: https://en.wikipedia.org/wiki/Kinnaur_district.

https://doi.org/10.52202/074664-0019

TOURISM AND HOSPITALITY IN SECONDARY EDUCATION

[1]Premlata Bourai
[1]Department of Education, Shri Guru Ram Rai University, Campus, Dehradun

Abstract

This research paper aims to explore the integration of Tourism and Hospitality in secondasry education, recognizing the evolving nature of education in a multicultural, global, and post-COVID world. Drawing inspiration from the National Curriculum Framework 2005 and the National Education Policy 2020, the research supports the claim that the inclusion of Tourism and Hospitality in the curriculum can enrich students' communication, critical thinking, and decision-making skills. A link between tourism, education, and societal values can be established after an examination of the historical, cultural, and economic significance of these fields. This research paper emphasizes the importance of practical skills, soft skill development, cultural awareness, environmental education, and overall enrichment in the secondary education experience. Advocating these skills would to nurture well-rounded individuals with a focus on sustainability. By incorporating these aspects, students are not only prepared for diverse career paths but also equipped to navigate the complexities of our ever-changing global landscape. As education transforms, investing in innovative programs becomes crucial to unlock the full potential of future generations.

Keywords: Tourism, Education, Hospitality, Skill development, Sustainability

1. Introduction

Education is not confined to four walls of a classroom anymore or to textbooks and exams. Education is being constantly modified and changed to incorporate the changing demands of a multicultural, global and most importantly a post covid world. Education being a sociological, physiological and developmental process is a complex sphere where a student acquires knowledge, expertise, skills and puts it to use in the real world. The National Curriculum Framework 2005 recommends the fusion of work and education into the domain of curricular as work gives experience, inculcates important social, cultural and emotional values in the students. Self-reliance, cooperation and creativity are other traits that are developed in students from a young age when practical aspects are introduced in their lives.

The introduction of Tourism and Hospitality in secondary education can be one such fusion of work and studies in the domain of curricular where an experience of involvement in productive work in an educational setting will enrich communication, critical thinking and decision making in different situations pertaining to different job roles.

The National Education Policy (NEP), 2020 also proposes to strengthen and promote the spirit of Ek Bharat Shreshtha Bharat (EBSB) among students. An activity related to it is the visit of students to 100 identified tourist destinations in the country. Under this programme students are encouraged to study the destinations and their history, scientific contributions, traditions, indigenous literature and knowledge, etc. as a part of augmenting their knowledge about different areas to understand the rich culture and diversity of the country.

Tourism and Hospitality are complex economic and labour-intensive fields. Tourism is derived from the French word 'Turisme' which means to tour and travel. Tourism is the process of temporary movement of humans from one place to another and returning to the point of origin of travel. According to the Tourism society of the United Kingdom it is the temporary movement of people to destinations outside of the places where they normally live and work, and their activities during their stay at these destinations. Hospitality on the other hand is the association between the guest and the host. It is a part of the tourism industry that provides services like accommodation, food

 https://doi.org/10.52202/074664-0020

etc to guests. Thus, tourism and hospitality are a complex web of different businesses. If these fields are introduced at the secondary level they will introduce the students to multifaceted areas of jobs, careers and will help in their personal and professional growth.

2. Tourism and Education: Establishing a Link

Whitehead, A. N. (1951) in his book *The Aims of Education*, states that Education is conceived primarily important for the development of future and that information or knowledge is worthy only if it contributes to the culture of the community. Education leads to proper understanding of all types of phenomena happening around and it gives the capability to look out for all the unanswered questions. "What we should aim at producing is men who possess both culture and expert knowledge in some special direction"[1].

According to Singh, tourism and education are in some ways intertwined. Both can contribute significantly to individual and societal values and beliefs. Practices in tourism arising from human, societal and environmental values contribute significantly to socio-cultural development. A perfect balance between the two contributes to the economy of the nation as well, it will also enable us to explore new avenues in tourism which are still unexplored due to lack of proper education.

Schriewer (2000) in his work states the importance of tourism as a subject. Tourism is an age old phenomenon. People have been traveling since times immemorial, but it took some time to understand tourism individually and later it was recognized by the government as a full-fledged industry. Since people started exploring the industry, a need for deep study of tourism was felt which led to the introduction of tourism as a subject.

Csikszentmihaly (1990), states how culture and education are linked. Culture has existed since mankind has evolved but a true understanding of the culture came when humans started to learn. Learning led to the building of the framework of the society. Tourism can serve as that gate that opens up the culture of a place for its students.

Education in India has been considered important since the existence of gurukuls. Institutes like Nalanda University, Taxila University explain how much the education was valued in the bygone era. Singh (2005) in his paper mentioned that post-independence India witnessed rapid expansion and reforms in higher education to evolve a system of education relevant to the life, needs and aspirations of the people. In the wake of educational developments in independent India, significant advisory and regulatory bodies specifically for the higher education were instituted.

With time, education also saw the evolution of new methods of education. New subjects were introduced; the prime example being the recent development in the New Education Policy which stresses on skill-based education, the foundation of hospitality. This enhances the individual potential and the creativity of the student and trains them for jobs and opportunities later in life from their early stages.

According to Daniel, Tourism affects the economy of a country. Since it has a multi-faceted character, as a consequence the number of tourism education courses has grown significantly to meet the demands of the qualified workforce. Dixit (2019) mentioned that the future economic growth rate and employment generation capacity of any nation depends on the knowledge and skill of its local workforce. In India, the tourism and hospitality industry continue to experience a critical shortage of a qualified and skilled workforce.

Cathy H.C Hsu (2018) opines that the increased adoption of blended learning and flipped classrooms, virtual learning and robot teaching assistants will change the nature of job of the educators. The integration of formal and informal learning spaces, with the use of augmented and virtual reality tools will further challenge educators' roles and their use of pedagogies. Thus, a vital link between tourism, hospitality and education exists whose importance will be discussed below.

Importance of Hospitality and Tourism in Secondary Education: The inclusion of tourism in the secondary education will equip students with practical skills by incorporating concepts like custom service, event management, culinary arts, hotel operations and meet and greet services. They will set foundational understanding amongst students to understand diverse facets relating to tourism. Career options would broaden for such students and they can make informed decisions regarding their academic and professional pathways.

This mode will help in nurturing crucial soft skills vital for personal and professional growth: through interactive learning experiences such as group projects, internships and role-playing scenarios. The students will develop effective communication, problem solving attitude, teamwork and adaptability to new situations and scenarios.

It will also help the students in developing a link with their culture. This education helps the student know the past, culture, beliefs of their country deeply. In a more global world, appreciation and respect for other cultures is vital. Topics like culture heritage, heritage tourism, travel geography and sustainable tourism practices give students insight into different societies and help them navigate cross-cultural interactions with sensitivity and empathy.

Tourism and hospitality education can augment the two topics, namely 'environmental education' and 'sustainability awareness'. Growing concerns about climate change and ecological conservation will help students learn about the importance of responsible tourism practices, resource management and sustainability.

Experiences like trips to destinations, guest lectures by professionals, participation in events and competitions will inspire students to pursue further studies or careers in these dynamics. It prepares students for the overall complexities of the world so that students can thrive in an ever-changing global world. It thus contributes in the overall enrichment of the educational experience.

Conclusion- Tourism and hospitality are two vital components where students will learn basic needs, services, and the demands of people of various cultures, ethnicities and regions. They will acquire skill sets which will help them not only in the tourism sector, but at large in life. Even if they don't establish a career in this field, they will be skilled enough to pursue other fields which is a desired objective of introducing this field in education by the government also. As educators and policymakers recognise the importance of these industries, investing in comprehensive and innovative educational programmes becomes imperative to unlock the full potential of the future generations.

The author has no competing interests to declare that are relevant to the content of this article.

References

1. Cathy H.C. Hsu.: "Tourism Education on and Beyond the Horizon". *Tourism Management Perspectives* (25), January (2018).
2. Csikszentmihaly, M.: Flow: *The Psychology of Optimal Experience.* Harper Perennial, New York (1990).
3. Daniel A.D., R. C. "Tourism Education: What about Entrepreneurial Skills". *Journal of Hospitality and Tourism*, Elsevier (2017).
4. Dixit S.K., Mawroh H. "Vocational Education and Training for Hospitality and Tourism Industry in India". In: Liu C., Schänzel H. (eds) *Tourism Education and Asia: Perspectives on Asian Tourism.* Springer, Singapore (2019).
5. Schriewer, J. "Forms of Externalization in Educational Knowledge". In: Seminar at Oxford University Day Conference on Theory, Method and Practice in Comparative Education (2000).
6. Singh, S. & Singh, T.V.: International Handbook of Tourism Education. Elsevier, India (2005)
7. Whitehead, A. N.: "The Aims of Education". North American Libraries Mentor Books. New York (1951).

 https://doi.org/10.52202/074664-0020

Gap in pagination due to formatting issues.

Pages 222-228

Impact of Electronic Word of Mouth (eWOM) on Consumer Behaviour and Attitude: A Comprehensive Analysis of Digital Influence in the Modern Marketplace

Kuldeep Kumar Choudhary[1], Monika Rani[2], and Aman Kumar Bhardwaj[3]

[1]Research Scholar, Chandigarh University, Mohali, India

[1&2]Assoicate Professor, UITHM, Chandigarh University

Abstract:

In the evolving landscape of digital marketing and consumer engagement, Electronic Word of Mouth (eWOM) has emerged as a pivotal influencer of consumer behavior and attitudes. This literature review provides an exhaustive analysis of the multifaceted impact of eWOM in the modern marketplace, synthesizing findings from a broad spectrum of research studies. The review delves into various domains, including tourism, online retail, social media, and e-hospitality, to offer a comprehensive understanding of how eWOM shapes consumer decision-making processes. A key aspect explored is the differential impact of eWOM based on the source, such as friend reviews versus general crowd reviews. Studies indicate that personal connections in eWOM, such as reviews from friends, exert a more substantial influence on consumer behavior compared to anonymous crowd reviews. This finding underscores the importance of personal trust and relationship in the efficacy of eWOM. In the context of tourism, research employing the theory of planned behavior reveals how eWOM significantly guides destination choices, highlighting its critical role in shaping travel decisions. Similarly, in online marketplaces, eWOM and digital marketing strategies are found to be significant determinants of purchasing decisions. This is particularly evident in platforms like Shopee, where improvement in online reviews and effective digital marketing strategies have a positive impact on consumer purchase decisions. The review also sheds light on the influence of eWOM in the realm of social media marketing. As consumers increasingly rely on information from their peers, eWOM becomes a crucial element for brands to leverage in building stronger customer-engagement relationships. Additionally, the impact of eWOM on consumer attitudes towards online travel packages during the new normal is evaluated, emphasizing factors such as online word of mouth, perceived risk, and destination trust. Cross-cultural analyses in the review reveal quantitative and qualitative differences in eWOM content across different regions, with variations in brand mentions and information-seeking behaviors. Furthermore, the emergence of tech-bloggers as influential actors in eWOM communication is explored, demonstrating how they establish their status through audience construction, content development, and network formation. The review addresses consumer confusion in the e-hospitality marketplace and its subsequent impact on negative eWOM and decision postponement. This aspect highlights the challenges and complexities in managing eWOM in various sectors. This literature review elucidates the profound and diverse impact of eWOM across multiple sectors, revealing its crucial role as a driving force in consumer behavior and attitude in the digital age.

Keywords: eWOM effectiveness, eWOM influence, eWOM marketing strategies, eWOM impact on consumer behavior.

https://doi.org/10.52202/074664-0022

1. Introduction:

In the digital era, the dynamics of consumer behavior and marketing have undergone a significant transformation, largely influenced by the advent and proliferation of Electronic Word of Mouth (eWOM) [1]. As a powerful form of social influence, eWOM has become an indispensable element in the modern marketplace, profoundly impacting consumer attitudes and decision-making processes. This literature review seeks to explore the extensive scope of eWOM, examining its multifaceted impact across various sectors such as tourism, online retail, social media, and e-hospitality. The concept of eWOM, essentially the digital counterpart of traditional word-of-mouth, encompasses the sharing of opinions, experiences, and information about products or services through online platforms. Unlike its traditional form, eWOM leverages the vast reach and immediacy of the internet, making it a more dynamic and far-reaching influencer [2]. The significance of eWOM lies in its ability to shape public perceptions and consumer behaviors in a more direct and immediate manner than conventional advertising methods. One of the pivotal aspects of eWOM is the source of the information. Personal connections, such as friend reviews, tend to have a more substantial impact on consumer behavior compared to anonymous reviews [3]. This distinction underscores the importance of trust and personal relationships in the effectiveness of eWOM. Consumers are more likely to be influenced by people they know and trust, highlighting the need for marketers to focus on building authentic and trustworthy sources of eWOM.In the tourism and Hotel industry, eWOM plays a crucial role in destination choice. Travelers increasingly rely on online reviews and social media posts to make informed decisions about their travel plans [4]. The theory of planned behavior applied in this context reveals that eWOM not only influences destination choices but also shapes the overall perception of the travel experience. The impact of eWOM is also prominently seen in online marketplaces. Studies have shown that eWOM, coupled with effective digital marketing strategies, significantly influences consumer purchasing decisions. In platforms like Shopee, the improvement of online reviews and the implementation of strategic digital marketing have been linked to positive consumer purchase behavior. Social media has become a fertile ground for eWOM, with consumers increasingly depending on peer information for brand interactions [5]. The role of eWOM in social media marketing highlights the shift from traditional advertising to more peer-driven, influencer-based marketing strategies. This shift necessitates a deeper understanding of how to leverage eWOM to build stronger customer-engagement relationships. The emergence of new influencers, such as tech-bloggers, in the eWOM landscape demonstrates the evolving nature of online communication. These influencers establish their authority through audience construction, content development, and network formation, playing a pivotal role in shaping consumer opinions and behaviors in specific domains [6]. Another critical area explored in this review is the impact of eWOM in the e-hospitality sector. The relationship between consumer confusion in online hospitality services and the prevalence of negative eWOM underscores the challenges faced by the industry in managing online reputations and customer expectations. Cross-cultural analyses included in this review reveal significant differences in eWOM content and consumer responses across different regions. These differences are essential for global brands and marketers to understand, as they indicate the need for tailored eWOM strategies that consider cultural nuances and consumer behaviors specific to each market [7].This literature review aims to provide a comprehensive understanding of the impact of eWOM on consumer behavior and attitudes. It delves into the intricacies of eWOM's influence across various sectors, highlighting its role as a crucial driver in the digital marketplace. The review underscores the need for businesses to

https://doi.org/10.52202/074664-0022

adeptly navigate the realm of eWOM, harnessing its power to build brand loyalty, influence consumer decisions, and stay competitive in the ever-evolving digital landscape [8].

Businesses and marketers working in the digital world must comprehend how eWOM affects consumer behaviour and attitudes. eWOM frequently outperforms traditional marketing and advertising in its ability to profoundly affect consumer perceptions, preferences, and purchase intentions. Customers are looking for genuine and trustworthy information from sources they consider to be reputable and relatable, and they are depending more and more on peer recommendations and evaluations to help them make decisions. As a result, companies must modify their marketing plans to take full advantage of eWOM and realise its potential for increasing sales, cultivating client loyalty, and establishing brand reputation [3]. This review seeks to shed light on the complex influences of electronic word-of-mouth (eWOM) on customer behaviour and attitudes in the contemporary marketplace, thereby assisting in the formulation of strategic decision-making.

2. Background of the Study:

In the digital age, electronic word-of-mouth, or eWOM, has become a major phenomena that is drastically changing how consumers communicate and make decisions. Due to the increasing use of social media platforms, review websites, and online forums, traditional word of mouth—which has long been acknowledged as a powerful force influencing consumer behavior—has given way to its digital equivalent, or electronic word of mouth, or eWOM. In contrast to traditional word-of-mouth, which was mainly limited to face-to-face encounters, electronic word-of-mouth (eWOM) crosses national borders and allows people to instantly communicate ideas, insights, and recommendations with a worldwide audience.

2. Objectives:

1. Investigate the influence of Electronic Word of Mouth (eWOM) on modern consumer behavior.
2. Analyze the impact of eWOM on consumer attitudes in the digital era.

3. Research Methodology:

The current study employs a research technique that entails a methodical approach to the collection and analysis of pertinent material pertaining to the influence of Electronic Word of Mouth (eWOM) on customer behaviour and attitudes in the modern marketplace. First, a thorough search approach will be used to locate academic books, papers, and articles from a variety of databases, including PsycINFO, Web of Science, and Google Scholar. To ensure the relevance and validity of the findings, the inclusion criteria will be limited to research that have been published within the last ten years. A thematic analysis will be carried out to find important themes, patterns, and theoretical frameworks connected to eWOM and its impacts on consumer behaviour and attitudes after relevant material has been chosen. This procedure will entail combining the results from the chosen research.

4. Literature Review:

4.1 Overview eWOM in the digital market landscape:

Electronic Word of Mouth, or eWOM, has become a powerful force in influencing consumer behaviour and attitudes in today's digital marketing environment. Social media, online review sites, and online forums have all proliferated, giving people unparalleled access to peer opinions and information on goods, services, and brands [9]. The term electronic word-of-mouth

 https://doi.org/10.52202/074664-0022

(eWOM) describes how customer experiences, views, and recommendations are shared electronically, bypassing geographic barriers and instantly reaching a large audience. This phenomena has completely changed how consumers research products, decide what to buy, and develop opinions about businesses.

Relationship between Electronic Word of Mouth (eWOM) and Consumer Behavior and Attitudes:

Businesses and marketers working in the digital world must comprehend how eWOM affects consumer behaviour and attitudes. eWOM frequently outperforms traditional marketing and advertising in its ability to profoundly affect consumer perceptions, preferences, and purchase intentions. Customers are looking for genuine and trustworthy information from sources they consider to be reputable and relatable, and they are depending more and more on peer recommendations and evaluations to help them make decisions [10]. As a result, companies must modify their marketing plans to take full advantage of eWOM and realise its potential for increasing sales, cultivating client loyalty, and establishing brand reputation [3]. This review seeks to shed light on the complex influences of electronic word-of-mouth (eWOM) on customer behaviour and attitudes in the contemporary marketplace, thereby assisting in the formulation of strategic decision-making.

The Mechanism of Influence

eWOM operates through various online platforms, including social media, forums, review sites, and blogs, where consumers share their experiences and opinions about products or services. This mode of communication is potent due to its perceived authenticity and relatability [11] unlike traditional marketing, where messages are unidirectional and crafted by the brands, eWOM is a consumer-generated form of marketing that offers a multi-dimensional view of products or services. Its influence is rooted in social proof, where consumers often rely on the opinions of others to make informed decisions. This phenomenon is especially potent in industries where the product experience is subjective, like hospitality, tourism, and fashion [12].

Impact on Consumer Trust and Credibility

Trust plays a pivotal role in the relationship between eWOM and consumer behavior. Consumers are more likely to trust and be influenced by reviews and opinions from real users, particularly when they come from within their social networks. Studies have shown that eWOM from friends or acquaintances carries more weight than anonymous reviews. The credibility of the source, therefore, becomes a key factor in determining the impact of eWOM. Brands that successfully encourage and leverage positive eWOM from credible sources can significantly enhance their trustworthiness and appeal to potential customers [13].

Influence on Purchasing Decisions

eWOM has a direct impact on consumers' purchasing decisions. Positive reviews and testimonials can significantly boost sales, while negative reviews can deter potential buyers. This influence is not just limited to the quantity of eWOM but also its quality, sentiment, and relevance [14]. For instance, in the online marketplace, detailed and authentic reviews can sway consumers' opinions more effectively than vague or generic praise. Consequently, companies are increasingly focusing on strategies to foster positive eWOM, recognizing its power to drive sales and build brand loyalty.

Role in Shaping Consumer Attitudes

Consumer attitudes towards brands and products are significantly molded by eWOM. This shaping of attitudes goes beyond mere product evaluation; it encompasses the entire brand perception, including values, identity, and reputation [15]. For example, when consumers frequently encounter positive eWOM about a brand's commitment to sustainability, their attitude towards the brand becomes favorably aligned with environmental consciousness. Likewise, a brand consistently associated with poor customer service in eWOM channels may struggle to overcome a negative perception, regardless of its marketing efforts.

Cultural and Contextual Variations

The impact of eWOM also varies across different cultures and contexts. In some cultures, community opinions are highly valued, making eWOM a more influential factor in consumer decisions. Additionally, the nature of the product or service in question can alter the effect of eWOM. High-involvement products, like cars or luxury goods, may see consumers engaging more deeply with eWOM, while low-involvement products might not elicit the same level of engagement [16].

Challenges and Complexities

Despite its significant advantages, eWOM presents challenges and complexities. Managing negative eWOM, for instance, requires tactful and responsive strategies. Moreover, the authenticity of eWOM is sometimes questioned, with incidences of fake reviews and paid endorsements muddying the waters. Brands must navigate these challenges carefully to maintain consumer trust and leverage eWOM effectively.

The relationship between eWOM and consumer behavior and attitudes is both profound and complex. eWOM acts as a powerful mediator in the modern marketplace, influencing consumer trust, purchasing decisions, and overall brand perception. Its impact is nuanced by factors such as source credibility, cultural context, and product nature. As digital platforms continue to evolve and integrate more deeply into consumer lives, understanding and harnessing the power of eWOM becomes increasingly crucial for brands aiming to thrive in the competitive landscape of consumer influence [17].

4.2 eWOM in Tourism:

Influence on Destination Choices:

In the travel business, eWOM is crucial in influencing consumers' destination preferences. Travellers are becoming more and more dependent on online evaluations, suggestions, and testimonies from other travellers to guide their own travel choices [18]. Travellers may be persuaded to consider or give preference to particular destinations over others by positive eWOM, which can improve impressions of locations, attractions, and lodging. Negative eWOM, on the other hand, has the power to turn off potential tourists, underscoring the significance of online reputation management for the travel industry.

https://doi.org/10.52202/074664-0022

4.3 eWOM in Online Retail:

eWOM is a significant factor in online shopping, influencing the decisions made by customers. Product impressions, trust, and buying intentions are influenced by online reviews, ratings, and recommendations [19]. Positive eWOM can increase conversion rates and sales by fostering a sense of trust and confidence in items. On the other hand, unfavourable eWOM might discourage customers and cause firms to miss out on prospects. Recognising the elements that affect eWOM's legitimacy and effect is crucial for merchants hoping to take advantage of its potential to boost sales and improve customer happiness.

4.4 eWOM in Social Media Marketing:

Role in Building Customer-Engagement Relationships:

Social media platforms have developed into effective avenues for the distribution of electronic word-of-mouth (eWOM), allowing companies to communicate and personalise their interactions with customers [20]. Through promoting community involvement, eWOM amplifies brand endorsement, and facilitates peer-to-peer communication, it plays a critical role in developing relationships with customers. Actively monitoring and engaging in eWOM conversations on social media can help brands build stronger customer relationships, boost brand loyalty, and encourage word-of-mouth recommendations [21].

4. Conclusion:

Electronic word-of-mouth (eWOM) has emerged as a potent force reshaping consumer behavior and influencing purchasing decisions in the digital age. The body of research highlighted in this discourse underscores the multifaceted nature of eWOM and its profound impact across diverse industries and contexts. As businesses navigate the complexities of the digital landscape, they must recognize the pivotal role that eWOM plays in shaping brand perception, driving customer engagement, and ultimately driving business growth. One of the key takeaways from the research is the importance of fostering a positive online reputation. Studies consistently demonstrate the significant influence of consumer reviews, social media advertising, and online brand presence on consumer attitudes and purchase intentions. By actively engaging with customers, soliciting feedback, and addressing concerns promptly, businesses can cultivate trust and credibility, thereby enhancing their eWOM effectiveness and bolstering brand loyalty. The integration of digital platforms and social media channels is essential for amplifying eWOM reach and engagement. Platforms like Instagram, TikTok, and online marketplaces provide fertile ground for eWOM dissemination, offering businesses vast opportunities to connect with consumers and drive brand advocacy. Marketers must adopt a multi-channel approach to eWOM marketing, creating compelling content that resonates with target audiences and encourages sharing and interaction across various digital touchpoints. Personalization emerges as another critical strategy for optimizing eWOM effectiveness. By leveraging data analytics and customer insights, businesses can tailor eWOM messages to align with individual preferences and behavior patterns. Personalized recommendations and offers not only enhance engagement but also increase conversion rates, ultimately driving ROI on eWOM initiatives. Moreover, integrating sustainability and social responsibility into eWOM strategies can foster deeper emotional connections with consumers, leading to long-term brand loyalty and advocacy. Technology plays a pivotal role in optimizing eWOM effectiveness and driving actionable insights. Advanced analytics tools and AI-driven solutions enable businesses to track and measure eWOM metrics, identify trends, and predict consumer behavior. By harnessing the power of data mining and predictive analytics, marketers can adapt their strategies in real-time, capitalize on emerging opportunities, and mitigate potential risks effectively. Transparency and authenticity emerge as fundamental principles underpinning successful eWOM marketing. Businesses must prioritize genuine and honest communication, avoiding deceptive practices or fake reviews that can erode trust and credibility. By being transparent about product attributes, pricing, and customer experiences, companies can build credibility and foster positive eWOM, ultimately driving sustainable growth and success. EWOM offers tremendous opportunities for businesses to engage with consumers, amplify brand awareness, and drive sales in an increasingly digital marketplace. By implementing strategies such as cultivating a positive online reputation, embracing digital platforms, personalizing campaigns, integrating sustainability, leveraging technology, and fostering transparency, companies can harness the full potential of eWOM to achieve their marketing objectives and gain a competitive edge. As eWOM continues to evolve and shape the digital landscape, businesses must remain agile and adaptive, continuously innovating and refining their strategies to stay ahead of the curve and capitalize on emerging opportunities.

5. Limitations:

While this review paper provides a comprehensive analysis of the impact of Electronic Word of Mouth (eWOM) on consumer behavior and attitudes, it is important to acknowledge several limitations. Firstly, the scope of the review is limited to specific domains such as tourism, online retail, and social media marketing, thereby excluding other potential areas of eWOM influence. Additionally, the review primarily focuses on empirical research studies, potentially overlooking valuable insights from theoretical perspectives and qualitative analyses. Furthermore, the review may be subject to publication bias, as it relies on existing literature that may prioritize positive findings over negative or null results. Moreover, given the rapidly evolving nature of digital technologies and consumer behavior, the review may not capture the most recent developments and emerging trends in eWOM research. Finally, while efforts have been made to provide a comprehensive overview, the depth of analysis on each topic may vary, potentially leading to gaps or oversights in certain areas. These limitations highlight the need for further research to address these gaps and provide a more nuanced understanding of the complexities of eWOM in shaping consumer behavior and attitudes.

https://doi.org/10.52202/074664-0022

8. References:

1. Cheung, C.M.K., Thadani, D.R.: The impact of electronic word-of-mouth communication: A literature analysis and integrative model. Decision Support Systems. 54, 461–470 (2012). https://doi.org/10.1016/j.dss.2012.06.008.
2. Hennig-Thurau, T., Gwinner, K.P., Walsh, G., Gremler, D.D.: Electronic word-of-mouth via consumer-opinion platforms: What motivates consumers to articulate themselves on the Internet? Journal of Interactive Marketing. 18, 38–52 (2004). https://doi.org/10.1002/dir.10073.
3. Litvin, S.W., Goldsmith, R.E., Pan, B.: Electronic word-of-mouth in hospitality and tourism management. Tourism Management. 29, 458–468 (2008). https://doi.org/10.1016/j.tourman.2007.05.011.
4. Lee, M.-R., Youn, S.: Electronic word of mouth (eWOM). International Journal of Advertising. 28, 473–499 (2009). https://doi.org/10.2501/s0265048709200709.
5. Hennig-Thurau, T., Gwinner, K.P., Walsh, G., Gremler, D.D.: Electronic word-of-mouth via consumer-opinion platforms: What motivates consumers to articulate themselves on the Internet? Journal of Interactive Marketing. 18, 38–52 (2004). https://doi.org/10.1002/dir.10073.
6. Jalilvand, M.R., Esfahani, S.S., Samiei, N.: Electronic word-of-mouth: Challenges and opportunities. Procedia Computer Science. 3, 42–46 (2011). https://doi.org/10.1016/j.procs.2010.12.008.
7. Dwyer, P.: Measuring the value of electronic word of mouth and its impact in consumer communities. Journal of Interactive Marketing. 21, 63–79 (2007). https://doi.org/10.1002/dir.20078.
8. Zhang, J.Q., Crăciun, G., Shin, D.: When does electronic word-of-mouth matter? A study of consumer product reviews. Journal of Business Research. 63, 1336–1341 (2010). https://doi.org/10.1016/j.jbusres.2009.12.011.
9. Meuter, M.L., McCabe, D.B., Curran, J.M.: Electronic Word-of-Mouth versus interpersonal Word-of-Mouth: Are all forms of Word-of-Mouth equally influential? Services Marketing Quarterly. 34, 240–256 (2013). https://doi.org/10.1080/15332969.2013.798201.
10. Hennig-Thurau, T., Walsh, G., Walsh, G.: Electronic Word-of-Mouth: Motives for and Consequences of reading customer articulations on the internet. International Journal of Electronic Commerce. 8, 51–74 (2003). https://doi.org/10.1080/10864415.2003.11044293.
11. Litvin, S.W., Goldsmith, R.E., Pan, B.: A retrospective view of electronic word-of-mouth in hospitality and tourism management. International Journal of Contemporary Hospitality Management. 30, 313–325 (2018). https://doi.org/10.1108/ijchm-08-2016-0461.
12. Abd-Elaziz, M.E., Aziz, W., Khalifa, G.S.A., Abdel-Aleem, M.: Determinants of Electronic word of mouth (EWOM) influence on hotel customers' purchasing decision. International Journal of Heritage, Tourism, and Hospitality. 9, (2015).
13. Jalilvand, M.R., Samiei, N.: The impact of electronic word of mouth on a tourism destination choice. Internet Research. 22, 591–612 (2012). https://doi.org/10.1108/10662241211271563.
14. Pourfakhimi, S., Duncan, T., Coetzee, W.: Electronic word of mouth in tourism and hospitality consumer behaviour: state of the art. Tourism Review. 75, 637–661 (2020). https://doi.org/10.1108/tr-01-2019-0019.
15. Fine, M.B., Gironda, J.T., Petrescu, M.: Prosumer motivations for electronic word-of-mouth communication behaviors. Journal of Hospitality and Tourism Technology. 8, 280–295 (2017). https://doi.org/10.1108/jhtt-09-2016-0048.
16. Jan, A., Khan, M., Ajmal, M.M., Patwary, A.K.: From traditional advertising to digital marketing: exploring electronic word of mouth through a theoretical lens in the hospitality and tourism industry. Global Knowledge, Memory and Communication. (2023). https://doi.org/10.1108/gkmc-08-2022-0199.
17. Manes, E., Tchetchik, A.: The role of electronic word of mouth in reducing information asymmetry: An empirical investigation of online hotel booking. Journal of Business Research. 85, 185–196 (2018). https://doi.org/10.1016/j.jbusres.2017.12.019.
18. Jalilvand, M.R., Samiei, N.: The impact of electronic word of mouth on a tourism destination choice. Internet Research. 22, 591–612 (2012). https://doi.org/10.1108/10662241211271563.

19. Roy, G., Datta, B., Basu, R.: Effect of eWOM valence on online retail sales. Global Business Review. 18, 198–209 (2017). https://doi.org/10.1177/0972150916666966.
20. Bilal, M., Zeng, J., Akram, U., Tanveer, Y., Sohaib, M., Raza, M.A.: The role of Motivational Factors for Determining Attitude towards EWOM in social media context. International Journal of Enterprise Information Systems. 16, 73–91 (2020). https://doi.org/10.4018/ijeis.2020040105.
21. Poturak, M., Turkyilmaz, M.: The Impact of eWOM in Social Media on Consumer Purchase Decisions: A Comparative Study between Romanian and Bosnian Consumers. Management and Economics Review. 138–160 (2018). https://doi.org/10.24818/mer/2018.12-02.

 https://doi.org/10.52202/074664-0022

Borderless Adventures: Empowering Tourism with Blockchain Passports

A Kazi Ashfaq Rouhan[1], Md Thouhedul Alam Tonoy[2][0009-0007-4374-4084]

[1]Airlines and Airport Management, [2]Computer Science and Engineering (Blockchain Technology), [1,2]Chandigarh University, Mohali, Punjab, India

Abstract.

A game-changing solution for the travel and hospitality industry: a Blockchain-based passport system. By leveraging Blockchain's decentralized, secure, and efficient framework, this system redefines passport management. It addresses challenges like fraud, inefficiency, and security breaches inherent in traditional systems. Through smart contracts, it automates processes, ensuring swift and accurate verification. Its decentralized nature ensures data integrity and resilience against cyber threats, fostering trust among stakeholders. Moreover, it promotes interoperability, inclusivity, and user-friendliness, promising a seamless travel experience. In summary, this system represents a paradigm shift, offering unmatched security, efficiency, and convenience for travelers worldwide. Furthermore, this innovative system could create opportunities for smart tourism, providing stakeholders with precise, real-time information. It could also enable governments to streamline immigration procedures, reducing queues and delays at borders. Additionally, the transparency offered by Blockchain technology can increase accountability in the industry, promoting ethical and sustainable tourism. Finally, with potential to simplify visa applications and other travel-related processes, this Blockchain-based passport system could revolutionize the way we travel, making it more accessible for all.

Keywords: Sustainable tourism; Decentralized; Security; B-passport; Smart tourism.

https://doi.org/10.52202/074664-0023

1. Introduction

In today's rapidly evolving landscape, where technology continually reshapes industries, the idea of using blockchain to transform tourism is sparking a lot of curiosity. One particularly intriguing concept gaining traction is the notion of a 'blockchain passport' for streamlined passport verification. This could mean big improvements in efficiency, security, and overall traveler experience. Blockchain passport information stored securely on a blockchain network, which is basically a decentralized and transparent digital ledger. This setup could potentially make verifying identities much smoother and more reliable. The beauty of blockchain lies in its decentralized nature, meaning no single authority has control over it. This decentralization, coupled with its transparency and immutability which refers once information is recorded, it can't be altered, forms the backbone of why it's so appealing for passport verification.

Taking reference from Satoshi Nakamoto, the mysterious figure behind the invention of blockchain, to understand how this technology can make passport verification more secure and efficient. Additionally, drawing on the insights of thought leaders like Tapscott and Tapscott, who believe blockchain has the power to revolutionize not just finance but various industries[1].The vibrant world of hospitality and tourism. Arriving at a luxurious hotel, where friendly staff welcome the guest with open arms and the promise of a cozy stay. By stepping out of the bustling streets, human senses come alive with the sights, sounds, and smells of a new destination. Whether flying high in the clouds, cruising along scenic routes, or taking a leisurely stroll through charming neighborhoods, each mode of travel adds its own flavor to the adventure. From famous landmarks to hidden gems off the beaten path, there's always something exciting to discover. With the help of expert travel guides or the sense of wanderlust, tailoring the journey to match interests perfectly. And along the way, people can meet locals and fellow travelers who'll greet them with genuine warmth and hospitality, enriching the experience with their stories and traditions. So, come join us on this extraordinary journey—a world of endless exploration, unforgettable moments, and meaningful connections awaits.
The paper is all about breaking new ground in tourism research by diving into how blockchain can shake up passport verification. Pushing the boundaries and inspiring others to join B-Passport system in tourism and hospitality exploring the promising application of blockchain technology.

2. Problem Statement:

In the realm of modern tourism, the persistent reliance on traditional passport systems poses a myriad of challenges that hinder the seamless travel experience for individuals worldwide. Despite the continuous evolution of technology, the shortcomings of physical passport issuance and verification processes persist, leading to inefficiencies, security vulnerabilities, and privacy concerns. Foremost among these challenges is the time-consuming nature of passport issuance, which can be particularly problematic for individuals facing urgent medical emergencies necessitating immediate travel abroad for life-saving treatment. Delays in passport delivery can exacerbate already stressful situations, underscoring the critical importance of streamlining and expediting the issuance process to meet travelers' urgent needs.

https://doi.org/10.52202/074664-0023

Additionally, the complex and often opaque visa processing procedures further compound travelers' woes, with lengthy wait times and potential inaccuracies in documentation posing significant barriers to seamless travel. Moreover, the inherent risk of data theft within government agencies responsible for passport issuance raises concerns about the security and privacy of travelers' personal information, highlighting the need for robust safeguards and authentication mechanisms[2].The risk of losing physical passports while traveling presents yet another logistical challenge, as individuals must navigate bureaucratic hurdles and administrative delays to obtain replacement documents, disrupting travel plans and causing undue stress. Furthermore, the proliferation of counterfeit passports underscores the need for enhanced authentication measures to detect fraudulent documents and prevent identity theft and security breaches[3].Moreover, the reliance on third-party entities for passport registration introduces elements of uncertainty and vulnerability, raising questions about the integrity and security of travelers' personal data. Additionally, the cumbersome border immigration authentication checks, while essential for ensuring national security, often result in long wait times and logistical challenges for travelers, detracting from the overall travel experience[4].

In light of these pervasive challenges, there exists a pressing need to explore innovative approaches to identity management and verification within the tourism sector. By leveraging emerging technologies and best practices in identity verification, this research seeks to identify solutions that enhance security, efficiency, and trust in the travel process, ultimately improving the overall experience for tourists and industry stakeholders alike

3. Literature Review:

In recent years, the travel and hospitality industry has faced numerous challenges related to passport management, including fraud, inefficiency, and security breaches inherent in traditional systems. To address these issues, researchers have explored the potential of blockchain technology as a game-changing solution. Three primary deficiencies within the current tourism landscape that a blockchain-based passport system could potentially address. Firstly, it emphasized the need for robust protection of sensitive data, such as booking information and travel itineraries, to address security concerns. Trust between tourists and industry players, advocating for the transparency and immutability of blockchain to establish authenticity and prevent fraud. Lastly, it criticized existing research for neglecting the perspectives of tourists on new technologies like blockchain, emphasizing the importance of a user-centric approach in system design and implementation[5]. The potential impact of blockchain technology on the tourism and hospitality industry as a foundational technology with transformative potential. While acknowledging the benefits of enhanced security, transparency, and efficiency offered by blockchain, and the need for a more comprehensive examination of how a blockchain-based passport system could specifically address challenges within the industry. These challenges include security concerns related to traveler identification, the need for improved trust and transparency between tourists and service providers, and the potential for streamlining travel processes. Addressing these research gaps would provide valuable insights into the feasibility and benefits of implementing blockchain-based solutions in the tourism and hospitality sector[6]. Current digital passport systems, particularly concerning privacy and security issues. Centralized data storage, privacy vulnerabilities, lack of transparency, and limited user control over personal data were identified as key concerns.

In response, the paper proposed a blockchain-based passport system as a potential solution. The decentralized nature of blockchain offers enhanced security against data breaches, while its encryption capabilities could bolster user privacy. Moreover, the transparency provided by blockchain enables users to monitor data processing activities, granting them greater control over their passport information. Although it did not propose a specific system, it emphasized the potential of blockchain to address critical issues in digital passport systems[7].

Passport literature covers a broad range of topics, diving into everything from the past of passports to how they're used in today's global travel scene. Researchers explore how governments issuing passports shape our ideas about belonging to a country, the identities of a person, and how freely a person can move across borders. They're also digging into the newer tech in passports, like biometrics and e-passports, and how these advancements affect privacy and safety. Legal experts are busy untangling the rules and regulations around passports, touching on big issues like who gets to decide and how this affects our rights as individuals. And there's a focus on the real-life experiences of passport holders, especially those who face obstacles because of who they are. Altogether, passport literature gives us a deep dive into how nationality, the world getting smaller, and the rules of the game set by governments all come together in shaping our journeys around the globe.

Table 3.1. Passport Revolution

Period	**Description**	**Key Development**
Ancient Era (450 BC - 202 BC)	Early precursors to passports emerged, functioning as travel permissions and safe passage guarantees.	- Nehemiah's letter from King Artaxerxes I of Persia (450 BC) - Chinese travel documents requiring information like age and height (202 BC - 9 AD)
Medieval Era (722 AD - 1548)	Diverse travel documents, often issued by local authorities and containing	- Arabic exit permit (722 AD) - Bulletta document from the Republic

 https://doi.org/10.52202/074664-0023

	specific details, appeared in various regions.	of Genoa (12th century) - Pass-Card Treaty among German states (1850)
Early Modern Era (1414 - 1914)	The concept of passports solidified, with England introducing a document resembling the modern passport in the 15th century.	- King Henry V's document (1414) - British Nationality and Status of Aliens Act (1914)
20th Century (1920 - 1980)	Standardization efforts commenced, with the League of Nations holding conferences and issuing passports to refugees. Machine-readable passports were introduced.	- Paris Conference on Passports (1920) - League of Nations Nansen passports (1922-1938) - Machine-readable passports (1980)
21st Century (Present)	Biometric passports incorporating features like chips and fingerprints became prevalent.	- Biometric passports (present)

4. Blockchain Transforms the Modern Passport System

4.1 A Passport's Evolution: Passport evolution has undergone significant transformation since their inception as simple identity markers. The 20th century witnessed a pivotal moment with the standardization of travel documents by the International Civil Aviation

Organization (ICAO). This standardization streamlined the travel process and enhanced security measures. However, persistent concerns such as forgery and identity theft persisted despite these advancements[6].

4.2 Revolution in Trust and Transparency: Blockchain technology has emerged as a groundbreaking solution, poised to disrupt the traditional passport system by introducing a myriad of unique features:

4.2.1 Blockchain Features:

4.2.1.1 Decentralization: One of the fundamental aspects of blockchain is its decentralized nature. Unlike centralized systems where a single entity controls data, blockchain operates on a distributed network where no single authority has control. This decentralization minimizes the risk of data manipulation and empowers individuals. A network where individuals, rather than central authorities, hold the key to their identity, enhancing privacy and security[8].

4.2.1.2 Transparency: Every transaction recorded on a blockchain is visible and verifiable by authorized parties. This transparency fosters trust within the system and acts as a deterrent to fraud. Individuals can access an immutable record of their travel history, knowing that any unauthorized changes will be immediately detected. An ecosystem where the integrity of passport records is upheld through transparent and auditable processes.

4.2.1.3 Autonomy: Blockchain technology enables individuals to have greater control over their personal data. Through cryptographic keys, individuals can grant or revoke access to their passport information as needed. This autonomy empowers individuals to manage their identities securely, reducing reliance on centralized authorities. The freedom to decide who has access to passport information, enhancing privacy and control.

4.2.1.4 Security: Cryptography plays a vital role in safeguarding passport data on a blockchain. Each piece of data is encrypted and linked to the previous transaction, making it virtually tamper-proof. This cryptographic security layer ensures that passport information remains secure and resistant to forgery and identity theft. A passport that is impregnable to fraudulent activities, enhancing confidence in the integrity of the travel document[4].

4.2.1.5 Immutability: Once recorded on a blockchain, data becomes immutable, meaning it cannot be altered or deleted. This permanence ensures the authenticity of an individual's travel history, eliminating the possibility of tampering or manipulation. Border authorities can rely on the veracity of blockchain records, streamlining the verification process and reducing unnecessary delays. Seamless border crossings facilitated by immutable and tamper-proof passport records.

4.2.1.6 Consensus Mechanisms: Blockchain operates on a consensus mechanism where validators collectively approve changes to the ledger. This ensures the integrity of the data and prevents malicious actors from manipulating the system. It is a secure ecosystem where everyone agrees on the validity of passport records, minimizing the risk of fraudulent activities[5].

4.3 Glimpse into the Future The integration of blockchain technology holds immense potential for the future of passport systems:

4.3.1 Visa applications are streamlined: The implementation of biometric verification and instant travel history checks. These measures expedite the visa approval process by

 https://doi.org/10.52202/074664-0023

swiftly authenticating applicants' identities and reviewing their travel records. As a result, administrative overhead is reduced, and efficiency is enhanced, benefiting both applicants and visa processing authorities. The integration of such technologies not only accelerates the approval process but also ensures a more thorough and accurate assessment of applicants' backgrounds, contributing to improved border security measures[6].

4.3.2 Border crossings are frictionless: Secure digital passports eliminate the need for cumbersome paper checks and long queues at border crossings. Authorities can quickly authenticate an individual's identity and travel history, facilitating smooth and efficient border crossings[6].

4.3.3 Data breaches are mitigated: The decentralized nature of blockchain minimizes the risk of large-scale data breaches, ensuring the security and privacy of individuals' passport information. Blockchain's cryptographic security measures make it extremely challenging for cybercriminals to compromise passport data, enhancing overall security in the passport system.

4.4 Tourism and Hospitality

4.4.1 New Tourism Management: A concept refers to contemporary approaches to managing tourism organizations, potentially involving the use of tools like the Balanced Scorecard.

4.4.2 Hospitality and Tourism Industry: The service sector encompasses diverse professions including hotel management, food and beverage service, event coordination, theme parks, travel, and tourism. These industries collectively generate billions of dollars in revenue, relying heavily on the discretionary income and leisure time of individuals. Their economic viability hinges on consumer spending habits and the availability of resources for leisure activities.

4.4.3 International Hotel Chains: Chains expanding their presence worldwide, although with varying degrees of growth and concentration across regions. For example, Latin America may experience concentrated growth in specific countries like Mexico and Brazil, while a tiny Gulf state like Dubai can witness a substantial increase in hotel supply within a short period.

4.4.4 Hospitality Industry: A diverse and complex sector encompassing various businesses such as lodging, foodservice, and entertainment. Despite assumptions of homogeneity and global trends, the industry is characterized by significant diversity, making it challenging to identify common issues and trends applicable to all operators universally.

4.4.5 Location Selection: A critical strategic decision influenced by factors such as demand, site accessibility, and operational suitability. The location choice differs between accommodation and foodservice sectors, with considerations varying based on consumer behavior and market dynamics.

Blockchain technology has the potential to revolutionize the modern passport system by introducing unprecedented levels of security, transparency, and efficiency. While challenges such as regulatory acceptance and interoperability remain, the benefits of blockchain in enhancing trust and revolutionizing international travel are undeniable. The future of passport systems is undoubtedly intertwined with this transformative technology, promising a more secure, transparent, and efficient travel experience for individuals worldwide.

https://doi.org/10.52202/074664-0023

5. Proposed Blockchain-Based Passport System Model

5.1 System Architecture:

The proposed blockchain-based passport system as a well-structured building, with two main components forming its foundation:

5.1.1 Main Chain (Root Chain):

There will be a main chain as the central pillar of the system, where all core passport data resides. It holds vital information like identity, nationality, and details about when the passport was issued. Each piece of data on this chain is locked down tight with cryptographic hashing, creating a unique and virtually unbreakable seal[8].

5.1.2 Sub-Blocks (Sub Chain):

These sub-blocks are like the building blocks surrounding the main chain. They store specific travel details, such as the stamps the individual gets when crossing the borders. By keeping this information separate from the main chain, we can manage updates more efficiently without compromising the integrity of the core passport data[8].

5.2 Security Features:

5.2.1 Cryptography:

Cryptography refers to the practice and study of techniques for secure communication in the presence of third parties. It involves the use of mathematical algorithms to encode data in such a way that it remains confidential and secure from unauthorized access or manipulation. In essence, cryptography acts as a shield, protecting sensitive information from being intercepted or deciphered by adversaries. Its applications span across various fields, including data security, authentication, and privacy protection, playing a crucial role in safeguarding digital communication and transactions in modern systems. Cryptography as the locks on the doors of the digital passport. It ensures that the data remains confidential, untampered with, and authentic. With public-key cryptography, who are in control of the private keys, which act as the personal set of master keys for accessing and sharing the passport information securely[7].

5.2.2 Digital Signatures:

Digital signatures serve as cryptographic assurances of the authenticity and integrity of digital data within a system. They are akin to traditional seals of approval, but in the digital realm, using mathematical algorithms to ensure that the data has not been tampered with and can be traced back to its rightful sender. Essentially, each piece of data carries its own unique seal, providing a means to verify its origin and verify its contents, thus enhancing security and trust in the system's communications and transactions. They confirm that any changes made in the passport information are legitimate and authorized. Only those with valid private keys can create these digital signatures, ensuring that any modifications to those data are traceable and accountable[5].

5.2.3 Consensus Mechanisms:

The consensus mechanism in blockchain is the process through which nodes within a decentralized network reach agreement on the validity of transactions and changes to the distributed ledger. It ensures the integrity of the blockchain by ensuring unanimous consensus among network participants, thus preventing instances of double-spending and maintaining the accuracy of the technology's current state. By using a method like Proof of Stake (PoS), the system ensures that everyone involved has a say in validating transactions and maintaining the integrity of the blockchain ledger. This distributed approach builds trust and eliminates the risk of a single point of failure[7].

5.3 System Functionalities:

https://doi.org/10.52202/074664-0023

5.3.1 Passport Issuance:

Approval for a passport by the government results in the secure addition of information to the main chain. This creates a unique digital identity within the system, akin to having a digital passport stamped with approval.

5.3.2 Visa Integration:

On the issuing of visa to travel to certain places, that information can be seamlessly linked to the passport data on the main chain. This makes the visa issuance and verification processes secure and transparent, ensuring smooth travels wherever a person wants to do so.

5.3.3 Border Crossing Management:

Every time crossing a border, entry and exit stamps are recorded in sub-blocks linked to the passport entry on the main chain. This creates a chronological record of the travel history, helping border officials verifying the movements with ease.

5.3.4 Data Access Control:

The gatekeeper in this system acts as a centralized authority responsible for regulating access to data, utilizing permissioned sharing mechanisms to determine which entities are authorized to view it. Analogously, it operates akin to border officials who grant or deny entry based on validation of passports. Just as a digital bouncer oversees access to a venue, this gatekeeper ensures that only authenticated and authorized parties can access sensitive information, enhancing security and control within the system[6].

5.4 Benefits of the Proposed Model:

5.4.1 Enhanced Security:

With robust security measures like the Keccak 256 cryptographic algorithm and digital signatures using SHA-256, data is fortified against unauthorized access, manipulation, or forgery. These technologies provide a strong shield, ensuring that digital identity and sensitive information remain protected from malicious actors. By employing such techniques, the system enhances its security posture, fostering trust and confidence among users regarding the integrity and authenticity of their digital identities and transactions[8].

5.4.2 Improved Efficiency:

Through the automation of processes and optimization of data management, the system facilitates expedited passport issuance, visa applications, and border crossing procedures. This automation leads to increased efficiency, reducing wait times and administrative burdens for both governments and travelers alike. By streamlining these operations, the system saves valuable time and resources while enhancing the overall experience of individuals navigating through immigration processes.

5.4.3 Increased Transparency:

All stakeholders have access to verifiable and immutable travel records, promoting trust and accountability within the system. Governments, travelers, and authorized institutions alike can rely on the transparency of the blockchain to ensure the integrity of passport data.

5.4.4 Reduced Costs:

Through the elimination of physical passports and simplification of administrative processes, the system holds promise in reducing costs for both governments and individuals. By transitioning to digital identification methods, expenses associated with passport production, distribution, and maintenance can be minimized. Additionally, streamlined administrative processes result in operational savings for government agencies. Ultimately, the accessibility of travel increases as financial barriers are lowered, potentially democratizing the experience for a broader range of individuals.

5.4.5 Enhanced User Experience:
Travelers experience the advantages of a convenient and secure system that provides streamlined procedures and empowers them with greater control over their data. This results in a travel experience characterized by efficiency, reduced hassle, and enhanced peace of mind. With simplified processes and robust security measures in place, travelers can navigate through checkpoints and transactions seamlessly, contributing to an overall more enjoyable journey. Moreover, the sense of control over their personal information fosters trust and confidence in the system, further enhancing the travel experience.

5.5 Challenges and Limitations:
5.5.1 Scalability Concerns:
The proposed blockchain-based passport system may encounter scalability challenges as it scales to accommodate a larger user base and increased transaction volume. Blockchain networks typically operate with finite processing capacities and face limitations in throughput and transaction speed. As the number of passport issuance, visa integrations, and border crossing records grows, the system may experience congestion, delays, or increased transaction fees. Scaling solutions such as sharding or layer-2 protocols could mitigate these issues, but implementing and testing these solutions effectively require careful planning and investment in infrastructure.
5.5.2 Regulatory Compliance:
Integrating the blockchain-based passport system with existing regulatory frameworks and international agreements presents a complex regulatory challenge. Governments and regulatory bodies may have divergent requirements regarding data privacy, security standards, and identity verification protocols. Achieving compliance with diverse legal and diplomatic requirements necessitates extensive coordination, standardization efforts, and potentially renegotiating international treaties. Ensuring the system adheres to data protection laws such as GDPR (General Data Protection Regulation) while facilitating seamless cross-border travel poses a significant regulatory hurdle that must be addressed through robust governance mechanisms and stakeholder engagement.
5.5.3 User Adoption and Education:
Transitioning from traditional paper passports to a blockchain-based system requires overcoming barriers to user adoption and providing comprehensive education and support to stakeholders. Resistance to change, particularly among individuals accustomed to traditional forms of identification, may impede the widespread adoption of the new system. Addressing concerns related to data privacy, security, and usability through targeted awareness campaigns, user training programs, and user-friendly interfaces is essential. Additionally, ensuring inclusivity and accessibility for individuals with limited technological literacy or access to digital infrastructure requires proactive measures to bridge the digital divide and promote equitable participation in the digital passport ecosystem.

6. Future scope

6.1 Improvements in Security and Privacy:
Blockchain technology's immutability and encryption features significantly enhance the security of passport data, effectively mitigating the risks associated with identity theft and fraudulent passport issuance. By storing passport information on a blockchain, data

https://doi.org/10.52202/074664-0023

integrity is maintained, making it nearly impossible for unauthorized parties to alter or manipulate records without detection.

Additionally, smart contracts, which are self-executing contracts with the terms of the agreement directly written into code, can further bolster security and privacy protections. Through the use of smart contracts, privacy measures can be enforced while facilitating safe and transparent transactions related to passport issuance, verification, or other processes. These contracts automate and validate the execution of agreements, ensuring that sensitive information remains confidential and transactions occur only between authorized parties, thereby enhancing security and trust in the system[8].

6.2 Straightforward Identity Verification:

Blockchain-based passport systems have the potential to streamline the identity verification process by offering a decentralized and impenetrable method. This can enable confirmed identification data to be shared effectively and securely across suitable parties, eliminating the need for additional verification steps[5].

6.3 Effective and Unchangeable Records:

Blockchain technology can provide a permanent and unchangeable record of passport-related transactions, confirming the validity of passport issuance, renewals, and other related processes.

6.4 Seamless and Interoperable Travel:

Blockchain technology revolutionizes identity verification processes at immigration checkpoints. By securely storing travelers' identity information on a blockchain, authorities can access necessary data swiftly and with high confidence, streamlining verification procedures. This efficiency leads to reduced processing times, allowing travelers to move through checkpoints seamlessly. Moreover, blockchain's interoperable nature enables different agencies or countries to access and authenticate identity data seamlessly, fostering smoother cross-border travel experiences for individuals while enhancing overall efficiency in immigration operations[2].

6.5 Individual Empowerment and User Control:

Individuals can have more control over their passport data, deciding when and how it's accessed and shared, thus enhancing personal privacy.

6.6 Reduced Fraud:

Blockchain's transparency and traceability are powerful tools in combating passport fraud, offering a means to easily verify the authenticity of passports and visa records. By storing such data on a blockchain, its tamper-proof nature ensures that records remain accurate and unaltered, bolstering the integrity of the information. This inherent security feature significantly reduces the risk of fraudulent activities related to passports and visas[3].

Moreover, the adoption of blockchain-based passport systems by more countries could lead to a standardized and widely accepted approach to identity verification and travel documentation. Such standardization enhances interoperability and simplifies processes across borders, promoting seamless travel experiences for individuals while simultaneously strengthening global security measures. As a result, blockchain technology holds the potential to revolutionize the way identity verification and travel documentation are managed, leading to greater efficiency, trust, and security in international travel.

6.7 Integration into Smart Cities and Smart Infrastructure:

As cities evolve into interconnected hubs of innovation, blockchain passports can seamlessly integrate with smart city infrastructure. From automated immigration checkpoints to intelligent transportation systems, these passports will play a pivotal role in enabling frictionless movement within smart urban environments.

6.8 Cross-Industry Applications:

Blockchain passports have applications beyond travel, such as healthcare, education, and finance. They streamline processes like patient registration in healthcare, academic credential authentication in education, and secure financial transactions. By leveraging blockchain's transparency and security, these passports offer efficient and secure solutions for identity verification and data management across diverse industries.

6.9 Environmental Sustainability:

Adoption of blockchain passports can contribute to environmental sustainability by reducing the reliance on paper-based documentation. Digital passports eliminate the need for physical document production and disposal, leading to significant reductions in paper waste and carbon emissions associated with traditional passport issuance processes.

6.10 Empowerment of Underserved Communities:

Blockchain passports have the potential to empower underserved communities by providing secure and verifiable identities. Through inclusive initiatives and partnerships, access to essential services such as healthcare, education, and financial inclusion can be extended to those traditionally marginalized or excluded from formal systems.

6.11 Streamlining cross-border data exchange:

Blockchain's secure and transparent framework has the potential to facilitate efficient and secure data exchange among countries regarding immigration and travel history. Through blockchain technology, nations can securely share and verify travel-related information without compromising data integrity or privacy. This streamlined exchange process not only enhances border security but also contributes to smoother international travel experiences for individuals[9].

6.12 Integration with loyalty programs:

Blockchain passports (B-passports) could be integrated with loyalty programs, providing personalized rewards and benefits to frequent travelers by leveraging their travel history stored on the blockchain. This linkage allows for seamless tracking and verification of travel activities, enabling loyalty programs to offer tailored incentives and perks to users. By utilizing blockchain technology, these rewards programs can ensure the integrity and security of user data while enhancing customer engagement and satisfaction. This integration not only incentivizes loyalty but also fosters a deeper connection between travelers and service providers[10].

7. Conclusion

The research paper explores the transformative potential of blockchain technology in redefining the traditional passport system. It proposes a blockchain-based passport system that could address issues such as fraud, inefficiency, and security breaches inherent in the current systems. This new system could revolutionize travel by providing unmatched security, efficiency, and convenience for travellers worldwide. It could also foster smart tourism, providing stakeholders with accurate, real-time information, and enabling governments to streamline immigration procedures. The future scope of this system includes improvements in security and privacy, seamless travel, reduced fraud, and integration into smart cities and infrastructure. In essence, the blockchain-based passport system could revolutionize travel, making it more accessible and promoting ethical and sustainable tourism.

https://doi.org/10.52202/074664-0023

References

1. Wright, C.: Bitcoin: a Peer-to-Peer electronic cash system. Social Science Research Network. (2008). https://doi.org/10.2139/ssrn.3440802.
2. Sarnacchiaro, P., Luongo, S., Sepe, F., Della Corte, V.: The role of blockchain technology in the tourism industry: analyzing the factors affecting its adoption. Quality and Quantity. (2024). https://doi.org/10.1007/s11135-024-01836-7.
3. Santos, L.L., Cardoso, L., Vila, N.A., Brea, J.A.F.: Sustainability Perceptions in Tourism and Hospitality: A Mixed-Method Bibliometric Approach. Sustainability. 12, 8852 (2020). https://doi.org/10.3390/su12218852.
4. Levis, D., Fontana, F., Ughetto, E.: A look into the future of blockchain technology. PloS One. 16, e0258995 (2021). https://doi.org/10.1371/journal.pone.0258995.
5. Sarnacchiaro, P., Luongo, S., Sepe, F., Della Corte, V.: The role of blockchain technology in the tourism industry: analyzing the factors affecting its adoption. Quality and Quantity. (2024). https://doi.org/10.1007/s11135-024-01836-7.
6. Sarhan, A.Y.: Innovative blockchain-based Agent digital passport solution. Procedia Computer Science. 225, 882–891 (2023). https://doi.org/10.1016/j.procs.2023.10.075.
7. Önder, I., Gunter, U.: Blockchain: Is it the future for the tourism and hospitality industry? Tourism Economics. 28, 291–299 (2020). https://doi.org/10.1177/1354816620961707.
8. Yang, Z., Yang, K., Lei, L., Zheng, K., Leung, V.C.M.: Blockchain-Based decentralized trust management in vehicular networks. IEEE Internet of Things Journal. 6, 1495–1505 (2019). https://doi.org/10.1109/jiot.2018.2836144.
9. Alzahrani, R.A., Herko, S.J., Easton, J.M.: Blockchain Application in Remote Condition Monitoring. IEEE. (2020). https://doi.org/10.1109/bigdata50022.2020.9377895.
10. Sönmeztürk, O., Ayav, T., Erten, Y.M.: Loyalty Program using Blockchain. IEEE. (2020). https://doi.org/10.1109/blockchain50366.2020.00074.

Comparing Technical Efficiency of Higher Education across Different States of India – DEA Application

Ms Harsha Sharma[1], Dr. Amit Srivastava[2]
[1]Research Scholar, Jaypee University of Information Technology, Waknaghat, Solan, HP, India
[2]Associate Professor, Jaypee University of Information Technology, Waknaghat, Solan, HP, India

Abstract.
The growth of an economy depends on the pillars of human resources and education. In a developing economy like India, approximately forty percent of the population is between the age of 18-25 years, hence it becomes even more important that the educational establishment is assessed in terms of its efficiency time and again. The training and education imparted to the higher education students in the field of innovation and intellectual growth prepare them for becoming a part of the working population. The national policies implemented for the upliftment of Indian higher education require a lot of human resource and therefore it is as important that the efficiency or inefficiency of those policies be analysed in terms of the qualitative efficiency of states. In this paper, we analyse the technical efficiency of states and union territories from 2016-2020 by employing Data Envelopment Analysis (DEA). Input oriented constant returns to scale model is used to find the efficiency of 34 DMUs using 2 inputs and 2 outputs, the study helps us to determine the efficiency of higher education systems in different states and union territories of India.

Keywords: Economics, Higher Education, Economics of Education, Human Resource, DEA

1. Introduction:

With the increasing pool of young population between the age of 18-25 years, the pressure on the Indian education system has been growing many folds. Today the number of graduates being enrolled per year is much more than the demand of graduates in the job market [60]. In such a scenario the focus falls on the quality of education being imparted by our states. Facing critical questions about its relevance and value, the higher education system stands at a crossroads, challenged to demonstrate its ability to prepare students for a rapidly evolving job market.

The governments and universities have been working tirelessly to make the higher education system more accountable through higher education regulatory institutions. Data collection agencies have been established to provide state wide and institution wise data on the public forum. There is an increasing demand of transparency and accountability in public education for which the government is working dedicatedly. Under-utilization of resources is a common problem when it comes to public education system. The definition of efficiency and optimality was given by Koopmans [44] and the measurement of efficiency was given by Debreu [28]. Data being released in the public forum helps us to understand whether the resources are being used effectively and efficiently by the Indian states. Data envelopment analysis developed by Charnes, Cooper and Rhodes [18] has been further modified and used extensively in the field of education [5][25]. The reasons for the popularity of this DEA specifically in the field of education are because it facilitates the researchers to develop models even in the absence of data on expenditures and finance on education. In the Indian context, there have been a lot of publications shedding light on the allocative and technical efficiency of the higher education system using DEA application [61] (Kaur, 2021).

In light of these considerations this paper seeks to assess the technical efficiency of higher education being imparted by the Indian states and Union territories. This paper offers a perceptive analysis of the varied efficiency scores achieved by individual states. For each state, we have an abundance of data represented by variables influencing efficiency scores derived from the formulations of Data Envelopment Analysis (DEA). Specifically, we include two inputs, the pupil-teacher ratio and the total number of colleges in each state, along with two outputs: total enrolment and the number of graduates across undergraduate, postgraduate, and PhD levels. It is noteworthy that enrolment can function both as an input and an output, depending on the context of educational institutions. The paper is structured into four sections; Section II covers the Literature Review, Section III addresses the methodology along

https://doi.org/10.52202/074664-0024

with data, results and discussion and Section IV provides conclusions along with policy implications.

2. Literature Review:

Measuring industrial efficiency is crucial for economic theorists and policymakers. This becomes exceedingly pertinent when seen in the context of assessing the competence of educational institutions. Prodigies like Adam Smith, Marshall, Marx have highlighted the correlation of economics with education, and in the process, emphasized the impact of effective educational outputs in economic growth and development [68]. Koopman defined technical efficiency [44] which was later built upon by Nobel Laureate Gérard Debreu who correlated it with the coefficient of resource utilization [28].

The term "Economics of Education" was coined by Schultz at the American Economic Association the 1960[63]. This opened the gates to several other economists like Michael James Farrell who attempted 1957 to provide a measure of productive efficiency using all inputs in the year 1957[31], and Abraham Charnes who proposed the use of a non-parametric approach for measuring the efficiency of decision-making units (DMUs) using the Charnes Efficiency score [17].

Following up with the idea, Charnes introduced data envelopment analysis (DEA), a robust system of evaluation for the relative efficiencies of multi-input multi-output production units. DEA aimed to identify best practices within comparable decision-making units, thereby developing an intrinsic correlation between them [17]. These models address the intricate structures of Decision-Making Units (DMUs), providing more insightful solutions for multi-criteria evaluations of homogeneous production systems.

Various parameters can serve as determinants for establishing the efficiency frontier, including but not limited to the number of employees, service quality, environmental safety, and fuel consumption. Educational accountability models have been explored by incorporating educational, statistical, and economic considerations. The role and potential improvements of quantitative models in assessing school and local education authority performance by integrating expenditure and socio-economic factors including both educational and economic effectiveness, recognizing inherent trade-offs through collaboration between quantitative models and qualitative assessments by bodies like HMI and educational advisers is imminent for educational performance assessment and accountability [40].

While DEA holds a key advantage in avoiding the imposition of an explicit functional form on the data, it becomes susceptible to data contamination, leading to potential warping of the calculated frontier [64]. In contrast, the econometric approach is adept at measuring empirical viewpoints but consequently imposes an overly restrictive approach.

The DEA models can be multifarious in nature. Over the years different ranking methods have been explored such as examining a cross-efficiency matrix and incorporating self and peer evaluations. Additionally, the super-efficiency method involves ranking through exclusion from the dual linear program, analyzing Pareto Frontier changes. Benchmarking ranks units chosen as valuable targets by potential prospects. Technical efficiency has been the subject of investigation across the board. Chakraborty evaluates technical efficiency across 40 school districts in Utah, employing both stochastic and non-stochastic estimation methods. Stochastic estimation reveals considerable variation in technical efficiency among districts, remaining consistent across distributional assumptions. The findings suggest that a majority of Utah's school districts exhibit technical efficiency, with mean scores of 85.8% and 89.7% under different parameters. The two-stage DEA model underscores the impact of socioeconomic and environmental factors on student success, irrespective of systematic variation among efficient and inefficient districts. Policy implications include recommendations for districts to enhance efficiency through effective management and tailored programs based on student demographics. Such examinations reveal essential parameters such as parental education which emerged as a pivotal factor influencing student performance. However, the study acknowledges limitations stemming from aggregated data, emphasizing the need for nuanced interpretations of district-level decisions impacting controllable inputs and outcomes [14]. Data Envelopment Analysis (DEA) has been employed to evaluate teaching efficiency in 2547 Economics graduates from UK Universities in 1993. Utilizing a methodology that decomposes individual efficiency into university and student components, the importance of the unit of analysis in DEA has been highlighted. Comparing institution-level and department-level DEA results, findings indicate that individual-level analysis provides insights into whether student efforts or institutional efficiency pose constraints on increased efficiency, emphasizing the significance of the chosen analytical strategy [41]. Global surveys have revealed divergences and lacunae in efficiency measurement in education. Efficiency scores for Canadian universities, obtained through stochastic and non-stochastic methods, exhibit substantial divergence in outcomes and rankings. However, an in-depth analysis of rankings reveals consistent relative positions across various efficiency rankings [49]. While recognizing the constraint of limited space and potential oversight, it is acknowledged that several crucial ancillaries in DEA, such as modeling integer variables, addressing congestion issues, managing missing data, allocating fixed inputs among Decision Making Units (DMUs), and understanding connections between DEA and general multiple criteria decision models warrant further study [25]. Subsequent research should address specific challenges in applying the DEA model to evaluate university performance, considering factors such as extended and fluctuating lead times between inputs and outputs. Moreover, there is a notable gap in studies on university performance measurement using DEA within the Asian region, making it intriguing to conduct comparative analyses across multiple Asian countries [45].

Measuring the efficiency of Spanish universities by incorporating knowledge transfer activities into their objectives employed through cluster analysis reveals distinct university behavioral pathways influenced by infrastructures, regional characteristics, and the adoption of entrepreneurial mindsets [10]. From an Italian perspective, utilizing Monte Carlo methods for estimating models with challenging likelihood functions has significantly expanded opportunities in applied economics. Estimating models with institution-specific parameters offers valuable insights into the sources of cost differentials across institutions, highlighting areas for improvement. In such an attempt to study Italian education institutions, notable variations in the cost of providing education to nonscience undergraduates and conducting research were uncovered, with some institutions appearing as outliers with higher costs, prompting the need for further qualitative investigation [1].Evaluation of the efficiency of post-graduate programs, and analysis of the NDEA models, considering both qualitative and quantitative variables, are proposed as an alternative to DEA models in education efficiency assessments. The NDEA model as an alternative is a valuable tool in educational evaluations, contributing to the ongoing discourse on efficiency assessment methodologies as it incorporates both productivity and quality indices, and considers internal program structures, treating each Decision-Making Unit (DMU) as a network of processes, taking into account the number of faculty as inputs, dissertations and theses as intermediate products, and scientific production as final outputs [51]. Over the past years, Data Envelopment Analysis (DEA) has experienced significant growth, particularly in the realm of network DEA (NDEA) models. Conventional methods and performance metrics, while effective in various scenarios, fall short of capturing the growing complexity of institutions. Traditional financial ratio analysis lacks an impartial combination of evaluation. Moreover, these models often treat decision-making units (DMUs) as black boxes, neglecting in-depth scrutiny of inputs and outputs. To overcome such shortcomings, evaluation methods are being modified to incorporate grey areas through novel measures like SBM DEA [48].

In light of technological developments, there have been initiatives to overcome the apparent limitations of the DEA models [77].The limitation of the deterministic DEA analysis lies in its challenging handling of extensive datasets due to time constraints, requiring recalculations for each new DMU added [58].To adjust to the needs of evolving DMUs and assessments required in measuring institutional efficiencies, Automated machine learning (Auto-ML), particularly utilizing meta-heuristics optimization like Genetic Programming, emerges as a promising solution to efficiently handle complex and extensive datasets, optimizing hyper-parameters for enhanced algorithm efficiency [58]. Among recent takes on the issue the significance of efficiency indicators in decision-making for production units, focusing on Data Envelopment Analysis (DEA) and its evolution, particularly the development of slack-based measure-network DEA (SBM-NDEA) have been considered [4]. There is a void about the efficiency analysis of the higher education system in the Indian states in light of the growing demands of the job market and the increasing enrolment numbers in the educational institutions. The idea should not be only to focus on the enrolment numbers but to make the passing out students ready for the job market as per the requirements of the hiring managers. This paper analysis the efficiency of the higher education system in the Indian states, using two inputs and two outputs.

3. Methodology:

There are various methods available for examining the efficiency of education systems like free disposal hull (FDH), regression analysis, order m- frontier method,

https://doi.org/10.52202/074664-0024

stochastic frontier analysis (SFA), data envelopment analysis (DEA) etc. Of these techniques DEA is the most widely used where efficiency analysis of public sector institutions is needed. DEA is preferred by most researchers in the field because of its ability to handle multiple inputs and outputs simultaneously without any mathematical formulations. DEA represents a technique of linear programming that empirically measures the comparative efficiency of multiple entities or DMUs that are similar. The DMU, responsible for transforming inputs into outputs, is considered to be a homogenous entity. DEA examines each DMU individually by solving a linear programming model (LP) with the objective of maximizing the efficiency (θ_o) at which the unit converts various inputs into outputs. To conduct a DEA study, it is necessary to have a matrix consisting of the inputs, outputs, and supplementary elements of the DMU sample. Once the DEA model has been formulated based on specific characteristics like metrics and orientation, the matrix is integrated into the model for solution, thereby yielding relative efficiency scores and operational benchmarks for each DMU as the primary outcomes.

General output maximization CCR DEA model can be represented as follows.

$$\max z = \sum_{j=1}^{J} v_{jm} y_{jm}$$

subject to

$$\sum_{i=1}^{I} u_{im} x_{im} = 1$$

$$\sum_{j=1}^{J} v_{jm} y_{jm} - \sum_{i=1}^{I} u_{im} x_{im} \le 0; n = 1, 2, K, N$$

$$v_{jm}, u_{im} \ge \varepsilon; i = 1, 2, I; \quad j = 1, 2, K, J$$

This program can be represented in matrix form as shown below

$$\max z = \underset{m}{VT} Y_m$$

Subject to

$$\underset{m}{UT} X_m = 1$$

$$\underset{m}{VT} Y - \underset{m}{UT} X \le 0$$

$$\underset{m}{VT} - \underset{m}{UT} > \varepsilon$$

(X - matrix of inputs, Y - matrix of outputs)

General input minimization CCR DEA model is represented as follows

$$\min z' = \sum_{i=1}^{I} u'_{im} x_{im}$$

subject to

$$\sum_{j=1}^{J} v'_{jm} y_{jm} = 1$$

$$\sum_{j=1}^{J} v'_{jm} y_{jn} - \sum_{i=1}^{I} u'_{im} x_{in} \le 0; \; n = 1, 2, K, N$$

$$v'_{jm}, u'_{im} \ge \varepsilon; \; i = 1, 2, K, I; \; j = 1, 2, K, J$$

The program can be represented in matrix form as shown below

$$\min z' = \underset{m}{U'T} X_m$$

Subject to

$$V'T_m Y_m = 1$$

$$V'T_m Y - U'T_m X \leq 0$$

$$V'T_m - U'T_m > \varepsilon$$

[57]

3.1. Data

Data was obtained from All India Survey of Higher Education (AISHE). AISHE collects state-wise data categorised into various variables like faculty, student enrolment, programs, examination results, infrastructure, and other parameters. The survey also calculates various indicators of educational development, such as Institution Density, Gross Enrolment Ratio, Pupil Teacher Ratio, and Gender Parity Index, based on the data collected. Together these account for a compounded annual growth rate of around 3.5% over 5 years from 2016 to 2020. In the year 2020, Bihar had the highest average enrolment per college. The total number of colleges has experienced a substantial growth of 1453 during the academic year 2020-21, resulting in a total count of 43,796 colleges, as compared to the previous year's count of 42,343 colleges in 2019-20. Among the various states, Uttar Pradesh, Maharashtra, Karnataka, Rajasthan, Tamil Nadu, Madhya Pradesh, Andhra Pradesh, Gujarat, Telangana, and Kerala are the top 10 states that boast the highest number of colleges. Furthermore, it is noteworthy to mention that the annual rate of increase in student enrolment has demonstrated a positive trend over the years. In particular, there has been a significant improvement in the increase of student enrolment in the academic year 2020-21, with a remarkable growth rate of 7.4%, surpassing the previous year's growth rate of 3% in 2019-20 and 2.7% in 2018-19. Moreover, when considering the overall increase in enrolment since the year 2014-15, the cumulative growth stands at an impressive 20.9%. In terms of student enrolment, the top 6 states that take the lead are Uttar Pradesh, Maharashtra, Tamil Nadu, Madhya Pradesh, Karnataka, and Rajasthan. These states collectively account for a significant portion, specifically 53.17%, of the total student enrolment. Such statistics highlight the prominence and influence these states have in the realm of higher education in the country [3].

This study uses 2 inputs and 2 outputs model. Three degrees have been included in the study – undergraduate, postgraduate and PhDs to cover a major part of higher education. Enrolment and out pass numbers have been kept as outputs whereas total number of colleges per state and pupil-teacher ratio has been considered as qualitative inputs. The total number of colleges includes private, government and government aided colleges. Variables such as pupil-teacher ratio are qualitative and hence can be described as efficiency-related. To make the efficiency judgment more comprehensive, the DEA approach has been used. The assessment of efficiency in the context of higher education has been a long-standing application of DEA. When it comes to specifying a DEA model, one must make a decision about the best combination of inputs and outputs to be included. This decision has been a topic of concern for quite some time, as the inclusion or exclusion of an input or an output can have an impact on DEA scores. Striking a balance between parsimony and information redundancy is the usual aim when dealing with DEA. On one hand, we strive for simple models that encompass all the relevant information within the system under investigation. However, we also worry about the exclusion of pertinent variables. On the other hand, we endeavour not to incorporate irrelevant variables into the model to avoid the problem of over-fitting. The literature in this realm agrees to treat human and financial resources as inputs, quantifying them, for example, by considering the number of academic personnel, expenditure per student, and similar factors. In this context, universities are recognized as multiproduct organizations that employ inputs collectively to generate outputs. Regarding the output aspect, the inclination is towards the number of graduates (in terms of teaching).

In this paper, we have opted for the number of total enrolments and the number of pass-outs as inputs, as the number of graduates is linked to the effectiveness of the teaching endeavour, with qualitative assessment also being possible through the ratio of students to teachers, hence incorporating it as an input. The DEA scores have been derived by employing input-oriented constant returns to scale (CRS) models.

3.2. Results and Discussion

We conducted the tests for the years under discussion, that is, 2016 to 2020. Firstly, the model was run for 2 inputs and 2 outputs

- O_1 - UG + PG+ PhD Enrolment
- O_2 - UG + PG+ PhD Out pass
- I_1 - Total No. of Colleges per state

https://doi.org/10.52202/074664-0024

- I_2–PTR

(All variables to be considered for 2016 - 2020 respectively)

Based on the inputs and outputs, the technical efficiency of higher education for states and UTs is estimated by DEA and an input-based approach and constant returns to scale are adopted.

Data shows that Delhi maintained its score of being 100% efficient state for all five years. Let us examine the reasons for such a result. The report showed that the number of admissions in colleges of Delhi were very regulated with very less increase in the year-on-year enrolment, whereas the out-pass numbers have been increasing consistently. It was seen that the number of colleges in the UT also did not increase much, the colleges from 2016 to 2020 have been moving in the range of 178-180, and the pupil teacher ratio also did not increase much with the increasing enrolments. This implies that one variable was increased to maintain the results, which is the number of faculty members. Assuming that the faculty numbers did not increase, then the pupil teacher ratio would have shown an increasing trend or the number of college campuses would have been increased to divide the increasing number of enrolments. Delhi has been known to be the most favourable destination for admission into undergraduate courses. Hence the consistent score of 1 is justified.

Now let us compare the numbers of Tamil Nadu that showed 100% efficiency in the DEA results. On analysing the variables of Tamil Nadu with other southern states it was found that other states like Andhra Pradesh and Kerala had a greater number of colleges but the enrolment and out-pass were the highest in Tamil Nadu. Almost all southern states had maintained their pupil teacher ratio, meaning the number of faculty was not a problem here rather a strong point for these states. But the enrolment in Kerala, Andhra Pradesh, Karnataka and Telangana were almost half of the number of students being enrolled in the colleges of Tamil Nadu. So, the high enrolment, high out-pass numbers and optimum pupil teacher ratio contributed to the efficiency score Tamil Nadu which was 1. Bihar had 100% efficiency in the year 2020. The average efficiency of the state has been between the range of 0.8 – 1.0. Bihar has had the highest pupil teacher ratio across all years. The average pupil teacher ratio in Bihar has been between 59 to 70 as compared to national average on 19-22. Such high pupil teacher ratio could be seen in Jharkhand but in this case the number of colleges is very less. We cannot compare Bihar with another neighbouring state UP, as it is the larger one both in terms of demography and population. The number of colleges in Bihar have risen but so has the number of enrolments. The PTR is also consistent in a particular range, hence we can understand that the pressure to deliver results falls on the faculty which is not increasing in the same proportion as the enrolment of students primarily in the undergraduate courses.

https://doi.org/10.52202/074664-0024

Table 3.1 – Showing the score (from 0 to 1) and rank of states from the year 2016 to 2020.

		2016		2017		2018		2019		2020	
No.	**DMU**	**Score**	**Rank**	**Score**	**Rank**	**Score**	**Rank**	**Score**	**Rank**	**Score**	**Rank**
1	Andaman & Nicobar Islands	0.2593	26	0.2431	28	0.2278	29	0.187	30	0.197	30
2	Andhra Pradesh	0.5605	12	0.4851	18	0.5045	19	0.5226	16	0.6131	11
3	Arunachal Pradesh	0.2591	27	0.2618	26	0.2127	30	0.2186	28	0.2231	28
4	Assam	0.5953	10	0.6363	10	0.6133	10	0.5761	10	0.5822	14
5	Bihar	0.891	4	0.8367	5	0.818	4	0.8514	4	1	1
6	Chandigarh	0.6841	8	0.6909	7	0.6414	9	0.6434	8	0.6538	10
7	Chhattisgarh	0.3553	22	0.3735	22	0.3735	22	0.3706	23	0.4004	21
8	Dadra & Nagar Haveli & Daman &Diu	0.0743	34	0.0693	34	0.0728	34	0.0758	34	0.0857	34
9	Delhi	1	1	1	1	1	1	1	1	1	1
10	Goa	0.1373	32	0.145	32	0.1497	32	0.1461	31	0.1477	32
11	Gujarat	0.4932	19	0.4585	20	0.4561	20	0.4733	20	0.5259	17
12	Haryana	0.5103	18	0.5706	15	0.5398	16	0.5307	15	0.5883	13
13	Himachal Pradesh	0.3197	23	0.3524	23	0.3294	24	0.322	24	0.3207	23
14	Jammu and Kashmir	0.3704	21	0.4149	21	0.4263	21	0.3946	22	0.3702	22
15	Jharkhand	0.5426	15	0.5737	13	0.5372	17	0.5618	13	0.5653	15
16	Karnataka	0.7491	7	0.6777	8	0.7299	7	0.7194	7	0.7291	8
17	Kerala	0.5476	14	0.5878	12	0.5768	13	0.572	11	0.6825	9
18	Madhya Pradesh	0.5588	13	0.5905	11	0.6509	8	0.6301	9	0.7313	7
19	Maharashtra	0.8371	6	0.849	4	0.8116	5	0.8187	6	0.8072	6
20	Manipur	0.2409	29	0.2069	30	0.2376	27	0.2395	27	0.2549	26

https://doi.org/10.52202/074664-0024

		2016		2017		2018		2019		2020	
No.	**DMU**	**Score**	**Rank**	**Score**	**Rank**	**Score**	**Rank**	**Score**	**Rank**	**Score**	**Rank**
21	Meghalaya	0.2243	30	0.2361	29	0.2335	28	0.209	29	0.2137	29
22	Mizoram	0.1783	31	0.1585	31	0.1616	31	0.1441	32	0.1598	31
23	Nagaland	0.1156	33	0.1256	33	0.1139	33	0.1119	33	0.1126	33
24	Odisha	0.54	17	0.5714	14	0.5455	15	0.5317	14	0.5112	20
25	Puducherry	0.2461	28	0.2599	27	0.2491	26	0.2446	26	0.2657	25
26	Punjab	0.5403	16	0.5646	16	0.5497	14	0.5082	18	0.5225	18
27	Rajasthan	0.573	11	0.6729	9	0.6047	11	0.5708	12	0.5182	19
28	Sikkim	0.2893	24	0.2759	24	0.3614	23	0.418	21	0.2497	27
29	Tamil Nadu	1	1	1	1	1	1	1	1	1	1
30	Telangana	0.4763	20	0.5107	17	0.5263	18	0.5033	19	0.6.35	12
31	Tripura	0.2749	25	0.2737	25	0.2598	25	0.5033	19	0.2736	24
32	Uttar Pradesh	0.8858	5	0.7421	6	0.7583	6	0.8214	5	0.8588	5
33	Uttarakhand	0.6829	9	0.4591	19	0.6011	12	0.5215	17	0.5432	16
34	West Bengal	0.974	3	0.9437	3	0.9246	3	0.9394	3	0.9938	4

Fig. 3.1. (Source: Authors' calculation) Graph shows the ranking of states listed in alphabetically descending order, with different colours showing the ranks achieved in respective years.

https://doi.org/10.52202/074664-0024

4. Conclusion:

The allocation of substantial funds by the central government towards the provision of education as a public good is an area of focus. Both the central and state governments have prioritized higher education in previous plans. To evaluate national policies, significant volumes of data, in the form of official statistics, are gathered by organizations such as AISHE. It is imperative to analyse this data in a manner that is comprehensible to individuals. This study explores whether the resources are being utilized adequately or if the governments are placing strain on teaching and infrastructural resources in order to accommodate the increasing number of enrolments. The results show that there is huge pressure on the teaching group in some states where the enrolment numbers are very high. It is imperative that the governing bodies analyse the increasing gap between the number of students being enrolled and the faculty recruitment at play. A low pupil teacher ratio gives scope to the faculty to groom individual student so that they can become an asset to the economic system. We selected two inputs and two outputs as qualitative variables to compare the scores of five years from 2016 to 2020. This paper will be helpful to administration authorities in improving specific areas for efficiency.

Reference

1. Agasisti, T. and Johnes, G.: Efficiency, costs, rankings and heterogeneity: the case of US higher education. Studies in Higher Education. 40(1), 60-82 (2015).
2. Ahn, T., Charnes, A. and Cooper, W.W.: Efficiency characterizations in different DEA models. Socio-Economic Planning Sciences. 22(6), 253-257 (1988).
3. AISHE.: All India Survey on Higher Education 2020-21. New Delhi: Ministry of Education, Department of Higher Education (2020).
4. Alves, C.G.M.D.F. and Meza, L.A.: A review of network DEA models based on slacks-based measure: evolution of literature, applications, and further research direction. International Transactions in Operational Research. 30(6), 2729-2760 (2023).
5. Athanassopoulos, A.D. and Shale, E.: Assessing the comparative efficiency of higher education institutions in the UK by the means of data envelopment analysis. Education economics. 5(2), 117-134 (1997).
6. Barra, C. and Zotti, R.: Measuring efficiency in higher education: an empirical study using a bootstrapped data envelopment analysis. International Advances in Economic Research. 22, 11-33 (2016).
7. Bauer, P.W.: Recent developments in the econometric estimation of frontiers. Journal of econometrics. 46(1-2), 39-56 (1990).
8. Beasley, J.E.: Comparing university departments. Omega. 18(2), 171-183 (1990).
9. Beasley, J.E.: Determining teaching and research efficiencies. Journal of the operational research society. 46, 441-452 (1995).
10. Berbegal-Mirabent, J., Lafuente, E. and Solé, F.: The pursuit of knowledge transfer activities: An efficiency analysis of Spanish universities. Journal of Business Research. 66(10), 2051-2059 (2013).
11. Bessent, A.M. and Bessent, E.W.: Determining the comparative efficiency of schools through data envelopment analysis. Educational Administration Quarterly. 16(2), 57-75 (1980).
12. Bessent, A., Bessent, W., Elam, J. and Clark, T.: Efficiency frontier determination by constrained facet analysis. Operations Research. 36(5), 785-796 (1988).
13. Burney, N.A., Johnes, J., Al-Enezi, M. and Al-Musallam, M.: The efficiency of public schools: the case of Kuwait. Education Economics. 21(4), 360-379 (2013).
14. Chakraborty, K., Biswas, B. and Lewis, W.C.: Measurement of technical efficiency in public education: A stochastic and non-stochastic production function approach. Southern economic journal. 67(4), pp.889-905 (2001).
15. Charnes, A. and Cooper, W.W.: Auditing and accounting for program efficiency and management efficiency in not-for-profit entities. Accounting, Organizations and Society. 5(1), 87-107 (1980).
16. Charnes, A. and Cooper, W.W.: The non-Archimedean CCR ratio for efficiency analysis: A rejoinder to Boyd and Färe. European Journal of Operational Research. 15(3), 333-334 (1984).
17. Charnes, A., Cooper, W.W. and Rhodes, E.: Measuring the efficiency of decision-making units. European journal of operational research. 2(6), 429-444 (1978).
18. Charnes, A., Cooper, W.W. and Rhodes, E.: Evaluating program and managerial efficiency: an application of data envelopment analysis to program follow through. Management science. 27(6), 668-697 (1981).
19. Charnes, A., Cooper, W.W. and Thrall, R.M.: A structure for classifying and characterizing efficiency and inefficiency in data envelopment analysis. Journal of Productivity Analysis. 2, 197-237 (1991).
20. Charnes, A., Cooper, W.W., Golany, B., Seiford, L. and Stutz, J.: Foundations of data envelopment analysis for Pareto-Koopmans efficient empirical production functions. Journal of econometrics. 30(1-2), 91-107 (1985).
21. Charnes, A., Cooper, W.W., Seiford, L. and Stutz, J.: A multiplicative model for efficiency analysis. Socio-Economic Planning Sciences. 16(5), 223-224 (1982).

 https://doi.org/10.52202/074664-0024

22. Charnes, A., Cooper, W.W., Niehaus, R.J. and Schinnar, A.P.: Measuring efficiency and trade-offs in attainment of EEO goals. In Work, Organizations, and Technological Change. 241-256 (1982). Boston, MA: Springer US.
23. Charnes, A., Rousseau, J.J. and Semple, J.H.: Sensitivity and stability of efficiency classifications in data envelopment analysis. Journal of productivity analysis. 7, 5-18 (1996).
24. Conroy, S.J. and Arguea, N.M.: An estimation of technical efficiency for Florida public elementary schools. Economics of Education Review. 27(6), 655-663 (2008).
25. Cook, W.D. and Seiford, L.M.: Data envelopment analysis (DEA)–Thirty years on. European journal of operational research. 192(1), 1-17 (2009).
26. Cook, W.D. and Zhu, J.: DEA cross efficiency. Data envelopment analysis: A handbook of models and methods. 23-43 (2015).
27. Cubbin, J. and Tzanidakis, G.: Regression versus data envelopment analysis for efficiency measurement: an application to the England and Wales regulated water industry. Utilities policy. 7(2), 75-85 (1998).
28. Debreu, G.: The coefficient of resource utilization. Econometrica: Journal of the Econometric Society. 273-292 (1951).
29. Debreu, G.: The Coefficient of Resource Utilization. Econometrica. 19(3), 273-292 (1951).
30. Diamond, A.M. and Medewitz, J.N.: Use of data envelopment analysis in an evaluation of the efficiency of the DEEP program for economic education. The Journal of Economic Education. 21(3), 337-354 (1990).
31. Farrell, M.J.: The measurement of productive efficiency. Journal of the royal statistical society: series A (General). 120(3), 253-281 (1957).
32. Firsova, A.A. and Chernyshova, G.Y.: Mathematical models for evaluation of the higher education system functions with DEA approach. 19(3), 351-362 (2019).
33. Friedman, L. and Sinuany-Stern, Z.: Scaling units via the canonical correlation analysis in the DEA context. European Journal of Operational Research, 100(3), 629-637 (1997).
34. Gershberg, A.I. and Schuermann, T.: The efficiency–equity trade-off of schooling outcomes: public education expenditures and welfare in Mexico. Economics of Education Review. 20(1), 27-40 (2001).
35. Giménez, V.M. and Martínez, J.L.: Cost efficiency in the university: A departmental evaluation model. Economics of Education Review. 25(5), 543-553 (2006).
36. Hanushek, E.A.: Some findings from an independent investigation of the Tennessee STAR experiment and from other investigations of class size effects. Educational Evaluation and Policy Analysis. 21(2), 143-163 (1999).
37. Hollingsworth, B. and Parkin, D.: Efficiency and productivity change in the English National Health Service: can data envelopment analysis provide a robust and useful measure? Journal of Health Services Research & Policy. 8(4), 230-236 (2003).
38. Hoxby, C.M.: How teachers' unions affect education production. The Quarterly Journal of Economics. 111(3), 671-718 (1996).
39. Jamasb, T. and Pollitt, M.G.: Electricity sector liberalisation and innovation: An analysis of the UK's patenting activities. Research Policy. 40(2), 309-324 (2011).
40. Jesson, D., Mayston, D. and Smith, P.: Performance assessment in the education sector: educational and economic perspectives. oxford Review of Education. 13(3), 249-266 (1987).
41. Johnes, J.: Measuring teaching efficiency in higher education: An application of data envelopment analysis to economics graduates from UK Universities 1993. European journal of operational research. 174(1), 443-456 (2006).
42. Johnes, J. and Li, Y.U.: Measuring the research performance of Chinese higher education institutions using data envelopment analysis. China economic review. 19(4), 679-696 (2008).
43. Khodabakhshi, M. and Asgharian, M.: An input relaxation measure of efficiency in stochastic data envelopment analysis. Applied Mathematical Modelling. 33(4), 2010-2023 (2009).

44. Koopmans, T.C.: An analysis of production as an efficient combination of activities. Activity analysis of production and allocation (1951).
45. Kuah, C.T. and Wong, K.Y.: Efficiency assessment of universities through data envelopment analysis. Procedia computer science. 3, 499-506 (2011).
46. Liu, J.S., Lu, L.Y., Lu, W.M. and Lin, B.J.: A survey of DEA applications. Omega. 41(5), 893-902 (2013).
47. Loganathan, M. and Subrahmanya, M.B.: Efficiency of entrepreneurial universities in India: A data envelopment analysis. Journal of the Knowledge Economy. 14(2), 1120-1144 (2023).
48. Mahmoudabadi, M.Z. and Emrouznejad, A.: Comprehensive performance evaluation of banking branches: A three-stage slacks-based measure (SBM) data envelopment analysis. International Review of Economics & Finance. 64, 359-376 (2019).
49. McMillan, M.L. and Chan, W.H.: University efficiency: A comparison and consolidation of results from stochastic and non-stochastic methods. Education economics. 14(1), 1-30 (2006).
50. McMillan, M.L. and Datta, D.: The relative efficiencies of Canadian universities: a DEA perspective. Canadian Public Policy/Analyse de Politiques. 485-511 (1998).
51. Meza, L.A., Mello, J.C.C.B.S.D., Gomes Júnior, S.F. and Moreno, P.: Evaluation of post-graduate programs using a network data envelopment analysis model. Dyna. 85(204), 83-90 (2018).
52. Miller, W.L.: The economics of education in English classical economics. Southern Economic Journal, pp.294-309 (1966).
53. Miller, W.L.: Education as a source of economic growth. Journal of Economic Issues. 1(4), 280-296 (1967).
54. Millimet, D.L. and Collier, T.: Efficiency in public schools: Does competition matter? Journal of Econometrics. 145(1-2), 134-157 (2008).
55. Mizala, A., Romaguera, P. and Farren, D.: The technical efficiency of schools in Chile. Applied Economics, 34(12), 1533-1552 (2002).
56. Norman, M. and Stoker, B.: Data envelopment analysis: the assessment of performance. John Wiley & Sons, Inc. (1991)
57. Ramanathan, R.: An introduction to data envelopment analysis: a tool for performance measurement. Sage, (2003).
58. Mohamad Razi, N.F., Baharun, N. and Omar, N.: Machine learning predictive model of academic achievement efficiency based on data envelopment analysis. Mathematical Sciences and Informatics Journal (MIJ), 3(1), 86-99 (2022).
59. Ruggiero, J. and Vitaliano, D.F.: Assessing the efficiency of public schools using data envelopment analysis and frontier regression. Contemporary Economic Policy. 17(3), 321-331 (1999).
60. Sadlak, J.: Efficiency in higher education—concepts and problems. Higher Education. 7(2), 213-220 (1978).
61. Sahney, S. and Thakkar, J.: A comparative assessment of the performance of select higher education institutes in India. Quality Assurance in Education. 24(2), 278-302 (2016).
62. Salas-Velasco, M.: The technical efficiency performance of the higher education systems based on data envelopment analysis with an illustration for the Spanish case. Educational Research for Policy and Practice. 19(2), 159-180 (2020).
63. Schultz, T.W.: Investment in human capital. The American economic review. 1-17 (1961).
64. Seiford, L.M. and Thrall, R.M.: Recent developments in DEA: the mathematical programming approach to frontier analysis. Journal of econometrics. 46(1-2), 7-38 (1990).
65. Sengupta, J.K.: Measuring economic efficiency with stochastic input-output data. International journal of systems science. 20(2), 203-213 (1989).
66. Sexton, T.R., Silkman, R.H. and Hogan, A.J.: Data envelopment analysis: Critique and extensions. New directions for program evaluation. 32, 73-105 (1986).

 https://doi.org/10.52202/074664-0024

67. Sinuany-Stern, Z., Mehrez, A. and Barboy, A.: Academic departments efficiency via DEA. Computers & Operations Research. 21(5), 543-556 (1994).
68. Adam Smith, A.: An inquiry into the nature and causes of the wealth of nations. Cannan Edition (New York: The Modern Library, 1937), 567-568 (1776).
69. Smith, P. and Mayston, D.: Measuring efficiency in the public sector. Omega. 15(3), 181-189 (1987).
70. Tavares, R.S., Angulo-Meza, L. and Sant'Anna, A.P.: A proposed multistage evaluation approach for Higher Education Institutions based on network Data envelopment analysis: A Brazilian experience. Evaluation and Program Planning. 89, 101984 (2021).
71. Tone, K. and Tsutsui, M.: Dynamic DEA: A slacks-based measure approach. Omega. 38(3-4), 145-156 (2010).
72. Tran, T.V., Pham, T.P., Nguyen, M.H., Do, L.T. and Pham, H.H.: Economic efficiency of higher education institutions in Vietnam between 2012 and 2016: a DEA analysis. Journal of Applied Research in Higher Education. 15(1), 199-212 (2023).
73. Veiderpass, A. and McKelvey, M.: Evaluating the performance of higher education institutions in Europe: a nonparametric efficiency analysis of 944 institutions. Applied Economics. 48(16), 1504-1514 (2016).
74. Viner, J.: Bentham and JS Mill: The utilitarian background. The American economic review. 39(2), 360-382 (1949).
75. Wolszczak-Derlacz, J.: An evaluation and explanation of (in) efficiency in higher education institutions in Europe and the US with the application of two-stage semi-parametric DEA. Research Policy. 46(9), 1595-1605 (2017).
76. Worthington, A.C. and Lee, B.L.: Efficiency, technology and productivity change in Australian universities, 1998–2003. Economics of education review, 27(3), 285-298 (2008).
77. Zhang, L., Du, X., Chiu, Y.H., Pang, Q. and Yu, Q.: Measuring industrial operational efficiency and factor analysis: A dynamic series-parallel recycling DEA model. Science of the Total Environment, 851, 158084 (2022).
78. Gao, Z.: Practical Knowledge Level Characteristics and Innovative Practice Achievement Transformation Mechanism of Vocational College Students. International Journal of Emerging Technologies in Learning (Online). 18(10), 232 (2023).

Evolution of Culinary Delights in Punjab: A Harmony of Tradition and Modernity

Sandeep Raheja[1], Kuldeep Kumar[2] and Rahul Garg[3]

[1]Assistant Professor, MMICT&BM (Hotel Management), Maharishi Markandeshwar (deemed to be) University

[2]Assistant Professor, UITHM, Chandigarh University

[3]Research Scholar, MMICT&BM (Hotel Management), Maharishi Markandeshwar (deemed to be) University

Abstract

This comprehensive review paper immerses readers in the rich culinary narrative of Punjab, India, meticulously tracing the intricate interplay between traditional gastronomic practices and the dynamic influences of the modern era. Beyond the historical roots, spanning from ancient civilizations to the influence of Sikh traditions, the exploration delves into the nuances of key ingredients like grains, dairy, and aromatic spices that have sculpted Punjab's unmistakable culinary identity. The examination extends to traditional cooking methods, offering a closer look at the iconic tandoori and dum cooking techniques, juxtaposed against the transformative force of globalization. This globalization, characterized by the infusion of international elements into local recipes, sparks the emergence of a vibrant fusion cuisine that redefines Punjab's culinary landscape. Emphasizing sustainability, the farm-to-table movement takes center stage, aligning traditional practices with contemporary values. Additionally, the application of molecular gastronomy techniques in urban settings adds a modern twist to age-old recipes. The burgeoning culinary tourism industry serves as a testament to Punjab's gastronomic diversity, attracting both locals and tourists to explore the region's innovative culinary offerings. In the face of challenges pertaining to authenticity and sustainability amid growing demand for diverse ingredients, this review elucidates how Punjab's culinary heritage adeptly navigates the delicate equilibrium between tradition and innovation in the ever-evolving global culinary panorama.

Keywords: Culinary Heritage, Punjab Cuisine, Sustainability, Globalization, Gastronomy

1. Introduction:

Punjab, the vibrant northwestern state of India, stands as a culinary haven with a rich tapestry of flavors that encapsulates its cultural diversity and historical legacy. This introduction embarks on a gastronomic journey, exploring the intricate fusion of tradition and modernity in Punjab's culinary landscape. Anchored in ancient civilizations and shaped by the Sikh traditions, Punjab's cuisine serves as a testament to the historical and cultural amalgamation that defines the region [1]. As a pivotal part of Indian culinary heritage, Punjab's gastronomy has evolved over centuries, navigating the influences of Mughal rulers and the transformative waves of globalization [2]. Key ingredients such as grains, dairy, and aromatic spices have become the pillars of Punjabi cuisine, contributing to its distinctive and robust flavors [3].

Traditional cooking methods, exemplified by the iconic tandoori and dum cooking techniques, carry the essence of Punjab's culinary traditions [4].

 https://doi.org/10.52202/074664-0025

However, in the contemporary era, the globalization of food markets has introduced a plethora of international influences, prompting chefs to infuse global elements into local recipes, fostering a dynamic fusion cuisine [5]. As Punjab grapples with the challenges of maintaining authenticity and sustainability amidst a rising demand for diverse ingredients, the culinary evolution is also marked by the emergence of a farm-to-table movement and the application of molecular gastronomy techniques [6]. This introduction sets the stage for a comprehensive exploration of how Punjab's culinary heritage navigates the delicate balance between preserving tradition and embracing innovation in a rapidly changing global culinary landscape.

2. Background of the study:

Punjab, historically renowned as the "Land of Five Rivers," boasts a culinary heritage as diverse and vibrant as its cultural tapestry. Rooted in ancient civilizations and enriched by subsequent historical influences, Punjab's cuisine stands as a testament to the region's rich cultural history. The historical legacy of Punjab's culinary traditions dates back to the ancient Indus Valley Civilization, where evidence of sophisticated culinary practices has been unearthed [7]. Subsequent centuries saw the region shaped by various rulers, including the Mauryas, Guptas, and Mughals, each leaving their mark on Punjab's gastronomic identity [8]. However, it was the advent of Sikhism in the 15th century that brought about a unique culinary influence, particularly through the institution of langar, where communal meals are served in Sikh Gurdwaras, promoting equality and community bonding [9].

Amidst this rich historical backdrop, Punjab's culinary landscape continues to evolve, influenced by globalization and changing dietary preferences. Traditional ingredients such as wheat, rice, dairy products, and aromatic spices remain central to Punjabi cuisine, reflecting the region's agrarian heritage [10]. Yet, the influx of international ingredients and culinary trends has led to the emergence of fusion cuisine, where traditional recipes are remained with a contemporary twist [11]. Against this backdrop, it becomes imperative to explore the nuanced interplay between tradition and innovation in Punjab's culinary scene, examining how the region's gastronomic heritage adapts to the demands of the modern era while preserving its cultural authenticity.

3. Gap of the Study:

The exploration of Punjab's culinary landscape has been a subject of in teaberry rest in various studies, delving into its historical roots, traditional ingredients, and cooking techniques. However, a noticeable gap exists in the literature regarding a comprehensive analysis of the recent culinary innovations in Punjab that navigate the delicate balance between tradition and modernity. While previous research has touched upon historical influences and traditional aspects, there is limited coverage of the contemporary fusion cuisine emerging from the integration of global elements into local recipes. Moreover, there is a dearth of studies that systematically investigate the challenges faced by Punjab's culinary heritage in terms of preserving authenticity and sustainability amidst the globalization-driven demand for diverse ingredients. The evolving trends in molecular gastronomy, the farm-to-table movement, and culinary tourism, although gaining prominence, lack in-depth scrutiny within the specific context of Punjab. This study aims to bridge these gaps by providing a nuanced understanding of how Punjab's culinary traditions have adapted to the dynamic culinary landscape, addressing the lacuna in current literature and contributing valuable insights to the broader discourse on the intersection of tradition and innovation in global culinary contexts.

https://doi.org/10.52202/074664-0025

4. Research Questions:

RQ.1 How have historical influences, particularly ancient civilizations and Sikh traditions, shaped the foundational aspects of Punjab's culinary traditions, such as key ingredients and cooking methods?

RQ.2 What is the impact of globalization on Punjabi cuisine, specifically in terms of the integration of international elements and the emergence of fusion cuisine?

RQ.3 What are the key characteristics and implications of recent culinary innovations in Punjab, including the farm-to-table movement, the application of molecular gastronomy, and the rise of culinary tourism?

5. Objectives of the study:

- To Investigate the Dynamics of Culinary Innovation in Punjab
- To Assess the Challenges and Opportunities for Punjab's Culinary Heritage

6. Research Methodology:

The research methodology involves an extensive review of existing literature on Punjab's culinary traditions, historical influences, and contemporary innovations. Primary data will be gathered through interviews and surveys with chefs, culinary experts, and industry stakeholders to provide qualitative insights into the challenges, opportunities, and innovations within Punjab's culinary landscape. Historical documents, culinary manuscripts, and traditional recipe books will be analyzed to understand the roots of traditional practices. Representative case studies of restaurants and culinary events will be examined to identify specific examples of culinary evolution. Surveys and questionnaires will be administered to consumers and chefs for quantitative data, complementing qualitative insights. Expert consultations with culinary specialists will offer additional perspectives. Data analysis will employ qualitative techniques to interpret information from interviews, surveys, and case studies, culminating in a synthesis of findings that explores the interplay between tradition and innovation in Punjab's culinary scene. The study will conclude with recommendations for sustaining and promoting Punjab's culinary heritage amidst evolving global culinary dynamics.

7. Literature Review:

7.1 Culinary Innovations in Punjab:

Farm-to-Table Movement:

The Farm-to-Table Movement has gained substantial traction in Punjab, reflecting a paradigm shift towards sustainable and locally-sourced produce. Traditionally an agrarian society, Punjab's embrace of this movement aligns with its deep-rooted connection to the land. Chefs and restaurants are increasingly prioritizing the use of fresh, locally-produced ingredients, establishing direct relationships with farmers to create a more direct and sustainable supply chain **[12]**. This not only ensures the freshness and quality of ingredients but also supports local farmers, contributing to the economic sustainability of the region.

The significance of the Farm-to-Table Movement lies in its impact on both the culinary and agricultural sectors. It revitalizes traditional farming practices, promotes biodiversity, and fosters a greater appreciation for seasonal produce.

https://doi.org/10.52202/074664-0025

This shift towards sustainability not only enhances the authenticity of Punjabi cuisine but also resonates with global trends emphasizing ethical and eco-conscious dining practices.

Molecular Gastronomy:

Punjab's culinary scene has witnessed a fascinating fusion of tradition and cutting-edge techniques with the introduction of Molecular Gastronomy. Traditionally known for its hearty and flavorful dishes, the application of scientific principles to cooking has brought a new dimension to Punjabi cuisine. Nitrogen-infused chaat, deconstructed sarson da saag, and innovative presentations of classic dishes are becoming increasingly common in upscale restaurants [13]. This innovative approach not only enhances the visual appeal of the dishes but also transforms the dining experience, captivating the senses in unexpected ways.

The integration of Molecular Gastronomy in Punjab reflects a desire to push the boundaries of traditional cooking while preserving the essence of local flavors. This culinary innovation serves as a creative expression for chefs, introducing an element of surprise and sophistication to familiar dishes. It has garnered attention not only locally but also on the global culinary stage, positioning Punjab as a hub for culinary experimentation and innovation.

Culinary Tourism:

Culinary Tourism has emerged as a dynamic force in Punjab, drawing food enthusiasts and tourists eager to explore the region's diverse gastronomic offerings. Punjab's culinary tourism is not confined to traditional restaurants; it extends to immersive experiences, culinary trails, and themed festivals celebrating the richness of Punjabi cuisine. This trend showcases the region's culinary diversity, providing a platform for local chefs to showcase their skills and for visitors to engage with the cultural nuances of Punjabi food [14].

The significance of Culinary Tourism lies in its potential to not only boost the local economy but also to preserve and promote culinary traditions [15]. By creating unique culinary experiences, Punjab leverages its rich gastronomic heritage to attract visitors, fostering cultural exchange and appreciation. This innovation not only showcases the region's culinary prowess but also enhances its cultural identity on a global scale.

Figure1. Factors Affecting Culinary Innovations

As mentioned in the figure1. Culinary innovation in Punjab is a dynamic interplay of historical influences, embracing ancient traditions and Sikh heritage. Globalization introduces diverse culinary elements, enriching the region's gastronomy. The availability and utilization of local ingredients like grains and spices fuel creativity. Culinary movements, including the Farm-to-Table approach, ensure sustainability and connect with local farmers. The preservation of cultural heritage acts as a guiding force, striking a balance between tradition and modernity. Together, these factors create a vibrant culinary landscape, where innovation thrives while honoring Punjab's rich history and agricultural bounty.

Challenges and Opportunities:

As Punjab's culinary scene undergoes a transformative journey, the preservation of authenticity and the pursuit of sustainability stand out as critical imperatives. This exploration delves into the challenges and opportunities inherent in maintaining the rich cultural heritage of Punjab's gastronomy while addressing the increasing demands for diverse ingredients and ethical practices.

7.2 Preservation of Authenticity:

Challenges:

Preserving the authenticity of Punjabi cuisine encounters multifaceted challenges, primarily stemming from the evolving global culinary landscape. The commercialization of traditional dishes for mass appeal often results in dilution and distortion of original recipes. Fast-food adaptations and standardized versions of iconic dishes risk overshadowing the nuanced flavors and cultural significance of authentic Punjabi cuisine.

Another challenge lies in the intergenerational transmission of culinary knowledge. With changing lifestyles and preferences, younger generations may not be as inclined to invest time in mastering traditional cooking techniques passed down through families. This poses a threat to the continuation of age-old practices that contribute to the distinctive taste and character of Punjabi dishes [16].

Opportunities:

Amidst these challenges, opportunities emerge to safeguard and celebrate Punjab's culinary authenticity. The rise of culinary documentation, both in print and digital media, provides a platform to preserve traditional recipes. Cookbooks, blogs, and online platforms crated by chefs and culinary enthusiasts contribute to the documentation and dissemination of authentic Punjabi recipes, ensuring they are accessible to a wider audience [17]. Campaigns for culinary education and awareness are also quite important. Campaigns that support storytelling sessions, culinary seminars, and traditional cooking classes can help close the generational divide and create a newfound appreciation for the creativity and cultural value found in real Punjabi food.

7.3 Sustainability in the Culinary Landscape:

Challenges:

The increasing demand for diverse ingredients, often driven by globalization and changing consumer preferences, poses challenges to sustainability in Punjab's culinary landscape. Traditional agricultural practices are disrupted as chefs seek exotic or out-of-season ingredients, leading to environmental strain and loss of biodiversity. Moreover, the demand for certain ingredients can

https://doi.org/10.52202/074664-0025

contribute to overexploitation and depletion of local resources, impacting the long-term viability of Punjab's agricultural ecosystem [18].

Waste management within the culinary industry is another pressing concern. The sheer volume of food waste generated by restaurants and households contributes to environmental degradation. The disposal of food scraps, packaging, and excess ingredients poses challenges in implementing sustainable practices within the culinary sector.

Opportunities:

Sustainability in Punjab's culinary landscape presents an array of opportunities that align with ethical, environmental, and economic considerations. Embracing the Farm-to-Table Movement, as discussed earlier, exemplifies an opportunity to support local farmers, reduce the carbon footprint associated with transportation, and promote seasonal, sustainable produce. Restaurants and culinary institutions can lead the way by adopting eco-friendly practices. Initiatives such as composting, recycling, and sourcing locally can contribute to minimizing the environmental impact of the culinary industry. Moreover, collaborations with local farmers and producers not only ensure the freshest ingredients but also contribute to the economic sustainability of the region.

Educational programs and awareness campaigns are instrumental in fostering a culture of sustainability. By raising awareness about the environmental implications of dietary choices and the importance of supporting local agriculture, consumers can be empowered to make informed decisions that contribute to the sustainability of Punjab's culinary landscape.

8. Discussion and Findings:

8.1 Synthesis of Findings:

In the discussion phase, the synthesis of findings involves weaving together the insights gathered from the review of literature, data collection, and analysis of various culinary innovations in Punjab. This synthesis aims to distill key patterns, recurring themes, and noteworthy observations that have emerged throughout the exploration of the Farm-to-Table Movement, Molecular Gastronomy, and Culinary Tourism.

Through a careful analysis of the literature, it becomes evident that Punjab's culinary evolution is a tapestry woven with threads of tradition and modernity. The synthesis encapsulates the essence of how historical influences, such as the Harappan civilization and Sikh traditions, laid the foundation for a unique culinary identity. Simultaneously, it encapsulates how contemporary influences, driven by globalization and technological advancements, have shaped the current gastronomic landscape.

8.2 Interplay between Tradition and Modernity:

Central to the discussion is the exploration of the intricate interplay between tradition and modernity in Punjab's culinary scene. Traditional cooking methods like tandoori and dum cooking, deeply rooted in the region's history, continue to coexist with innovative culinary techniques such as Molecular Gastronomy. The interplay is not merely juxtaposition but a dynamic fusion, where chefs leverage modern approaches to reimaging traditional dishes without compromising their authenticity.

The discussion delves into how the Farm-to-Table Movement represents a return to the roots, emphasizing sustainability and local sourcing, aligning with

https://doi.org/10.52202/074664-0025

Punjab's agrarian heritage. Meanwhile, Molecular Gastronomy introduces a contemporary flair, demonstrating that tradition and innovation are not mutually exclusive. Instead, they converge to create a culinary experience that respects the past while embracing the present.

8.3 Implications for Punjab's Culinary Heritage:

The implications drawn from the discussion are far-reaching, influencing not only the gastronomic landscape of Punjab but also the cultural and economic aspects of the region. By embracing the Farm-to-Table Movement, Punjab reinforces its commitment to sustainability, supporting local farmers and preserving the authenticity of its agricultural practices. The integration of Molecular Gastronomy reflects a forward-looking approach, positioning Punjab as a hub for culinary experimentation and innovation, contributing to the global discourse on gastronomy.

Furthermore, the discussion delves into the implications of Culinary Tourism, emphasizing how it serves as a cultural ambassador for Punjab. The region's diverse culinary offerings become a source of pride and a means of global engagement, fostering cultural exchange and economic growth. Culinary Tourism not only brings economic opportunities but also acts as a catalyst for the preservation and promotion of Punjab's culinary heritage on an international stage.

In essence, the discussion section serves as the analytical core of the review paper, offering a nuanced interpretation of the findings and drawing meaningful connections between the various facets of Punjab's culinary innovations. It sheds light on how the delicate balance between tradition and modernity shapes the region's gastronomic identity and explores the broader implications for the sustainability, cultural richness, and global recognition of Punjab's culinary heritage.

9. Conclusion:

In concluding this exploration into the culinary innovations of Punjab, a synthesis of key findings illuminates the complex yet harmonious interplay between tradition and modernity. The historical roots of Punjab's culinary heritage, enriched by the Harappan civilization and Sikh traditions, have seamlessly merged with contemporary influences, reflecting in practices such as the Farm-to-Table Movement, Molecular Gastronomy, and Culinary Tourism. The preservation of authenticity amidst globalized influences, the integration of sustainable practices, and the embrace of innovative culinary techniques collectively contribute to the vibrant tapestry of Punjab's gastronomic identity.

10. Recommendations for Sustaining Culinary Heritage:

Sustaining Punjab's culinary heritage demands a concerted effort to balance innovation with preservation. To safeguard the authenticity of Punjabi cuisine, there is a need for continued documentation of traditional recipes, culinary techniques, and historical influences. Culinary education initiatives, both formal and informal, should be encouraged to ensure the transmission of knowledge across generations, bridging the gap between traditional practices and contemporary lifestyles.

The Farm-to-Table Movement offers a sustainable pathway forward, emphasizing the importance of local sourcing and seasonal produce. Recommending policies that support local farmers, incentivize sustainable agricultural practices, and promote awareness about the environmental impact

https://doi.org/10.52202/074664-0025

of culinary choices can further fortify the foundation of Punjab's culinary heritage.

In addition, fostering collaborations between culinary institutions, chefs, and local communities can create a network that actively works towards preserving and promoting authentic Punjabi cuisine. Initiatives such as culinary festivals, workshops, and cultural exchanges can contribute to a more sustainable and inclusive culinary heritage.

11. Future Directions for Research:

As we look toward the future, several avenues beckon for further exploration within the realm of Punjab's culinary innovations. Future research endeavors could delve deeper into the socio-cultural aspects of culinary practices, investigating how changing demographics and cultural dynamics influence the evolution of Punjabi cuisine. A longitudinal study tracking the impact of culinary innovations on local economies, farmer livelihoods, and cultural identity would provide valuable insights into the sustained effects of these culinary trends.

Moreover, as technology continues to play a significant role in shaping culinary landscapes globally, understanding the role of digital platforms and social media in promoting and shaping the narrative around Punjab's culinary heritage would be a fruitful area of research. Exploring the potential of incorporating sustainable practices in both large-scale and small-scale culinary establishments can contribute to a more comprehensive understanding of the challenges and opportunities in sustainable gastronomy.

In conclusion, this exploration into Punjab's culinary innovations not only highlights the richness of its gastronomic identity but also underscores the imperative of embracing sustainable practices while navigating the delicate balance between tradition and modernity. The recommendations and future research directions outlined aim to inspire a continued dialogue, ensuring the preservation and evolution of Punjab's culinary heritage for generations to come.

References:

1. Tamang, D.D.: Culinary tradition and Indigenous knowledge in the Nepal Himalayas. In: Springer eBooks. pp. 27–65 (2015). https://doi.org/10.1007/978-3-662-48159-2_3.
2. Dhillon, C.P.S., Bhinder, H.S.: Exploring Sensory and Cultural Experiences of Tourists with Local Gastronomy Case of Amritsar, Punjab, (India). Turkish Online Journal of Qualitative Inquiry, 12(9). 12, 971–981 (2021).
3. Banerji, C.: Eating India : exploring the food and culture of the land of spices. (2009).
4. Jacob, R., Sharma, A.: Naming food and creating identity in transnational contexts. In: Springer eBooks. pp. 99–113 (2018). https://doi.org/10.1007/978-981-10-8468-3_6.
5. Berry, K.: Progress and emerging issues in culinary tourism: a study with special reference to Punjab. JOHAR, 9(1). 9, 28–47 (2014).
6. Antani, V., Mahapatra, S.: Evolution of Indian cuisine: a socio-historical review. Journal of Ethnic Foods. 9, (2022). https://doi.org/10.1186/s42779-022-00129-4.
7. Appadurai, A.: How to make a national Cuisine: Cookbooks in Contemporary India. Comparative Studies in Society and History. 30, 3–24 (1988). https://doi.org/10.1017/s0010417500015024.
8. Miller, R.E.: Mappila Muslim Culture: How a historic Muslim community in India has blended tradition and modernity. (2015).

9. Gupta, V.: Sustainable tourism: learning from Indian religious traditions. International Journal of Contemporary Hospitality Management. 11, 91–95 (1999). https://doi.org/10.1108/09596119910250751.
10. A history of cooks and cooking. Choice/Choice Reviews. 38, 38–2143 (2000). https://doi.org/10.5860/choice.38-2143.
11. Latour, B.: Why Has Critique Run out of Steam? From Matters of Fact to Matters of Concern. Critical Inquiry. 30, 225–248 (2004). https://doi.org/10.1086/421123.
12. Fine, G.A.: Justifying work: Occupational rhetoric's as resources in restaurant kitchens. Administrative Science Quarterly. 41, 90 (1996). https://doi.org/10.2307/2393987.
13. Henderson, J.C.: Food tourism reviewed. British Food Journal. 111, 317–326 (2009). https://doi.org/10.1108/00070700910951470.
14. Chhabra, D., Lee, W., Zhao, S., Scott, K.D.: Marketing of ethnic food experiences: authentication analysis of Indian cuisine abroad. Journal of Heritage Tourism. 8, 145–157 (2013). https://doi.org/10.1080/1743873x.2013.767816.
15. Bagale, A., Branger, S.H.: Food tourism: An effective marketing tool for the Indian tourism industry. International Journal of Science Technology & Management, 4(2), 2394-1529. 6, 227–230 (2020).
16. Ashley, C., Maxwell, S.: Rethinking rural development. Development Policy Review. 19, 395–425 (2001). https://doi.org/10.1111/1467-7679.00141.
17. Suryanarayana, P., Saraswat, M., Mrudula, T., Krishna, T.P., Krishnaswamy, K., Reddy, G.B.: Curcumin and turmeric delay Streptozotocin-Induced diabetic cataract in rats. Investigative Ophthalmology & Visual Science. 46, 2092 (2005). https://doi.org/10.1167/iovs.04-1304.

https://doi.org/10.52202/074664-0025

Disruptive Innovation and Digital technology in Human Resources for Hospitality

Dr Praveen Kumar[1[0000-0003-0294-3830]] and Mr.Harpreet Singh[2[0009-0001-1296-5366]],

[1,2] UITHM, Chandigarh University,
India

Abstract: In the twenty-first century, every organizational activity is becoming reliant on digital technology, and it is impossible to envision a company disruption occurring today without embracing some sort of technical innovation. Customers may receive robotic assistance due to technological disruption in HR. particularly in the hospitality sector, as the human element of an experience—such as meal service in a restaurant—will be lost if technology takes the place of human resources. As a result, the service eventually becomes mechanical. As an example, consider the replacement of a meal service by a robotic machine or robot. Although human resources are neither a product nor a service, the innovation theory seems to apply more to them than to any other kind of product or service. The phrase "human resource" refers to both the individuals who work for the company and the department in charge of maintaining the supplies needed to enable employees to do their jobs. In 1960, the phrase "human resource" was initially used as people began to recognize the importance of labor relations, motivation, conduct, and production.

Keywords: Digital disruption, cognitive, affective, and human capital

1 Introduction

In the field of business administration, disruptive innovation is defined by Christensen, Raynor, and Mc Donald (2015) as an innovation that first creates a new market and value network before eventually upending an established market and value network and dislodging well-established market-leading companies, goods, and partnerships. These academics developed the disruptive innovation theory, which has grown to be the most influential concept of the contemporary era. The theory states that not all breakthroughs are seen as disruptive. Using the example of the vehicle from the late 19th century, for instance, this is not an easy illustration of disruption because these were pricey, luxury-class goods that didn't affect the horse-drawn vehicle industry. The mass-produced automobile, which defied expectations by changing the whole transportation market in contrast to the first three years of the automobile, altered the transportation market, which had been stable until the launch of the cheaper Ford Model in 1908. Instead of the established market leaders, entrepreneurs and fresh entrants have developed disruptive innovation. Because their development may divert resources from sustaining breakthroughs, it is initially exceedingly difficult for the firms to pursue disruptive innovation because they are not well-funded. The disruptive process takes time to establish, but once it does, the effects spread quickly throughout the market, upending the established marketplaces in the process. It's also claimed that several businesses, like Seagate Technology, U.S. Steel, and Bucyrus, have been identified as victims of disruption and are currently stagnating in the industry. There is an issue that arises because it seems that the idea has been incorrectly used to explain everything in all areas of life, not only business, but also public institutions and education. Source:

https://doi.org/10.52202/074664-0026

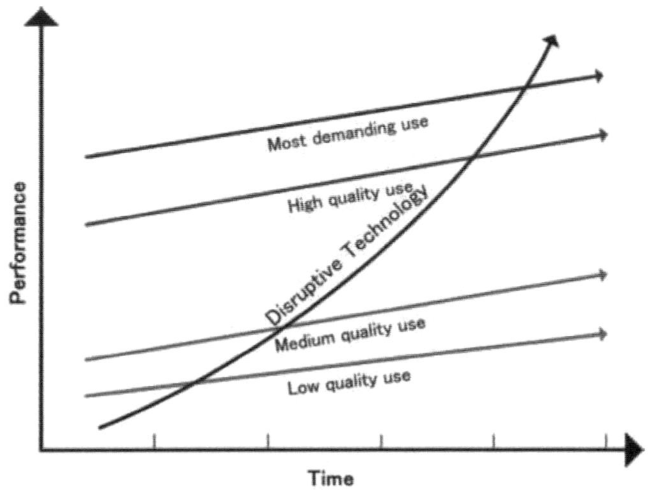

Source: Clayton M., Michael R. and R. McDonald, What is Disruptive Innovation, December**(2015)**

For a product or service to be considered truly disruptive, it must provide an answer to the following queries.

1. Are the newcomers aiming for a market that the incumbents find unappealing financially? When this inquiry is connected to the disruptive innovation in hospitality human resources, it takes on the following form: Are the newcomers focusing on the organization's human resources department, which the prior management settings saw as unproductive and insignificant?
2. Are the newcomers' different business models allowing them to serve this market segment profitably? We might ask the following question in relation to the hospitality human resource: might the entrants professionally and effectively inspire the human resource utilizing various tactics, or service this segment?
3. If the answer to (2) is "yes," then what we really need to know is if the smaller companies have access to "enabling technology" that will enable them to use their new business model to take on bigger, more complex projects without sacrificing the cost advantages and performance that allowed them to successfully complete smaller projects. That is, is it possible for the smaller EPCM businesses to "move upmarket" without just growing into the bigger firms that already dominate the market?

2 Literature review

In an interview with the hospitality channel, Santen R. and Kirk J. (2016) stated that disruptive forces are present in the hospitality industry. It goes without saying that new players are entering the market and taking a different route. For instance, how visitors locate rooms, how hotel staff is ready for that, and how hotels handle their human resource concerns.
We must comprehend and identify the methods via which the disturbance is occurring. Thus, it also requires a shift in perspective to realize that the way we conduct business is evolving and that we must be prepared to accept this development. Thus, there is also the human element, and we give this a lot of attention. According to Christensen, Raynor, and McDonald (2015), a disruptive invention is not a game-changer. Innovation is what separates a historically significant and unaffordable product or service into a new market niche. It is challenging to apply his theory to any business's human resource management, or what is more widely known as human capital management, particularly in the hospitality sector.

 https://doi.org/10.52202/074664-0026

The innovation idea appears to be more applicable to any kind of good or service, yet a human resource is neither of these things. The phrase "human resource" refers to both the individuals who work for the company and the department in charge of maintaining the supplies needed so that staff members can do their jobs more easily. In 1960, the word "human resource" was initially used when individuals began to recognize the importance of labour relations, motivation, conduct, and production. In contrast to an organization's financial and material resources, Tracey (2004) defines human resources as "the people that staff and operate an organization."

The example we are reviewing is a disruptive innovation if we can say "yes" to each of the three questions; if we can't answer any of them, the procedure isn't actually disruptive innovation-related. A single individual or employee in your company is a human resource. The word "disruptive human resource" is defined by Thomas (2016) in two ways: first, it refers to proactive HR, while the second definition is reactive HR. A new term, "disruptive HR," is being used to describe how the disruptive world is posing worldwide issues. The talent pool, employment market, and corporate goals all require disruptive HR to think creatively and creatively to bring value to both the organization and itself. But according to his dissemination of this idea throughout the globe, it may be applied to human resources and used as a catalyst to aid in the process of disruption that occurs in Having said that, it is crucial to recognize the disruptive changes that the human resources component is undergoing. Organizations should view their HR departments in the same light as Ritz Carlton Hotels worldwide, which manages its HR department under the motto "We take care of those who take care of our valuable customers." This phrase highlights the HR component of the hospitality industry as a crucial element that has the potential to influence all of the strategic changes that our company undertakes.

In the twenty-first century, as all organizational functions are shifting to a digital technology foundation, it is impossible to envision a business disruption occurring today without incorporating some sort of technical innovation. Customers may experience robotic service as a result of technological disruption in HR, particularly in the hospitality sector. This is because, when technology replaces human resources, such as in the case of restaurant meal service, the customer will not recognize the human touch component of the experience, and as a result, the service eventually becomes mechanistic.

Reinke (2016) asserts that the human element is the reason restaurants succeed. They are edible museums and works of living art that are only possible because of the considerate service that individuals provide. This is the reason why other so-called "disruptive" technologies—which aim to eliminate human interaction by using iPads, tables, or ordering kiosks—have not proven successful in the market. However, these models will only be successful at first in fast-casual environments when customers are eager to get their food. Technology won't be much use to customers seeking a real dining experience. However, Cloth bound's model, which modifies the system in secret, is starting to become noticeable. An internet portal called Clothbound links restaurants with possible staff. It offers a practical, straightforward employment board tailored specifically for the restaurant business, much like monster.com. Kelly Daigle, co-founder of Clothbound, was most recently honored by Zagat for her creative thinking and market influence when they included her in their 30 under 30 list of Boston's finest. With the help of Cloth bound's technology platform, the restaurant network is now available online, effectively connecting the two networks of employers and potential employees. Reinke (2016) emphasizes the disruptive nature of human resources once more. Disruption thrives on increasing the effectiveness of current systems, as demonstrated by the success in other areas of the hospitality sector, such as accommodation (Airbnb) and transportation (Uber). Due of their systems' emphasis on passenger ease and freedom, Uber and Airbnb have seen tremendous success. By using any of the two models, customers can be in charge of their own destiny and book their ideal stay without having to pay any third-party costs.

This eliminates the need for them to rely on systems run by other parties, such as hotels, hostels, cabs, etc. These cutting-edge models are used in an even more practical way thanks to online interactions and smart technology.
Despite significant disagreements over Uber and Airbnb, But the restaurant can't be found online, and a successful restaurant relies just as much on experiences as it does on goods. Pre-ordering, iPads on tables, and other restaurant model adaptations have not endured long since they dilute the core essence of a dining experience. Any self-respecting restaurant should serve authentic, sensorial, entertaining, and tech-free meals.

The theory of disruptive innovation is not without flaws. It has proven successful in certain industries, but it hasn't been able to connect with many others. The first aim is to relate this theory to the human resources of the hospitality industry. After reviewing numerous written publications, published conference recordings, write-ups, and interviews with numerous individuals in the field, the researcher discovered that this type of research had never been done before, particularly in the hospitality industry. Since it was the first of its kind, it was extremely difficult to apply disruptive innovation theory to hospitality human resources because there aren't many resources available in this context. This is the reason the researcher also included his more than ten years of expertise managing thousands of people in the hospitality industry, from operations to academics, while working in five-star hotels in Nepal and overseas. The study also includes interviews, group discussions, talks, and input from a large number of people in Nepal's hospitality business.

3 Methods and methodology:

Study selection process and methods

This study used an SLR technique to conduct a literature review. To establish the requirements for articles' inclusion and the techniques for analysis, a protocol was created. The first table contains the inclusion criteria. A typical procedure advised for a PRISMA SLR was followed in the articles (Liberati et al., 2009). Three steps make up the PRISMA: inclusion, screening, and identification. The identification stage includes the use of keywords or searched terms, search criteria, databases, and data extraction. "Electronic Human Resource Management" (E-HRM), "Disruptive Technology For Human Resource Management," "Digital Human Resource Management," and "Human Resource Management Information Systems" (HRIS) , predicators, and results were the typical search phrases employed. The search criteria were created with the operators "AND" between the antecedents and outcomes and "OR" for terms that are similar to disruptive technologies for HRM. The search parameters were entered into LENS.ORG and Scopus.

Article screening can be divided into two categories: screening and eligibility verification. Both human and automatic screening are used in the process. The articles were included by adhering to the inclusion criteria from one to five listed in Table and using the databases' default automatic screening feature. Section 3.1 explains the amount of articles that have been excluded thus far along with the reasons behind them. Additionally, each publication was manually evaluated by two of the author's independent colleagues in accordance with the first, second, third, and fourth inclusion criteria listed in Table . The evaluation of methodological reporting is necessary for the eligibility checking of the papers (Meline, 2006; Priyashantha et al., 2021; Priyashantha and Dilhani, 2022). Five articles that used quantitative research approaches were included in the study as inclusion criteria; the remaining publications were removed. Five studies that used empirical research methodology were included. Two associates of the The author completed the assignment on their own, and debate settled any differences over inclusion. At this point, the amount of articles excluded and their justifications are stated.

 https://doi.org/10.52202/074664-0026

The remaining articles were then exported to an MS Excel file, which was used to give VOSviewer with the data it needed for the required analysis. This sheet had the title, abstract, keywords, names and affiliations of the authors, journal name, cited numbers, and year of publication of the articles.

3.1Disruptive Human Resource Approach

The phrase "disruptive innovation theory" can be applied to two distinct stages of the development of human resources: disruption of the HR system and disruption of HR performance.

3.2 Disruption in Human Resource System

Th omas (2016), drawing from an article by Josh Bersin in Forbes (2014), lists the innovations in the HR system that have altered the organizational development roadmap.

1. *Engagement of the Workforce System*: The system of old office records is being replaced by modern managers with an employee-friendly engagement plan. in creating their own documentation. An employee can directly access a well-integrated system called the employee-employer interface. Both management and the individual employee have constant access to all of the employee's HR records. This interface makes it easier to manage staff attendance, notices, financial information, performance reviews, feedback (both positive and negative), leave records, warnings, and appreciations automatically.

   The approach has advantages for companies with qualified staff members and disadvantages for those with averagely qualified staff members. Mobile devices are a key component of this project.
2. *Human resource management approach:* The development of international networks and connections has made the process of outsourcing people considerably simpler. HR managers hire people directly through Linked Inn, Glassdoor, Indeed, and other social media platforms. The more than 130 Ritz Carlton hotels worldwide have created an online hiring system via their registered websites. Any deserving graduate in hospitality or worker from anywhere in the world can participate in this process by registering in their hiring area and sending their information for a position that is open for review. There will be a Skype interview with the panel following the evaluation. Following a successful interview and other formalities, the candidate will be hired on the spot.
3. *Employers' and employees' job platforms that are easy to use:* HR in the modern day is IT-friendly. It is not necessary for employers or employees to look for one another. These are the internet employment portals that are accessible for both purposes. The company may also access their necessary Human Resources online through these job portals, where employees can enter their personal information.
4. *Mobile Programs:* Kleiner Perkins and Caufeild & Byers (KPCB) (2016) reports that there are 1.6 billion smart phone users, 5.2 billion mobile users, 789 million laptop users, and 745 million desktop users worldwide.
5. *Data-driven choices:* Only 4% of large organizations are able to assess, predict, and model their human resources, compared to 90% who can do the same with their financial structure, according to Deloitte data from 2014. Data analysis for hiring, choosing, onboarding, and training and development of employees is now a science-based solution rather than a product. (Thomas (2016)

According to Thomas (2016), 85% of private Indian enterprises are prepared to invest in their employees' professional development. Thomas conducted research on 42 of these companies. Because it wasn't required, 43% of employees chose not to participate in training and development. The companies' inability to manage change was deemed to be their top obstacle. The majority of respondents—59%—cite HR managers as their primary role models for implementing such reforms in HR.

4 Disruption in the Performance of Human Resources

- It is crucial that every innovation in human resources that is considered disruptive also affects employee performance. Any newcomer should concentrate on the current workforce and discover methods to change employees' mindsets to create the most innovative products. The qualities and aptitudes that allow a job holder to carry out the tasks listed in a task statement that outlines what the job holder does are known as Knowledge, Skill, and Attitude (KSA) (Quinones and Ehrenstein, 1997). KSA's equivalents in a learning environment are cognitive, psychomotor, and affective; these three domains define the final stages of training (objectives). There are two different ways to interfere with employees' performance levels.

We refer to individuals who are career-driven, modern, and aware of changes, innovations, and technology as explorers. They also recognize external competition. At this staff level, the disruption is referred to as exploratory disruptive innovation.

1. We refer to the employees—who are technically sound but unfamiliar with the competition—as sustainers because they are dedicated, specialized, and driven by their families. This kind of disruption is known as persistent disruptive innovation.
   - Let's examine Kraiger et al.'s (1993) taxonomy of learning outcomes. Three regions of Saudi Arabia are defined as follows, per the researchers.

- Cognitive (knowledge): This is founded on declarative and factual principles, and it teaches staff members how to mentally organize concepts and information. The definition of this also includes metacognitive Allocating and regulating cognitive resources is a strategy.
- Psychomotor (skill-based): This involves the employees' acquisition of automation and compilation. There is a procedural linkage and routine development in compilation. The capacity to carry out a task with other tasks and without conscious supervision is known as automaticity.
- Affective (concepts). To put it simply, this is an attitude. There are two distinct sorts of attitudes in general understanding: positive attitudes and negative attitudes. Here, attitude refers to one's perspective toward learning, self-efficacy, performance perception, and goal-setting. The impact of this is associated with the employees' motivating state.

Rural tourism provides a distinctive insight into India's culture and economy[1]. Rural communities possess a variety of cultures, craftsmanship, and natural beauty that have not been fully utilized. Rural tourism has not achieved its full cultural and economic potential due to many difficulties[2]. Indian rural areas possess diverse cultural traditions and distinct ethnic identities, although they are often disregarded by tourists from urban centers.

5 Cloud-based modern human resources:

2015 saw Oracle and its affiliates Write something that disturbs One recent and developing digital standard is human capital management. But the integration of big data, social media, cloud computing, mobile, and digital consumer experiences with HRM technology is changing how businesses interact and manage their workforce. This innovative approach to people management is being used by HR managers in the modern workplace to enable an agile workplace where employees can easily access information and data to complete tasks with purpose. Sixty percent of DEOs said that cloud computing technology is strategically important for their company. The cloud is radically altering how organizations will do business going forward. By 2020, ninety percent of those over the age of sixty will be utilizing mobile phones. The entire generation will have grown up in a digital world by 2020. For them, using computers, cellphones, tablets, messaging, and social networking will come naturally.

Organizations that can swiftly and readily adjust to shifting business conditions will outperform their competitors as the world becomes more unpredictable and chaotic.

 https://doi.org/10.52202/074664-0026

6 Case studies from Ritz Carlton and Kempinski Hotel

Whether or not the Ritz-Carlton corporation, which is most renowned for treating its employees well, is aware of the disruption theory, it is employing a very sophisticated approach to manage disruption in human resources. In contrast to the disruptive innovation notion, this establishment is not a recent arrival; rather, it is a long-standing hospitality establishment that employs current staff management techniques. One of the top providers of luxury accommodations in this category is Ritz-Carlton, a hotel company renowned for its own standards. According to Training Magazine, it is the only service provider in the United States to hold the Malcolm Baldridge National Quality Award (MNQA). as the sole business most renowned for its staff training and development. According to Reiss R. (2009), the motto of the organization is "We are ladies and gentlemen serving ladies and gentlemen," which further emphasizes its distinctive culture. One of its unique regulations allows each staff to spend up to $2,000 on making every guest feel satisfied. This is case-specific rather than monthly.

"The 12 Service Values," "The Credo," "The Three Steps of Service," "The Sixth Diamond," and other proprietary statements that are taught to all 38,000 workers across 73 hotels in 24 countries codify Ritz-Carlton's expectations about service. Simon Cooper, the company's CEO for the last eight years, describes what makes Ritz-Carlton unique as essentially emphasizing people, product, and location in relation to investment. They invest more than $500,000, roughly. $1 million, for instance, each room. Select the most ideal and practical location for their customers to ensure that they have simple access to the hotel. In a similar vein, the company's ladies and gentlemen serving ladies and gentlemen statement demonstrates how much emphasis is placed on the moral principles and professional growth of its staff.

The Ritz Carlton is known for its "line up" custom, which dates back to the early days of French restaurants and involves the chef calling his staff each evening at 5:30 p.m. Every Ritz Carlton hotel in the world has a round table every day where each member has the chance to share a "wow story," or something amazing that one of their staff members has accomplished. This demonstrates a unique and inventive strategy that many hotels overlook, yet what Ritz Carlton provides for its employees is an illustration of disruptive innovation. Reiss R. (2009) elaborates in a Forbes interview that at the Ritz Carlton

The Ritz Carlton is known for its "line up" custom, which dates back to the early days of French restaurants and involves the chef calling his staff each evening at 5:30 p.m. Every Ritz Carlton hotel in the world has a round table every day where each member has the chance to share a "wow story," or something amazing that one of their staff members has accomplished. This demonstrates a unique and inventive strategy that many hotels overlook, yet what Ritz Carlton provides for its employees is an illustration of disruptive innovation. Reiss R. (2009) elaborates on that further in an interview with Forbes at the Ritz Carlton.

1. It gives each and every Ritz-Carlton employee the authority to spend up to $2,000 on a guest without the general manager's consent. And it's not yearly. According to the occurrence, yes. This demonstrates that the hotel's ability to instill trust in each and every employee is a significant accomplishment made possible by managing employee disruptions at the motivation level, which in turn fosters a positive attitude and promotes stability, devotion, and productivity.
2. For a single night, a carpenter is hired to build a shoe rack of the guest's choosing. In a similar vein, a laundry manager from Puerto Rico would fly to New York to personally return a dress if the stain was still present after two attempts. This demonstrates the level of staff motivation.
   This is a novel approach that Ritz uses to empower staff members to go above and beyond the call of duty in order to delight guests and feel successful. Whether intentional or not, it is an illustration of disruptive innovation strategy in the HR domain.

https://doi.org/10.52202/074664-0026

3. Usually, they hire 2% of those who submit job applications. Turnover can be minimized by hiring the proper women and gentlemen, developing them, and giving them career chances. Since it advances the careers of our ladies and gentlemen, training is quite vital. Naturally, it's harder to keep employees satisfied during a difficult economic period, but it's still crucial.
4. The voluntary turnover rate is a gauge of the development and training of new talent that Ritz Carlton tracks. In order to ensure that they hire the proper kind of ladies and gentlemen and then train them to the finest available career alternatives so that the turnover is very low, Ritz Carlton usually hires roughly 20% of the applicants. The low turnover rate and training and development plan at Ritz Carlton demonstrate the company's ability to use disruptive innovation in HR, something its rivals are unable to do.
5. The foundation of the Ritz Carlton culture is trust. And the quickest way to destroy an organization's culture is for its leadership to not uphold the ideals that the organization demands of them. If a culture is not embodied at the highest levels of the organization, it will not last. Building a culture takes an incredibly long period. Each year, Kempinski invests much in the development of its human resources. The organization plans to keep improving staff development. It has put in place training programs for all staff levels that are aimed at developing managerial acumen, improving technical proficiency and operational skills, and fostering cross-departmental training. The foundation of the Ritz Carlton culture is trust. And the quickest way to destroy an organization's culture is for its leadership to not uphold the ideals that the organization demands of them. If a culture is not embodied at the highest levels of the organization, it will not last. Building a culture takes an incredibly long period. Each year, Kempinski invests much in the development of its human resources. The organization plans to keep improving staff development. It has put in place training programs for all staff levels that are aimed at developing managerial acumen, improving technical proficiency and operational skills, and fostering cross-departmental training.

Kempinski offers staff involvement, growth, and training throughout the entire year. It firmly feels that in order to achieve maximum engagement from the personnel, contemporary methods and characteristics are required. Kempinski (2010) reported that their success in opening the properties was attributed to staff who are regularly trained and taken off the floor to make them competent enough to handle the competition. Karen Thorburn, Regional Director of Training at Kempenski and 2009 Winner of the Hotelier Middle East Awards, said as much. They pay equal attention to the study's explorers and sustainers. This demonstrates that staff growth and training at both the explorer and sustainer levels can cause disturbance.

Additionally, Kempinski hired experts to traverse the area and provide skill-building workshops on how to make the greatest bread and tea, as well as coffee. These classes were held at Kempinski hotels all over the world.In this area, Kempinski put these ideas for traveling experts into practice in Tanzania, Djibouti, and the United Arab Emirates, culminating in the famous Emirates Palace in Abu Dhabi.
The company's strong emphasis on people was further reasserted by the Kempinski Hotel Ajman Employee Satisfaction Survey pilot. The poll was carried out under the slogan "People are everything." The poll, which was held in combination with a full program of enjoyable events throughout the day, was open to all employees. a buff et lunch, massages, cooking workshops, camel rides, prize competitions, bowling, beach access for staff members and their families, and much more. Kempinski has the highest completion rate (99%) of any facility in the region for the Employee Satisfaction Survey.

Additionally, Kempinski has piloted multiple management programs including an English e-learning program. "The feedback has been encouraging," the company says. Additionally, Kempinski has created a housekeeping-specific skills-enhancement program targeted at supervisors and assistant housekeepers looking to advance their careers.

 https://doi.org/10.52202/074664-0026

There are numerous things, such staff costing, creating budgets, business planning, marketing concepts, and leadership/management abilities, that one simply does not have the opportunity to learn on the job due to the long hours and intense hands-on nature of the hospitality sector. Offering employees organized means to absorb this knowledge advances their professional development and increases a company's service competence and industry competitiveness.

- Figure:1The Four Areas' Example Lodging Firms can be viewed as disruptive innovation strategies in the HR domain.

Methods of Selective Hiring and Retention	Initiatives for Leadership Development and Competency Training	Benefits and incentive schemes to Raise Efficiency Cut Expenses	Changing the Work Process to Increase Employee Involvement
Dahlmar Properties: Work Preview Sessions, Orientation, Tests, and Quizzes	Coastal Hotel Group: Employee Loan Buddy System	Accor: 360° feedback The Houstonian: Bonus system for all employees	Ritz Carlton-Tyson's Corner: Independent Work Groups
Ramada Franchise Systems: Qualification exams Interactive Training via CD	Select Hotels: Establishing Leadership Capabilities	Roadway INN - Orlando: $100 monthly budget; prizes, pins, and gifts	The Boulders: Independent, Self-Directed, Three-Person Housekeeping Teams With the Authority to Select Work Areas and Take Own Inspections
Day Hospitality Group Mandatory 90 Day Paid Sabbatical Leave Program for GM's	Ashley Hotels- Keswick Hall: Service Quality Training Guest Say	Day Hospitality Group: Mandatory 90 Day Paid Sabbatical Leave Programs for GMs	Inn at Essex: Empowering employee decision making Promise Hotels: Guaranteeing 100% Guest Satisfaction
Four Seasons and Regent Hotels:??? Greenbrier: Culinary Trainings Hyatt Regency Scotsdale:High School Sponsored Classes Sheraton-Denver West: Job Sharing in Sales	Simpson House Inn: Testing/Training Language Skill Building Staff Meeting (Powwows) Motel 6: Employee Development Program Cross Training Management on Duty General Management Training	Tainar Inns: Cost Controls Inn at the Market: Outsourcing Human Resource Director	Hilton Hotels-Minneapolis: Line Level Employees are Empowered to Make Guest Satisfaction Decisions

Source: Adapted from, Kate Walch and Cathy A. Innovation in Hospitality Human Resource

7 Findings and conclusions from the study

In this 21st-century world, where all organizational activities are shifting to a digital technology foundation, any business disruption is unimaginable without incorporating a technological innovation. Technological disruption in human resources (HR) has the potential to result in a robotic customer experience, particularly in the hospitality sector, where replacing human resources with technology—for example, replacing a restaurant's meal service with a robotic machine or robot—will prevent the customer from realizing the human touch component of that experience, making the service ultimately mechanistic. In a Cornell University poll of 243 managers, 64% of the participants emphasized that the managers' top concerns include hiring, retaining, training, and morale. According to this analysis, it is difficult to motivate today's human resources unless hospitality organizations concentrate solely on innovations—something that their rivals are unable to do.

The discovery of the connection between disruptive innovation and hotel HR practices is the primary contribution of this study. There aren't many studies on hospitality innovation, particularly radical innovation (Chan et al., 1998; Rodgers, 2007), especially when it comes to examining how disruptive HRM practices affect the hotel sector. It has been discovered that two key HRM practices—hiring dynamic core customer-contact workers, or what we refer to as multi-skilled workers, and providing training to these workers for a variety of skills—have improved both incremental and disruptive innovation in the hotel industry. It is debatable if "hire for skill and train for skill" would be a superior approach to boost disruptive innovation in hospitality HRM, despite the long-standing management tenet of "select for attitude and train for skill."The results run counter to those of Tracey et al. (2007), who concluded that consciousness and general mental capacity are significant predictors of front-line personnel job performance.

According to Thomas A.'s research, 85% of private Indian enterprises are prepared to train new staff. The research covered 42 different companies. Since training and development were optional, 43% of employees chose not to participate, which indicates that management needs to focus on instilling a sense of importance in employees regarding training and development in order to deal with disruptions. This demonstrates how crucial it is for HR managers to manage disruptive innovation in the field. Additionally, it is discovered that hotels are not openly using the disruptive innovation theory since they are unable to connect it to their regular business operations, particularly in human resources.

Two distinct aspects of human resources—exploratory disruptive innovation and sustaining disruptive innovation—are the subject of this study. According to research, skilled staff members with lesser levels of academic qualification who engage in disruptive innovation require approaches for maintaining disruptive innovation, which aligns with Dr. Christensen's theory. This demonstrates that an innovation may only be considered disruptive if it originates from a highly unique market that competitors have never entered or have chosen to ignore, and is expertly managed by new entrants. Dr. Christensen and colleagues introduced the concept of disruption at this level, which we refer to as sustaining disruptive innovation in this study.

 https://doi.org/10.52202/074664-0026

References

1. Singh, S. et al.: India's new-found love for wine tourism: a decanter of expectations and change. International Journal of Wine Business Research. 33, 3, 313–335 (2020). https://doi.org/10.1108/ijwbr-05-2020-0021.
2. Rosalina, P.D. et al.: Rural tourism: A systematic literature review on definitions and challenges. Journal of Hospitality and Tourism Management. 47, 134–149 (2021). https://doi.org/10.1016/j.jhtm.2021.03.001.
3. Zhang, X.: RETRACTED ARTICLE: Impact of rural tourism on residents' well-being in traditional ancient villages: a case of North Guangxi. Heritage Science. 9, 1, (2021). https://doi.org/10.1186/s40494-021-00616-y.
4. Dey, B. et al.: Influence of destination attractiveness factors and travel motivations on rural homestay choice: the moderating role of need for uniqueness. International Journal of Culture, Tourism and Hospitality Research. 14, 4, 639–666 (2020). https://doi.org/10.1108/ijcthr-08-2019-0138.
5. Ćurković, M.D.: The role of the traditional crafts as intangible heritage on the global tourist market. SHS Web of Conferences. 92, 06005 (2021). https://doi.org/10.1051/shsconf/20219206005.
6. Ćurčić, N. et al.: The role of rural tourism in strengthening the sustainability of rural areas: the case of Zlakusa Village. Sustainability. 13, 12, 6747 (2021). https://doi.org/10.3390/su13126747.
7. Nicolaides, P.A.: Sustainable Ethical Tourism (SET) and rural community involvement, https://doaj.org/article/fbaed36d9328469c8526fbe3dcab3ba7.
8. Gößling, S.: Integrating e-scooters in urban transportation: Problems, policies, and the prospect of system change. Transportation Research. Part D, Transport and Environment. 79, 102230 (2020). https://doi.org/10.1016/j.trd.2020.102230.
9. Kumar, S., Shekhar, S.: Technology and innovation: Changing concept of rural tourism – A systematic review. Open Geosciences. 12, 1, 737–752 (2020). https://doi.org/10.1515/geo-2020-0183.
10. Brouder, P.: Tourism development against the odds: The tenacity of tourism in rural areas. Tourism Planning & Development. 9, 4, 333–337 (2012). https://doi.org/10.1080/21568316.2012.726259.
11. Brovarone, E.V., Cotella, G.: Improving Rural Accessibility: a multilayer approach. Sustainability. 12, 7, 2876 (2020). https://doi.org/10.3390/su12072876.
12. Dłużewska, A., Giampiccoli, A.: Enhancing island tourism's local benefits: A proposed community-based tourism-oriented general model. Sustainable Development. 29, 1, 272–283 (2020). https://doi.org/10.1002/sd.2141.
13. Cowan, T.: The village as urban infrastructure: Social reproduction, agrarian repair and uneven urbanisation. Environment and Planning. E, Nature and Space. 4, 3, 736–755 (2019). https://doi.org/10.1177/2514848619868106.
14. Apostolopoulos, N. et al.: Sustaining rural areas, rural tourism enterprises and EU Development Policies: A Multi-Layer Conceptualisation of the obstacles in Greece. Sustainability. 12, 18, 7687 (2020). https://doi.org/10.3390/su12187687.
15. Qi, J. et al.: Spatial distribution characteristics of the rural tourism villages in the Qinghai-Tibetan plateau and its influencing factors. International Journal of Environmental Research and Public Health (Online). 19, 15, 9330 (2022). https://doi.org/10.3390/ijerph19159330.
16. Javaid, M. et al.: Artificial Intelligence Applications for Industry 4.0: A Literature-Based Study. Journal of Industrial Integration and Management. 07, 01, 83–111 (2021). https://doi.org/10.1142/s2424862221300040.
17. Iyer, L.: AI enabled applications towards intelligent transportation. Transportation Engineering. 5, 100083 (2021). https://doi.org/10.1016/j.treng.2021.100083.
18. Zhou, X. et al.: Smart Tour route planning algorithm based on naïve Bayes interest data mining Machine learning. ISPRS International Journal of Geo-information. 9, 2, 112 (2020). https://doi.org/10.3390/ijgi9020112.
19. Manfreda, A. et al.: Autonomous vehicles in the smart city era: An empirical study of adoption factors important for millennials. International Journal of Information Management. 58, 102050 (2021). https://doi.org/10.1016/j.ijinfomgt.2019.102050.
20. Forkan, A.R.M. et al.: Mobile IoT-RoadBot. Proceedings of the 28th Annual International Conference on Mobile Computing and Networking. (2022). https://doi.org/10.1145/3495243.3558271.
21. Mhlanga, D.: Industry 4.0 in Finance: The impact of Artificial intelligence (AI) on digital financial inclusion. International Journal of Financial Studies. 8, 3, 45 (2020). https://doi.org/10.3390/ijfs8030045.
22. Mogaji, E. et al.: The implications of artificial intelligence on the digital marketing of financial services to vulnerable customers. Australasian Marketing Journal. 29, 3, 235–242 (2020). https://doi.org/10.1016/j.ausmj.2020.05.003.
23. López-Sanz, J.M. et al.: Sustainable development and rural tourism in depopulated areas. Land. 10, 9, 985 (2021). https://doi.org/10.3390/land10090985.
24. Khan, N. et al.: Factors affecting tourism industry and its impacts on global economy of the world. Social Science Research Network. (2020). https://doi.org/10.2139/ssrn.3559353.

25. Soliman, M. et al.: Sensory marketing and accessible tourism: An AI-generated article, https://journal.robonomics.science/index.php/rj/article/view/53.
26. Zhang, Q. et al.: Location identification and personalized recommendation of tourist attractions based on image processing. Traitement Du Signal/TS. Traitement Du Signal. 38, 1, 197–205 (2021). https://doi.org/10.18280/ts.380121.
27. Arora, M., et al.: Ai-Driven Personalized Travel Planning: Enhancing Tourist Experiences In Uzbekistan. *2023 3rd International Conference On Technological Advancements In Computational Sciences (Ictacs)* 1317–1322 (2023) Doi:10.1109/Ictacs59847.2023.10390025.
28. Harish, P., Thomas, T.K.: Shaping and reshaping of tourism: The influence of IR 4.0 for modifying the tourist behaviour. In: Emerald Publishing Limited eBooks. pp. 119–132 (2023). https://doi.org/10.1108/978-1-80455-157-820231008.
29. Skublewska-Paszkowska, M. et al.: 3D technologies for intangible cultural heritage preservation—literature review for selected databases. Heritage Science. 10, 1, (2022). https://doi.org/10.1186/s40494-021-00633-x.
30. Fararni, K.A. et al.: Hybrid recommender system for tourism based on big data and AI: A conceptual framework. Big Data Mining and Analytics. 4, 1, 47–55 (2021). https://doi.org/10.26599/bdma.2020.9020015.
31. Kheiri, J.: Tourists' engagement in cultural attractions: an exploratory study of psychological and behavioural engagement in indigenous tourism. International Journal of Anthropology and Ethnology. 7, 1, (2023). https://doi.org/10.1186/s41257-023-00087-y.
32. Khrais, L.T.: Role of artificial intelligence in shaping consumer demand in E-Commerce. Future Internet. 12, 12, 226 (2020). https://doi.org/10.3390/fi12120226.
33. Qin, Y. et al.: Artificial Intelligence and Economic Development: An Evolutionary investigation and Systematic review. Journal of the Knowledge Economy. (2023). https://doi.org/10.1007/s13132-023-01183-2.
34. Hossam, A.T.A.: Evaluating the effectiveness of AI-Driven customer segmentation in enhancing targeted marketing strategies, https://vectoral.org/index.php/JCSD/article/view/50.
35. Ajiga, D.I. et al.: Ai-Driven Predictive Analytics In Retail: A Review Of Emerging Trends And Customer Engagement Strategies. International Journal of Management & Entrepreneurship Research. 6, 2, 307–321 (2024). https://doi.org/10.51594/ijmer.v6i2.772.
36. Eddyono, F. et al.: Optimization model: the innovation and future of e-ecotourism for sustainability. Journal of Tourism Futures. (2021). https://doi.org/10.1108/jtf-03-2021-0067.
37. Shafiee, S. et al.: Developing sustainable tourism destinations through smart technologies: A system dynamics approach. Journal of Simulation. 17, 4, 477–498 (2022). https://doi.org/10.1080/17477778.2022.2030656.
38. Abdallah, M. et al.: Artificial intelligence applications in solid waste management: A systematic research review. Waste Management. 109, 231–246 (2020). https://doi.org/10.1016/j.wasman.2020.04.057.
39. Wang, M. et al.: Multi-Objective optimization of envelope design of rural tourism buildings in southeastern coastal areas of China based on NSGA-II algorithm and Entropy-Based TOPSIS method. Sustainability. 15, 9, 7238 (2023). https://doi.org/10.3390/su15097238.
40. Rocha, H.R.O. et al.: An Artificial Intelligence based scheduling algorithm for demand-side energy management in Smart Homes. Applied Energy. 282, 116145 (2021). https://doi.org/10.1016/j.apenergy.2020.116145.
41. Samara, D. et al.: Artificial intelligence and big data in tourism: a systematic literature review. Journal of Hospitality and Tourism Technology. 11, 2, 343–367 (2020). https://doi.org/10.1108/jhtt-12-2018-0118.
42. Wall, P.J. et al.: Artificial intelligence in the Global South (AI4D): potential and risks, https://arxiv.org/abs/2108.10093.
43. Robinson, K.P. et al.: Authente-Kente: Enabling Authentication for Artisanal Economies with Deep Learning. Social Science Research Network. (2021). https://doi.org/10.2139/ssrn.3769151.
44. Phillips, R. et al.: Introduction to the special issue: Intersections of Artificial Intelligence and Community Well-Being. International Journal of Community Well-Being. 3, 4, 425–435 (2020). https://doi.org/10.1007/s42413-020-00099-y.
45. Kaur, P. et al.: Unlocking the potential of artificial intelligence (AI) for healthcare. In: Artificial intelligence. (2023). https://doi.org/10.5772/intechopen.111489.
46. EmiRoğlu, B.D.: Times Factors Affecting Interactions Between Tourists. Journal of Gastronomy, Hospitality and Travel. (2022). https://doi.org/10.33083/joghat.2022.184.
47. Wilson, C., Van Der Velden, M.: Sustainable AI: An integrated model to guide public sector decision-making. Technology in Society. 68, 101926 (2022). https://doi.org/10.1016/j.techsoc.2022.101926.
48. Yiğitcanlar, T. et al.: Green Artificial Intelligence: towards an efficient, sustainable and equitable technology for smart cities and futures. Sustainability. 13, 16, 8952 (2021). https://doi.org/10.3390/su13168952.
49. Gupta, A., Singh, S.: Application of industry 4.0 technologies in sustaining supply chain. 2022 3rd International Conference on Intelligent Engineering and Management (ICIEM). (2022). https://doi.org/10.1109/iciem54221.2022.9853157.

Sustainability and Cultural Tourism: Preserving the Past for the Future With Special Reference to Manipur

Wahengbam Priyalakshmi[1][0009000415962138] and Dr. Sanjeev Kumar Saxena[2][00000221399255]

[1]Research Scholar, Department of Hospitality & Tourism Management, Assam University, Silchar. India

[2]Associate Professor, Department of Hospitality & Tourism Management, Assam University, Silchar. India.

Abstract

Cultural tourism is a special form of tourism practice that presents the unique opportunity to celebrate, preserve, and sustain the rich culture and heritage of the destination. Manipur, being a state with rich and diverse cultural and heritage resources is a perfect destination for cultural tourism. It is a state nestled in the northeastern part of India surrounded by nine hills. With different communities living together, the state does not lack fairs and festivals, rituals, celebrations etc, throughout the whole year. The study delves into various sustainability practices that apply to cultural tourism and help in balancing tourism development along with the conservation of its cultural heritage. This ensures a harmonious coexistence between tradition and progress. It also examines the values of education and awareness programs in instilling a sense of pride and responsibility among visitors and tourists alike. Also, the study underscores the importance of various stakeholders like the government, non-governmental organizations, tour operators, and local communities. By embracing sustainability, Manipur can unlock its cultural tourism potentiality for economic growth, social cohesion, and conservation. The data for the paper is collected through secondary sources.

Keywords: Sustainability, Cultural, Heritage, Local communities, Potentiality

1. Introduction

Tourism is becoming the third largest economy in the world (ICOM and WFFM/ FMAM, 2007). According to the Macrothema Review 3[7], 180-189, 2014, sustainable development has become one of the most usual concepts mentioned in the business world for the last three decade. In this era where the world is focusing on sustainable practices in every aspect, tourism is also one of the industries where sustainable practices are in full swing. Cultural tourism, which is also one of the trending forms of tourism in this generation, is now taking up preservations so as to crucially safeguard the culture and heritage for future generations. As the tourism industry evolves globally, it strikes a balance between the economic development and cultural conservation. It has also become imperative in regions like Manipur, where the past has a significant cultural value. Manipur is in the north eastern part of India, is a beacon for rich culture where traditions are deeply ingrained in the fabric of daily life.

Cultural tourism is an economic booster for the local community and it also serves for the purpose of safeguarding the culture and heritage of the area. There are even challenges to ensure that the tourism activity does not compromise the integrity of the cultural sites or they do not exploit the indigenous cultures for commercial gain. This brings in the need for promoting sustainable practices in tourism to safeguard the culture of the state and the legacy so as to simultaneously foster economic growth of the local communities.

Manipur has the potentiality to pave the way for a model that combines economic growth with cultural preservation through sustainability practices in cultural tourism. Engaging local communities in decision-making processes, implementing eco-friendly tourism practices and promoting responsible travel among the visitors are some of the practices that can be engage with the host community and the tourists. A path towards a successful sustainable cultural tourism can be pave through collaborative efforts among government, non-governmental organizations, stake holders and the local communities. This will help in preserving the past for the future generations.

2. Literature Review

According to EM Istoc 2012, Cultural tourism plays an important role in shaping the environment to meet the needs of consumers. The beauty of a place is now defined from the physical, tangible assets and also from the series of experiences developed around those assets, giving a living culture vibe to the local environment. A special attention should be paid to the relationship between tourism and cultural heritage (tangible and intangible), because beyond certain limits, the tourism activity can become a negative element in terms of conservation and use of this heritage. The need to integrate cultural projects with intervention programs for socio - economic development and local promotion is also an important element.

Aranburu B Plaza et al 2016, in his article mentions that policy makers and developers must understand the behavior of the tourist and also how they utilize the destination and resources available so as to set the place as a sustainable cultural tourism destination. When a tourist visits a destination they consume not only the space but also the image of the destination as they make a mental representation of the destination. Economic sustainability of the destination depends on the centrality of the tourism resources. Tourists who opted for cultural tourism practices are mostly educated so they are highly selective with their cultural destinations, time and resources.

According to F Nocca, 2017, sustainable development has become the main topic for many conferences. If the cultural landscape can play a role in sustainable development, it could be positive only if the local communities are able to produce empirical evidence about its contribution to improve economic, social, and environmental productivity of the city. It has also become important to provide empirical evidence to make sure that the conservation is an investment and not a cost. Tools to evaluate the contribution of cultural tourism for the achievement of sustainable development and the identification of new effective model for sustainable management of cultural heritage are necessary in today's era.

R Fistola et al, 2018 mentions that cultural and heritage tourism represents the physical infrastructures for supporting the sustainable and slow tourism. Tourism is in fact the facilitator for territorial development. In this generation, tourists prefer tourism forms that take care of the local population, culture and history and also characterized by the tourists who are willing to spend on the destination and its resources.

Mahna Nikbin et al 2012, in his article mention that sustainability is what the modern tourism industry should strive for. Sustainability has become an attractive topic in this era is promising a sustainable growth for present and also for the future. It has also become very important to make sustainability in

https://doi.org/10.52202/074664-0027

socio-cultural advantages for the development of tourism and for the host community.

3. Objectives

1. To explore the intersection of sustainability and cultural tourism in Manipur.

2. To discuss the challenges faced in balancing tourism development with the preservation of cultural identity in Manipur.

3. To provide recommendations to ensure long term sustainability of cultural tourism in Manipur.

4. Research Methodology

The data is collected truly from secondary sources like websites, journals, books, and articles due to time constraints. It was analyse through qualitative and content analysis. The study is exploratory and descriptive in nature.

5. Discussion

5.1 Cultural tourism in Manipur

Cultural tourism is a form of sustainable tourism if it is managed responsibly. Promotion of local cultures, traditions and heritage can foster the support for preserving the culture for generations to come. Some of the key to ensure cultural tourism benefits both the local community and the tourists is to go for sustainable practices like minimizing the environmental impact, respecting local communities and promoting economic benefits for the locals.

Manipur is a state with rich cultural resources that includes traditional dances, music, handlooms and handicrafts and many other traditional practices. When one brings out sustainability and cultural tourism in the state, preserving the past for the future generations is very important. Sustainable practices are very essential for the preservation of the state's culture. Development of tourism initiatives that respect and support the local communities, minimizing the impact on the environment, contribute to the development of socio-economic for the locals without compromising the cultural identity are some of the initiatives that can be taken up for conserving the culture of the state.

One way of achieving the balance of promoting as well as conserving is the culture through tourism is by initiating the community based tourism that empowers the local community and their involvement in the activities related to tourism. This ensures that the benefits for the local community is distributed equally and believes decisions and values are being respected. Moreover, environmental impact and preservation of the state's natural beauty for years to come can be done by practicing waste management, energy conservation and responsible wildlife viewing.

Furthermore, education and awareness programs can help in fostering a sense of pride and ownership to the local communities towards their cultural resources. This will not only help the present generation but will also helps the future generation to develop a sense of responsibility towards their cultural resources and ensure longevity and relevance in the modern world. Manipur can preserve the past culture for the future while benefiting the community in a sustainable way by practicing holistic approach.

5.2 Challenges faced by Manipur cultural heritage

Promoting sustainable cultural tourism can bring about some challenges in Manipur. One of the main challenges is the risk for exploitation of the culture and commodification of culture. Rather than respecting the culture and heritage of the state for identity and attraction, there arise the danger for depletion of cultural practices for mere attractions for the tourists. This leads to the depletion of cultural practices for commercial gain understanding the authenticity and for long term viability.

The threat of environmental degradation is another challenge faced by Manipur cultural heritage. With increase in the number of tourism activities there raise the risk for straining the ecologically sensitive areas like the scenic landscapes and biodiversity hotspots. If there is no proper management and unchecked tourism development, it could lead to destruction of the habitat, pollution and other forms of environmental harm, destroying the very resources that make Manipur an attractive cultural tourism destination.

Also, an obstacle to sustainable cultural tourism arises with the socio-economic disparities within the state. Different communities can exert tensions among themselves as they get unequal access to resources available and opportunities and this can hinder inclusive development initiatives. Proactive measures are required to promote equal participation of the communities and avoid disparities by ensuring that all the residents get access in the benefits of tourism related income and opportunities.

Moreover, a challenge can be posed for preserving the traditional cultural practices with the rapid growth in globalization and modernization. There is risk for cultural erosion as the new generation is getting exposed to external influences and alternative lifestyles and loses the heritage knowledge. It has become an important task to balance the preservation of cultural heritage with the need for adaptation and innovation to meet the contemporary demands for the engagement of the local community.

Hence, efforts to promote sustainable cultural tourism can be slowed down with the hindrance of inadequate infrastructure and institutional support. Also the implementation of effective conservation and management of the strategies may be impeded by insufficient funding, limited access to training and capacity building programs and weak governance structure. To overcome these challenges, a coordinated approach that involves the government organizations, local communities, civil societies and other stakeholders are required to strengthen the institutional frameworks, improve infrastructure and promote sustainable cultural tourism practices that can benefit both the present and the future generations to come.

5.3. Sustainability practices on Cultural Tourism

Sustainability practices for cultural tourism encompasses a wide range of strategies that aimed to minimize the negative impact on the environment, support the local community and also preserve the cultural heritage of the region. One of the sustainability practices is to promote responsible tourism practices among the tourists that includes respecting the local customs and traditions, minimizing generation of waste and conserve the natural resources. Also, it is important to foster community engagement and empower the local community so that the tourism practices benefits the local community and enhance their lifestyles.

Preservation and restoration of cultural heritage sites and artifacts is another important sustainability practice for cultural tourism. It involves implementing

https://doi.org/10.52202/074664-0027

conservation measures to protect historical buildings, monuments and artifacts from degradations and destructions that may be caused from tourism related activities. Moreover, the use of eco-friendly infrastructure and transportation options should be priorities for sustainable tourism development so as to reduce the carbon emissions and minimize ecological footprint.

Furthermore, equitable distribution of economic benefits should be prioritizing to benefit the host community in the sustainable cultural tourism practices. This can be achieved through community based tourism projects that involve the local community in decision making processes and also in providing opportunities for the locals to participate in business and activities related to tourism. Education and training programs should also be invested upon to empower the local community with skills and knowledge that are required to understand tourism opportunities in a sustainable way.

Thus, it has become very essential to integrate sustainability practices in cultural tourism practices for long term viability and resilience of the destination while also preserving the unique and rich cultural resources for future generations. Cultural tourism can therefore become a powerful force for a positive change and inclusive growth by prioritizing environmental conservation, supporting the community development and promoting responsible travel behavior.

5.4 Sustainability emerging trends and opportunities

The world is developing at a rapid speed and with this there emerge trends and opportunities for sustainability in cultural tourism that offers promising pathways for addressing the present environmental and social challenges while promoting economic growth and resilience. The increase in focus on circular economy principles is one notable emerging trend. It emphasizes the importance of designing products and systems that minimize waste and maximize resource efficiency. This helps the businesses to innovate and develop sustainable products and services and in the meanwhile reduce the environmental carbon footprint.

The growing recognition of the importance of biodiversity conservation and restoration of ecosystem is also another trend that is growing very fast. Almost everyone is concern about the climate change and loss of habitat. Emphases are given on restoring the natural ecosystem by safeguarding the biodiversity and services of the ecosystems. This helps in providing opportunities for investing in solutions related to nature-based like the reforestation, sustainable agriculture and conservation-based tourism that deliver multiple benefits for both the local people as well as the tourist.

Moreover, sustainable and ethical consumption across various sectors that includes food, fashion and energy are rising in demand. Products and services that align with the values like organic, fair trade and cruelty free goods are increasingly seek by the consumers. This helps in creating opportunities for businesses to differentiate by adopting various sustainable practices through their supply chain and by offering transparent and ethically sourced products to meet the ever evolving preferences of the consumers.

Furthermore, new possibilities are being opened up by the digital revolution for promoting sustainability through technologies like AI, blockchain and renewable energy solutions. This helps in enabling more efficient resource management, decentralized energy system and transparent supply chain that helps in accelerating progress towards sustainability goals. Also the role of digital platforms and social media is very crucial in raising awareness, mobilizing action and also fostering collaboration among stakeholders in the movement of sustainability.

Hence, the emerging trends and opportunities in sustainability offers an immense potential for positive change and also creates a more sustainable and resilient future for the people and the planet. A more sustainable world for the present and future generations can be harness by embracing innovation, collaboration and responsible leadership, business, governments and civil society.

6. Conclusion

In conclusion, it is essential to ensure longevity and vitality by preserving the cultural and heritage of Manipur through sustainable tourism practices for future generations. The state can safeguard the unique cultural assets by prioritizing sustainability while simultaneously fostering economic development and protection of the environment. Also the state can strike a harmonious balance between the preservation of the past while embracing the opportunities of the future through initiatives like community based tourism, responsible behavior of the tourists and the conservation of natural and cultural resources.

Moreover, local communities are empowered and their socio-economic well beings are also enhanced by sustainable cultural tourism. Manipur can promote a sense of ownership and pride among the people by involving the local community in decision-making processes of tourism and also by creating opportunities for them to showcase the cultural heritage. This will not only benefit the local community but will also enhance the experiences of the tourists by fostering mutual respect and understanding between the cultures.

Eventually, the success of sustainability and cultural tourism in Manipur depends on the collaboration and commitment of the various stakeholders, government organizations, civil societies and even the tourists. The state can ensure that the cultural heritage remains a source of pride and inspiration for future generations by working together to implement sustainable practices and thus contributing to the prosperity and well being of the whole region.

Reference

1. Girard, L.F., Nijkamp, P.: Cultural tourism and sustainable local development. (2016).https://doi.org/10.4324/9781315258720.

2. Iştoc, E.M.: Urban cultural tourism and sustainable development. Urban Cultural Tourism and Sustainable Development. 1, 38–57 (2012).

3. Aranburu, I., Plaza, B., Esteban, M.: Sustainable cultural tourism in urban destinations: Does space matter? Sustainability. 8, 699 (2016). https://doi.org/10.3390/su8080699.

4. Zhang, G., Chen, X., Law, R., Zhang, M.: Sustainability of Heritage Tourism: A Structural Perspective from Cultural Identity and Consumption Intention. Sustainability. 12, 9199 (2020). https://doi.org/10.3390/su12219199

5. Korez-Vide, R.: PROMOTING SUSTAINABILITY OF TOURISM BY CREATIVE TOURISM DEVELOPMENT: HOW FAR IS SLOVENIA? Innovative Issues and Approaches in Social Sciences. 6, (2013). https://doi.org/10.12959/issn.1855-0541.iiass-2013-no1-art05.

 https://doi.org/10.52202/074664-0027

6. Nocca, F.: The role of Cultural Heritage in Sustainable Development: Multidimensional Indicators as Decision-Making Tool. Sustainability. 9, 1882 (2017). https://doi.org/10.3390/su9101882.

7. Casagrandi, R., Rinaldi, S.: A theoretical approach to tourism sustainability. Conservation Ecology. 6, (2002). https://doi.org/10.5751/es-00384-060113.

8. Astawa, I.P., Triyuni, N.N., Santosa, I.D.M.C.: Sustainable tourism and harmonious culture: a case study of cultic model at village tourism. Journal of Physics. Conference Series. 953, 012057 (2018). https://doi.org/10.1088/1742-6596/953/1/012057

9. Tyrrell, T.J., Johnston, R.J.: Tourism Sustainability, Resiliency and Dynamics: towards a more Comprehensive perspective. Tourism and Hospitality Research. 8, 14–24 (2008). https://doi.org/10.1057/thr.2008.8.

10. Ko, T.G.: Development of a tourism sustainability assessment procedure: a conceptual approach. Tourism Management. 26, 431–445 (2005). https://doi.org/10.1016/j.tourman.2003.12.003.

11. Fistola, R., La Rocca, R.A.: Slow mobility and cultural tourism. walking on historical paths. In: Green energy and technology. pp. 301–322 (2018). https://doi.org/10.1007/978-3-319-77682-8_18.

12. Chhabra, D.: Sustainable marketing of cultural and heritage tourism. (2010). https://doi.org/10.4324/9780203855416

13. Richards, G., Hall, D.: Tourism and sustainable community development. (2002). https://doi.org/10.4324/9780203464915

14. Hall, D.: Tourism development and sustainability issues in Central and South-Eastern Europe. Tourism Management. 19, 423–431 (1998). https://doi.org/10.1016/s0261-5177(98)00039-9.

RAISA in Tourism and Hospitality Industry: A Study on Technology Implementation and Implications in Indian Tourism and Hospitality Sector

Chanchreek sharma[1], Deepakshi Garg[2] & Kabeer Dubey[3]
[1,2,3]Postgraduate scholar, NCHM-Institute of Hospitality, Noida

Abstract.
Human touch has always been an influential aspect in the hospitality industry. With increasing globalization and fast paced changes in the trends and technology. Robotics, AI, and Service automation have become a hot topic for discussion in tourism and hospitality industry. RAISA has emerged to be the flag bearer of the new and innovative technological trends being adapted by the tourism and hospitality industry globally. The study focuses on RAISA in tourism and hospitality industry: A study on technology implementation and implications in Indian tourism and hospitality sector. The advent of automation in the tourism and hospitality industry is effectively supported by economy and goal of providing better customer service. This automation of the hospitality industry highly relying on the human interactions has been highly prioritized in area of R&D post 2016. India being a country of diverse culture, highly relies on the tourism and hospitality sector for both economy and employment. Many of the Indian hotel chains and restaurants have started to incorporate AI, Robotics and IOT in their work force, like the Yellow House Robo restaurant, The Oberois, Biriyani by Kilo (in South, India). The study is exploratory in nature, using quantitative research methodology. The primary data collection was done using a self-administered survey questionnaire, while the secondary data was collected from a wide study of 24 research papers on RAISA focusing both global and India aspects. The sample size of target was 150 and, in the end, only 133 respondents participated. The collected data was analysed using M.S Excel and R&R studio to derive some concrete insights on the technology implementation and implications in Indian tourism and hospitality sector. Models were built for finding significant relationship to support the research. The study finding can be an aid to the Indian hotel chains, startups and QSR to plan the incorporation of RAISA in the tourism and hospitality sector and capitalize on the technological innovations and maintain a balanced human-technology relationship to efficiently improve the productivity and enhance the customer experience. An open eye should be kept for all the implication and challenges and strategies can be formulated to improve the incorporation and overcome the short comings of incorporating RAISA in the Indian tourism sector.

https://doi.org/10.52202/074664-0028

.

Keywords: RAISA, UNTWO, Human interactions, Tourism and Hospitality industry, Tourism demand.

1. Introduction

An invention of Thomas Cook, Thomas Shelby and Alfred Hitchcock, tourism is basic need in day-to-day life. Having several definitions, commonly, tourism is the movement of people either inland or outland for any purpose ranging from leisure activities to culinary experience or business, etc. The type of movement is temporary, and the duration can vary but range less than a year. [3]

The tourism industry comprises of all the sectors that cater to travellers. But to classify broadly there are 4 main sectors, namely: - Transportation (conveyance), Accommodation (hotels, motels, hostels, B&Bs, etc), Ancillary Services (travel insurance, foreign exchange, passport, and visa services, etc), and Sales and distribution (travel agency, tourism organizations, touring companies).

Hospitality or accommodation is a very important part of tourism. The hotel industry with all its aspect is directly proportional to economic growth and stability of the destination. Competitiveness in hotel industry is measured by its market share. The main aim of any hotel is to maintain and increase its productivity that encompasses an all-around approach and consolidate efficiency, quality, predictability, and effectiveness in terms of performance standards [19].

The tourism and hospitality industry are no exception when it comes to successfully conveying emotions to the client. It has always been about the human touch, or the homely feeling and warmth provided. Human touch was once seen as a competitive advantage, but Morden times have seen a shift from that human touch to technology and artificial intelligence (AI) [18].

This is an era of rapid growth and change in a very dynamic and highly competitive business environment. The customer preferences change, new technologies are introduced almost every alternate day, thus forcing organizations in all sectors to reinvent, reorganize, and continuously restructure to keep up in this competitive environment [20].

The tourism and hospitality industry are also accustomed to these changing trends, with RAISA being the differential factor in present and future of this sector. IOT has been in the tourism sector for past 66 years and has now become an important aspect of this Morden tourism and hospitality sector.

The advent of automation in hospitality industry is fuelled by economy, efficiency, and goal of providing better customer experience. It has numerous benefits, but also certain downsides, out of which a major downside that can be taken into consideration is the need to balance automation with personalised human interactions.

India has been a country of rich culture and is known for its invaluable treatment and facilities to all the tourists. Tourism has been a plus point for India in terms of economy and employment both. Tourism industry is a very important source of foreign exchange earnings in India. Recent times have seen India rise as a popular tourist attraction in the world, creating value for the tourists [12].

The covid-19 pandemic in 2020 has severely affected the tourism sector in India in general and globally as whole. given the important role tourism sector plays in Indian economy, the pandemic has had a devastating effect on the Indian economy. According to Ministry of Tourism (MOT), post covid, the number of foreign tourists decreased to 1.52 Mn in 2021 from 10.93 Mn in 2019. But the year 2022, again saw increase to 6.19 Mn. The 2013 report of World Economic Forum's Travel and Tourism Competitiveness, India secured 11th position in the Asia-Pacific region and out of 140 economies globally, it ranked 65th on the index [12]. As of 2020, according to WTTC, Indian tourism industry's contribution to GDP increased from 958.17 Bn to 2190.24 Bn in just 7 years. Studies showed this as a fruitful contribution of proper HRM strategies implemented in the Indian tourism and hospitality industry, thus showing the power of human labour in this sector.

https://doi.org/10.52202/074664-0028

2. Literature Review

Tourism has a composite nature of its product as it has tourists purchase, experience, and sum of all services inclusive. All the products and sub-products are inter-dependent. This shows that if one sub-product is unavailable then the other will not be of any use, also bad experience of one sub-product can hamper or badly influence the impression of another sub-product. Where, accommodation is seen as the most influential sub-product in all aspects [3].

Human interactions are vital to everything that occurs in the tourism and hotel sector. Goods being provided in industry are intangible, but as it is service-based sector, it becomes very important to have highly skilled and trained labour to carry out the operations. the human resource forms the base for any hospitality organization [4]. But we cannot unsee the fact that machines and advancement is technologies may greatly benefit the experience and service provided in the industry. The human factor is essential for efficient operations of all divisions of the tourism sector. however, in the wake of pandemic (Covid-19), RAISA has gained popularity in the hospitality sector.

Tourism and hospitality industry is catching up the change in trend and use of technology at a very fast rate. The over-crowded market and fear of losing has made different hospitality organizations to use innovation and technology as a tool to win the race, and trying to meet the current guest expectations, the hotel industry is adopting RAISA.

Innovation in the modern times is seen as the most crucial element in the tourism sector. It is a tool to hit the bullseye in today's fiercely competitive market, which is created by the surrounding community providing excellent quality which is well worth the cost. RASIA brings about the change and improvements by which the tourism and hospitality sector can enhance their service and serve their patrons, that have not been talked about earlier. [14].

First literary mention of hotels using technology, robotic applications was in early 1990's [23], but the first ever use of technology in the sector was the automated reservation system introduced by Sheraton in 1958. This forces us to note that automation is not only use of robots and AI or service automation, but any use of technology or IOT, which can be automatic reservation system, authorized key card elevators, Check-ins & check-outs via mobile, maintenance and housekeeping, etc.

The use of RAISA in the tourism sector primarily benefits the enhancement of the visitor experience, proved by the previous research in the scope of technology in the tourism and hospitality sector. The hotel operations accustomed to the specific requirements of the clients mainly emphasises the customer experience. Its main aim is to covert the happy customers into devoted customers and improve customer retention in the sector, thus improving the word-of-mouth promotions. The application of RAISA in the guest cycle can be seen in the following:

- **Application of RAISA at the Pre-arrival stage:** -Mobile technologies, Virtual reality, Chatbots
- **Application of RAISA at the Arrival stage:** -Digital kiosks, Mobile check-in, Smartphone Room Keys/Non-stop check-in (NSCI)
- **Application of RAISA during the Occupancy stage:** -IOT, and service automations, F&B service by the robots.
- **Application of RAISA during the Departure stage:** -Travel assistant, Express Checkout, Digital kiosks

Thus, RAISA can be of great help to hotel operations [11].

Some recent studies have categorised the roles for service automation into Support and Substitute. Also, these studies have constantly pressed concern on human touch and supported the upliftment of RAISA in tourism and hospitality industry post Covid-19 [6].

Tourism & hospitality industry, like every other industry requires certain perquisites for proper functioning, like: - raw material, finance and fundings, machinery, etc. But, most importantly it requires a skilled labour to work in different departments focused in HRM of a company. It involves managing manpower to improve individual, group, and organizational effectiveness in total. It helps to access the human touch in every aspect of Tourism and hospitality sector, thus significantly impacting the service quality and touching the emotional aspect of guest satisfaction [10].

The hotel industry is ever evolving, the changes seen are not in its nature but keeping in view the ask of the hour. This has led to the evolution of hotels into Boutique hotels, smart hotels, green hotels, and adventure hotels, leaving behind the concept of old classification on basis of its clients or guests like business hotels, resort, etc [8].

Studies post-2016 have prioritized the use of automation in the tourism industry and have attributed certain reasons for the current influence of automation. The reason is broadly linked to resources and quality management. Another reason not focused but of importance is the shortage of skilled labour [20]. Some of the automatable tasks in the hospitality industry include heavy-duty work, data processing, repetitive tasks, statistical calculations, etc. These tasks can be automatized using computers, mobiles, chatbots, kiosks, autonomous pickups, and robots [5].

RAISA can improve operational efficiency and revolutionize guest experience by integrating smart devices and AI-driven systems to increase guest satisfaction and comfort, also contributing to decreased costs [4]. The artificial intelligence is entirely changing the corporate world. The involvement of chatbots, and intelligent service bots are leading the race as important customer support avenues. they provide an all-around solution for customer support, and communication on all platforms. The potential use of robots in different domain areas of tourism and hospitality industry include, hotel industry, meetings and events, restaurants, theme and amusement parks, museums and galleries, and airports [16].

India is a country of diverse culture. Tourism and hospitality industry form a major pillar of the Indian economy. Indian economy is a volatile economy, with constant changes happening to attain the title among the great nations. The tourism and hospitality industry are ranked among the top 5 sectors in India, when viewing in the aspect of economy [7].

The recent trends have seen advancement in Engineering and sciences due to heap of unprecedented amounts of data. There has been an outbreak of robotics from confined areas of factories to the dynamics of human environment [21]. These developments have also

https://doi.org/10.52202/074664-0028

shattered the boundaries of the tourism and hospitality sector, which can be seen as different robotic applications in cooking and service. Some examples of such robot interventions in tourism and hospitality industry are, Burger robot in Troition opened in 2018 with the capacity to cater to 120 orders/hour, developed in California; Robot baristas at Cafe X in Canales that came into working in 2018 that can produce 3 beverages in 40 seconds, and many other examples from Japan, England, India, and other parts of the world.

The utilisation of technology by hotels and other businesses to enhance guest experiences and boost operational productivity has led to a steady development in automation in the Indian hotel sector. In a survey Indian hotel have more than 300 installations of robots used hotels, Oberois, ITC, Taj, and Marriott being a few of them [10]. "Demand for such products in the hospitality segment is coming back. Some examples of use of robotics in India are: - Robo-chefs in Chennai & Bangalore, Ruby- the Yellow House Robo restaurant (Jaipur & Noida), Robot SM Post test restaurant, Chennai, and Robot restaurant, Bangalore.

The review of study specifically focused on guest attitude towards automation in hotels in Delhi, by [15], showed mixed customer reviews on use of RAISA in hotel industry. Their data findings showed an inclination towards use of robots as they have a faster work rate with more accuracy. But, on the other hand the respondents feared the lack of social skills putting a question mark on their problem-solving capability. The main conclusion of the study focused on the need for hoteliers to maintain balance between the robotics and human employees which would thus help cater to the guest needs.

Economic viability of RAISA in India, depends on how well the companies can weigh the pros and cons of RAISA. it may boost productivity and competitiveness, but its repercussions on society and economy (job displacement inclusive). It is upon the stakeholders and management to not only become tech savvy and incorporate technology, but also find ways to combine advanced technology with the human touch and keep human resource handy and skilled to utilize RAISA fully for better productivity and outcome.

Several factors influence the adoption of automation in the Indian hospitality industry, including the target market, the institution's size and kind, and financial considerations. While larger hotel chains may invest in comprehensive automation systems, smaller hotels may focus on certain areas that align with their objectives.

RAISA being introduced as substitution of human employees raises the fear of job loss and unemployment [5]. Now, service organizations have started shifting to AI technology and robotics to handle simple to complex tasks, replacing humans. Although RAISA has yet not affected employment negatively, but the current situation forces to predict that in future, technology will overtake the human touch [17]. Although RAISA's use till date is sporadic and adoption is minimal, it still presents an influential support to the tourism and hotel industry around the world, but there must be a careful implementation to cater the wishes of customer and fears of the employees [1].

Incorporation of technology-based initiatives in many organizations has become common now-a-days, but a negative perception of customers can also be seen in accordance with these technological developments [2]. The everyday life now is seeing AI and Robotics become crucial in the social discourse. Intelligent services using AI accurately predict key tasks allowing individuals, and organizations to deliver quick responses and solutions to all problems from major to minor [9].

The need of the hour asks luxury hotel brands to integrate smart technology with their services. the rapid innovation in technology has influenced the guest expectations and challenged the regular working strategies in the tourism and hospitality sector [8].
Major Implications in India for inclusion and adaption of RAISA can be divided into huge implementation costs, automation-employee balance, job conservation for employees, and enhancing customer satisfaction by creating a better understanding of RAISA in the tourism and hotel industry. The major costs of adopting RAISA can be divided into: - Financial costs (initial costs), non-financial costs (resistance and barriers costs). these costs can lead to financial implications and hence should be properly catered to before adoption of RAISA [6].

https://doi.org/10.52202/074664-0028

3. Research Methodology

The research study on RAISA in tourism and hospitality industry: A study on technology implementation and implications in Indian tourism and hospitality sector aims: -

- To find the Viability of RAISA in Indian tourism and hospitality industry.
- To analyse the influence of RAISA on guests, employees & suppliers in tourism and hospitality.
- To identify business implications due to adoption of service automation and robotics.

Owing to the dearth of research publications on the Indian tourism and hospitality sector within the framework of RAISA, we conducted the study utilising an exploratory approach and a quantitative research technique.

To provide the foundation to the study and impart a fundamental knowledge of RAISA in the tourism and hospitality sectors, a formalised literature review was conducted with the assistance of secondary data sources. A substantial body of research was available on RAISA's effects on employment and customer happiness in the tourism and hospitality sectors globally, as well as its growth during and post the Covid-19 epidemic. But relatively few scholars have specifically addressed the Indian tourism and hospitality sector when discussing about RAISA. This research paper's main goal is to close this information gap and offer insightful commentary on the implementation and implications of RAISA in the Indian tourism and hospitality industry.

Self-administered survey questionnaires were created to gather primary data, or firsthand knowledge, to support our goals. 28 questions in total (27 required and 1 conditional) were broken down into several sections of the questionnaire. The initial segment of 6 inquiries aimed at obtaining the respondents' demographic data. Insight into respondents' understanding of RAISA and how they view it in relation to the tourism and hospitality industry globally sought for by the 6-question segment that followed. In the third phase of the comprehensive questionnaire, there were 10-questions designed to elicit responses regarding employees, suppliers, and customer satisfaction (9 were compulsory and 1 conditional). The last 6-question section was designed to gather respondents' opinions of RAISA in the travel and hospitality industry, with an emphasis on India.

With 1.42 billion people and a diversified population, India is a nation of thought leaders. Considering this, the target audience consisted of a combination of Indian suppliers, employers, employees, and visitors in the tourism sector. When the focus of the research is on the local population, sampling presents a challenge for tourism researchers [13]. The study aims to collect information about the implementation and consequences of RAISA in the Indian tourist and hospitality industry. Consequently, non-probability and convenience sample techniques were employed in compliance with normal sampling methods. A sample size of 150 respondents was targeted, but the actual number of responses collected was 133. Collected data was then analysed using MS Excel to help formulate the interpretations and consolidate the conclusions of the research.

https://doi.org/10.52202/074664-0028

4. **Findings**

The survey was filled by 133 respondents. the survey was formed considering all roles in the tourism and hospitality industry, that is the guests, tourism and hospitality industry employees and employers, and the raw material suppliers. out of the 133 respondents, 42 were guests, 34 were raw material suppliers, 28 employees of the industry and 29 were employers or from the management. A code book (Table 4.1) was created. Firstly, a summary statistic test (Table 4.2) was carried out on the data collected followed by a correlation test was conducted on the collected data using MS Excel, a heat map (Fig4.1) was generated using the R&R Studio. Different models were created based on the correlation coefficient (Table 4.3) and regression analysis was done to formalize the statistical significance of the models. Based on data analysis the following interpretations were formulated regarding the role of RAISA in the tourism and hospitality industry in India: -

- The data was divided into models with positive correlation (Table 4.4) and negative correlation (Table4. 5)
- The first 4 models, that is M1-M4 highlight strong positive correlation. the model M1 and M2 previewed the positive influence of familiarity with RAISA on its role in the tourism and hospitality industry and enhanced customer satisfaction. The 3rd model supported a good impact of RASIA's importance in the industry and its role in the industry. M4 denoted strong positive correlation among the negative impact of RAISA on employee enthusiasm and job displacement showing the potential risks of replacement and employee feelings about the job security.
- The Models M5-M7 depicted moderate positive correlation among the positive impact of RAISA on customer satisfaction, visitor experience and how it can influence the human resource of the industry.
- The models, M8 and M13 showed weak positive correlation between the importance of RAISA for hotels and its ability to nullify the human error, previewing RAISA as a boon to reduce the human error in daily operations.
- the models, M9, M10, M14, M17 covering RAISA's importance in the industry and its influence on customer satisfaction and supplier efficiency was also supported by weak positive correlation. this correlation suggested the significance of RAISA to enhance the service quality and operational productivity, giving and edge of easy work options for the suppliers as well.
- The models, M15 and M16 preview the weak positive correlation in the economic viability of RAISA in India and its role in the industry. This justifies the need to focus on future implications of the economic viability of adoption of RAISA technologies.
- The models, M18-25 denote weak negative correlation among the variables of familiarity with RAISA, previous experiences with RAISA, influence of RAISA on customer satisfaction, ease of work for suppliers, job displacement due to RAISA, and its impact on the enthusiasm of the employees pressing the potential concerns regarding job security and the negative impact on the human touch that forms the base for the homely feeling in the tourism and hospitality industry. Previous encounters of the guests with RAISA can be very influential in the customer satisfaction.
- The last model, M26 reveals a strong negative correlation between familiarity with RAISA and previous experience with RAISA, showing that the familiarity with RAISA is strongly dependent on previous encounters with RAISA in the truism and hospitality industry. it also emphasises on the gap between theoretical knowledge and practical application the RAISA in the hospitality sector.

 https://doi.org/10.52202/074664-0028

Table 4. 1. Code Book

Categories	Variable Name	Question	Data Type	Variable Value 1	Variable value 2	Variable value 3	Variable value 4	Variable value 5
Demographic Information	Age	Age	Interval	< 18	19-24	25-40	41-55	>55
	Gender	Gender	Ordinal	Male	Female	Others		
	Occupation	Occupation	Ordinal	Student	Tourism and hospitality professional	Other industry professionals	Unemployed	
RAISA	Awareness about RAISA	Are you aware about Robotics, Artificial Intelligence & Service Automation (RAISA)	Nominal	Yes	No			
	Familiarity with RAISA	How familiar are you with RAISA?	Ordinal	Not at all familiar	Not familiar	moderately fmailiar	famillliar	totally familiar
	Previous experience with RAISA	Have you encountered RAISA in any tourism sector?	Nominal	Yes	No			
General Satisfaction	Importance of RAISA in hotels	How Important would you rate RAISA for hotels?	Ordinal	Not at all important	Not important	Moderately Important	Important	Very important
	Economic viability of RAISA	RAISA is economically viable in Indian tourism and hospitality industry.	Ordinal	Strongly Disagree	disagree	neutral	agree	strongly agree
	RAISA role in tourism and hospitality industry	On a scale of 1-5 how helpful is RAISA in tourism and Hospitality Industry in India.	Ordinal	Not at all helpful	Not helpful	Moderately helpful	Helpful	Very helpful
RAISA and Employees	RAISA nullify human error	Does RAISA nullify the human error?	Ordinal	Strongly Disagree	disagree	neutral	agree	strongly agree
	Negative impact of RAISA on enthusiasam	Will RAISA negatively impact the level of enthusiasm and dedication a worker feels toward their job.	Ordinal	Strongly Disagree	disagree	neutral	agree	strongly agree
	Job displacement due to RAISA	Do you think RAISA endangers the job opportunities of employees in the organization?	Ordinal	Strongly Disagree	disagree	neutral	agree	strongly agree
	RAISA replace human touch	Can Robotics, Artificial Intelligence, and Service automation replace the human touch in India.	Nominal	Yes	No			
	Impact of RAISA on employees	On a scale of 1-5 how will RAISA impact employees and job opportunities in India.	Ordinal	Severily negative impact	slightly negative impact	Neutral	Slightly positive impact	Greatly positive impact
RAISA and Suppliers	RAISA for suppliers	Is RAISA helpful or problematic for suppliers.	Ordinal	Problematic	Can't Say	Helpful		
	RAISA ease the work of suppliers	RAISA ease the work of the supplier.	Ordinal	Strongly Disagree	disagree	neutral	agree	strongly agree
RAISA and Customer satisfaction	RAISA positively affect your	RAISA positively affect your visit.	Ordinal	Strongly Disagree	disagree	neutral	agree	strongly agree
	Prior interaction with robots in hospitality industry.	Have you visited any hotel or restaurant using robots in cooking or service?	Nominal	Yes	No			
	Influence of RAISA in customer satisfaction	On a scale of 1-5 how influential is RAISA in customer satisfaction in India.	Ordinal	Not at all influential	slightly influential	Moderately influential	Influential	Very influential

Table 4. 2. Summary Statistics

	Mean	Standard Error	Standard Deviation	Sample Variance	Skewness
Age	3.15	0.07	0.78	0.61	0.40
Gender	1.28	0.04	0.47	0.22	1.23
Occupation	2.44	0.06	0.70	0.49	-0.58
Awareness about RAISA	1.01	0.01	0.09	0.01	11.53
Familiarity with RAISA	4.19	0.07	0.75	0.56	-1.20
Previous experience with RAISA	1.05	0.02	0.22	0.05	4.05
Importance of RAISA for hotels	4.42	0.05	0.63	0.40	-0.99
RAISA nullify the human error	3.86	0.06	0.75	0.56	-1.11
Negative impact of RAISA on enthusiasam	3.24	0.09	1.00	1.00	-0.50
Job displacement due to RAISA	3.11	0.11	1.28	1.63	-0.57
RAISA for suppliers	4.11	0.06	0.73	0.53	-0.87
RAISA ease the work of the supplier	4.23	0.06	0.74	0.54	-1.44
RAISA positively affect your visit.	1.80	0.03	0.40	0.16	-1.55
Prior interaction with robots in hospitality industry.	4.19	0.07	0.76	0.58	-1.28
Economically viablitiy	4.43	0.06	0.71	0.50	-1.48
RAISA role in tourism and hospitality industry	3.63	0.10	1.12	1.25	-0.85
Impact of RAISA on the employees	4.40	0.06	0.75	0.56	-1.13
Influence of RAISA in customer satisfaction	3.03	0.12	1.39	1.92	-0.28
Can RAISA replace humans	3.03	0.12	1.39	1.92	-0.28

Table 4.3. Pearson's Correlation Coefficient

	Age	*Gender*	*Occupation*	*Awareness about RAISA*	*Familiarity with RAISA*	*Previous experience with RAISA*	*Importance of RAISA for hotels*	*RAISA nullify the human error*	*Negative impact of RAISA on enthusiasam*	*Job displacement due to RAISA*	*RAISA for suppliers*	*RAISA ease the work of the supplier*	*RAISA positively affect your visit.*	*Prior interaction with robots in hospitality industry.*	*Economically viablitiy*	*RAISA role in tourism and hospitality industry*	*Impact of RAISA on the employees*	*Influence of RAISA in customer satisfaction*	*Can RAISA replace humans*
Age	1.000	-0.095	0.446	0.206	0.235	-0.261	0.270	0.139	0.069	0.119	0.035	0.012	0.070	-0.099	0.093	0.196	-0.049	0.078	0.100
Gender	-0.095	1.000	-0.119	0.323	0.044	0.076	0.037	-0.239	-0.242	-0.193	-0.153	-0.042	-0.080	-0.195	-0.061	0.026	-0.180	0.006	-0.107
Occupation	0.446	-0.119	1.000	0.070	0.030	-0.099	0.044	0.070	0.152	0.156	-0.125	0.013	0.007	-0.072	0.009	0.033	0.023	0.071	0.025
Awareness about RAISA	0.206	0.323	0.070	1.000	0.095	-0.021	-0.058	-0.218	0.154	0.129	-0.319	-0.132	-0.383	-0.177	-0.252	0.070	-0.206	0.070	0.124
Familiarity with RAISA	0.235	0.044	0.030	0.095	1.000	-0.600	0.552	0.208	-0.202	-0.212	0.282	0.364	0.468	-0.180	0.291	0.559	0.038	0.581	-0.115
Previous experience with RAISA	-0.261	0.076	-0.099	-0.021	-0.600	1.000	-0.373	-0.183	-0.023	0.006	-0.216	-0.265	-0.396	0.116	-0.292	-0.428	-0.134	-0.442	-0.176
Importance of RAISA for hotels	0.270	0.037	0.044	-0.058	0.552	-0.373	1.000	0.235	-0.402	-0.408	0.333	0.462	0.619	-0.032	0.232	0.728	-0.187	0.493	-0.301
RAISA nullify the human error	0.139	-0.239	0.070	-0.218	0.208	-0.183	0.235	1.000	0.105	-0.024	0.256	0.290	0.319	0.063	0.228	0.196	0.194	0.423	0.019
Negative impact of RAISA on enthusiasam	0.069	-0.242	0.152	0.154	-0.202	-0.023	-0.402	0.105	1.000	0.814	-0.207	-0.273	-0.323	0.062	-0.118	-0.285	-0.130	-0.149	0.557
Job displacement due to RAISA	0.119	-0.193	0.156	0.129	-0.212	0.006	-0.408	-0.024	0.814	1.000	-0.072	-0.159	-0.334	-0.076	-0.169	-0.296	-0.178	-0.269	0.657
RAISA for suppliers	0.035	-0.153	-0.125	-0.319	0.282	-0.216	0.333	0.256	-0.207	-0.072	1.000	0.529	0.434	-0.058	0.208	0.279	0.115	0.219	-0.010
RAISA ease the work of the supplier	0.012	-0.042	0.013	-0.132	0.364	-0.265	0.462	0.290	-0.273	-0.159	0.529	1.000	0.587	-0.033	0.203	0.496	0.122	0.435	-0.003
RAISA positively affect your visit.	0.070	-0.080	0.007	-0.383	0.468	-0.396	0.619	0.319	-0.323	-0.334	0.434	0.587	1.000	0.002	0.377	0.618	0.059	0.544	-0.148
Prior interaction with robots in hospitality industry.	-0.099	-0.195	-0.072	-0.177	-0.180	0.116	-0.032	0.063	0.062	-0.076	-0.058	-0.033	0.002	1.000	0.027	-0.023	0.092	-0.067	0.038
Economically viablitiy	0.093	-0.061	0.009	-0.252	0.291	-0.292	0.232	0.228	-0.118	-0.169	0.208	0.203	0.377	0.027	1.000	0.332	0.431	0.336	0.017
RAISA role in tourism and hospitality industry	0.196	0.026	0.033	0.070	0.559	-0.428	0.728	0.196	-0.285	-0.296	0.279	0.496	0.618	-0.023	0.332	1.000	-0.048	0.617	-0.075
Impact of RAISA on the employees	-0.049	-0.180	0.023	-0.206	0.038	-0.134	-0.187	0.194	-0.130	-0.178	0.115	0.122	0.059	0.092	0.431	-0.048	1.000	0.095	0.046
Influence of RAISA in customer satisfaction	0.078	0.006	0.071	0.070	0.581	-0.442	0.493	0.423	-0.149	-0.269	0.219	0.435	0.544	-0.067	0.336	0.617	0.095	1.000	-0.063
Can RAISA replace humans	0.100	-0.107	0.025	0.124	-0.115	-0.176	-0.301	0.019	0.557	0.657	-0.010	-0.003	-0.148	0.038	0.017	-0.075	0.046	-0.063	1.000

https://doi.org/10.52202/074664-0028

	Age	*Gender*	*Occupation*	*Awareness about RAISA*	*Familiarity with RAISA*	*Previous experience with RAISA*	*Importance of RAISA for hotels*	*RAISA nullify the human error*	*Negative impact of RAISA on enthusiasam*	*Job displacement due to RAISA*	*RAISA for suppliers*	*RAISA ease the work of the supplier*	*RAISA positively affect your visit.*	*Prior interaction with robots in hospitality industry.*	*Economically viablitiy*	*RAISA role in tourism and hospitality industry*	*Impact of RAISA on the employees*	*Influence of RAISA in customer satisfaction*	*Can RAISA replace humans*
Age	1.000	-0.095	0.446	0.206	0.235	-0.261	0.270	0.139	0.069	0.119	0.035	0.012	0.070	-0.099	0.093	0.196	-0.049	0.078	0.100
Gender	-0.095	1.000	-0.119	0.323	0.044	0.076	0.037	-0.239	-0.242	-0.193	-0.153	-0.042	-0.080	-0.195	-0.061	0.026	-0.180	0.006	-0.107
Occupation	0.446	-0.119	1.000	0.070	0.030	-0.099	0.044	0.070	0.152	0.156	-0.125	0.013	0.007	-0.072	0.009	0.033	0.023	0.071	0.025
Awareness about RAISA	0.206	0.323	0.070	1.000	0.095	-0.021	-0.058	-0.218	0.154	0.129	-0.319	-0.132	-0.383	-0.177	-0.252	0.070	-0.206	0.070	0.124
Familiarity with RAISA	0.235	0.044	0.030	0.095	1.000	0.600	0.552	0.208	-0.202	-0.212	0.282	0.364	0.468	-0.180	0.291	0.559	0.038	0.581	-0.115
Previous experience with RAISA	-0.261	0.076	-0.099	-0.021	0.600	1.000	-0.373	-0.183	-0.023	0.006	-0.216	-0.265	-0.396	0.116	-0.292	-0.428	-0.134	-0.442	-0.176
Importance of RAISA for hotels	0.270	0.037	0.044	-0.058	0.552	-0.373	1.000	0.235	-0.402	-0.408	0.333	0.462	0.619	-0.032	0.232	0.728	-0.187	0.493	-0.301
RAISA nullify the human error	0.139	-0.239	0.070	-0.218	0.208	-0.183	0.235	1.000	0.105	-0.024	0.256	0.290	0.319	0.063	0.228	0.196	0.194	0.423	0.019
Negative impact of RAISA on enthusiasam	0.069	-0.242	0.152	0.154	-0.202	-0.023	-0.402	0.105	1.000	0.814	-0.207	-0.273	-0.323	0.062	-0.118	-0.285	-0.130	-0.149	0.557
Job displacement due to RAISA	0.119	-0.193	0.156	0.129	-0.212	0.006	-0.408	-0.024	0.814	1.000	-0.072	-0.159	-0.334	-0.076	-0.169	-0.296	-0.178	-0.269	0.657
RAISA for suppliers	0.035	-0.153	-0.125	-0.319	0.282	-0.216	0.333	0.256	-0.207	-0.072	1.000	0.529	0.434	-0.058	0.208	0.279	0.115	0.219	-0.010
RAISA ease the work of the supplier	0.012	-0.042	0.013	-0.132	0.364	-0.265	0.462	0.290	-0.273	-0.159	0.529	1.000	0.587	-0.033	0.203	0.496	0.122	0.435	-0.003
RAISA positively affect your visit.	0.070	-0.080	0.007	-0.383	0.468	-0.396	0.619	0.319	-0.323	-0.334	0.434	0.587	1.000	0.002	0.377	0.618	0.059	0.544	-0.148
Prior interaction with robots in hospitality industry.	-0.099	-0.195	-0.072	-0.177	-0.180	0.116	-0.032	0.063	0.062	-0.076	-0.058	-0.033	0.002	1.000	0.027	-0.023	0.092	-0.067	0.038
Economically viablitiy	0.093	-0.061	0.009	-0.252	0.291	-0.292	0.232	0.228	-0.118	-0.169	0.208	0.203	0.377	0.027	1.000	0.332	0.431	0.336	0.017
RAISA role in tourism and hospitality industry	0.196	0.026	0.033	0.070	0.559	-0.428	0.728	0.196	-0.285	-0.296	0.279	0.496	0.618	-0.023	0.332	1.000	-0.048	0.617	-0.075
Impact of RAISA on the employees	-0.049	-0.180	0.023	-0.206	0.038	-0.134	-0.187	0.194	-0.130	-0.178	0.115	0.122	0.059	0.092	0.431	-0.048	1.000	0.095	0.046
Influence of RAISA in customer satisfaction	0.078	0.006	0.071	0.070	0.581	-0.442	0.493	0.423	-0.149	-0.269	0.219	0.435	0.544	-0.067	0.336	0.617	0.095	1.000	-0.063
Can RAISA replace humans	0.100	-0.107	0.025	0.124	-0.115	-0.176	-0.301	0.019	0.557	0.657	-0.010	-0.003	-0.148	0.038	0.017	-0.075	0.046	-0.063	1.000

Table 4. 4. Positive Correlation Models with correlation coefficient, R Square, and p-Value showing statistical significance.

No.	Moderate Positive Correlation (Between 0.5 and 0.8)	R (Correlation Coefficient)	R Square	p-Value
M1	RAISA role in tourism and hospitality industry and Familiarity with RAISA	0.559	0.31	2.82E-12
M2	Influence of RAISA in custromer satisfaction and Familiarity with RAISA	0.581	0.34	2.31E-13
M3	RAISA role in tourism and hospitality industry and Importance of RAISA for hotels	0.728	0.53	3.23E-23
M4	Negative impact of RAISA on enthusiasam and Job displacement due to RAISA	0.814	0.66	1.04E-32
M5	RAISA positively affect your visit and RAISA ease the work of suppliers	0.587	0.34	1.17E-13
M6	Influence of RAiSA on customer satisfaction and RAISA positively affect your visit	0.544	0.30	1.25E-11
M7	Influence of RAISA on the employees and RAISA role in tourism and hospitality industry	0.617	0.38	2.54E-15
No.	**Weak Positive Correlation (Between 0 and 0.5)**	**R (Correlation Coefficient)**	**R Square**	**p-Value**
M8	Importance of RAISA for hotels and RAISA nuliffy the human error	0.235	0.06	0.007
M9	Importance of RAISA for the hotels and RAISA ease the work of suppliers	0.462	0.21	2.13E-08
M10	Importance of RAISA for hotels and Influence of RAISA on customer satisfaction	0.493	0.24	1.65E-09
M11	RAISA positively affect your visit and RAISA nullifies the human error	0.319	0.10	0.000
M12	Influence of RAISA on customer satisfaction and RAISA nullifies the human error	0.423	0.18	3.97E-07
M13	Economically viable in India and RAISA nullifies the human error	0.228	0.05	0.008
M14	Influence of RAISA on customer satisfaction and RAISA ease the work of suppliers	0.435	0.19	1.64E-07
M15	Economical viability in India and RAISA role in tourism and hospitlaity industry	0.332	0.11	9.55E-05
M16	Economic viability and impact of RAISA on the eomployees	0.431	0.19	2.22E-07
M17	Influence of RAISA in customer satisfaction and Economic viability in India	0.336	0.11	7.67E-05

https://doi.org/10.52202/074664-0028

Table 4. 5. Negative Correlation Models with correlation coefficient, R Square, and p-Value showing statistical significance.

No.	Weak Negative Correlation (0 to -.05)	R (Correlation Coefficient)	R Square	p-Value
M18	RAISA nullify human error and Awareness with RAISA	-0.218	0.05	0.012
M19	RAISA positively affect your visit and Awareness about RAISA	-0.383	0.15	5.34E-06
M20	Economic vialbility in India and Awareness with RAISA	-0.252	0.06	0.003
M21	Job displacement due to RAISA and Familiartiy with RAISA	-0.212	0.04	0.014
M22	RAISA ease the work of suppliers and Previous experience with RAISA	-0.265	0.07	0.002
M23	Influence of RAISA on customer satisfaction and Previous experience with RAISA	-0.442	0.20	9.83E-08
M24	Importance of RAISA for hotels and Negative impact of RAISA on enthusiasm	-0.402	0.16	1.62E-06
M25	RAISA positively affect your visit and Job displacement due to RAISA	-0.334	0.11	8.62E-05
	Strong Negative Correlation (Close to -1)			
M26	Familiarity with RAISA and Previous experience with RAISA	-0.600	0.36	2.35E-14

5. Discussions

The findings of the data analysis rise the following points: -

- The research provides a better understanding of the relationship among the different variables of RAISA's implementation in the Indian tourism sector.
- The insights from the findings can be of use to various organizations in India to make informed decisions, build strong policies to support the incorporation of RAISA in the sector.
- The findings help to focus on the economic viability – focusing on all the pros and cons of initial implementation costs, maintenance costs, training costs and befits post implementation.
- The findings of the data analysis show the productivity that can be boost by involvement of RAISA in the tourism and hospitality industry.
- The study findings also highlight the negative impact of RAISA on the current employees of the industry and the potential risk of total job displacement in the tourism and hospitality sector in India.
- The findings also give clarity to the challenges and area to maximize the productivity and output of the organization.
- The findings also press note on the better understanding required in technology for both the employees of the organization and the guests being catered to by RAISA.

These topics of discussion form the full base of the study and are the major aspects that can be seen and encountered with while implementing RAISA in the Indian tourism and hospitality industry. The implications formulated in the findings should also be taken into consideration while incorporating RAISA in the industry and strategies should be implemented to overcome these implications.

https://doi.org/10.52202/074664-0028

6. Conclusions

Post the data analysis, the undermentioned conclusions can be formulated in accordance with the objectives of the research:

- The models, M3, M4, M5, and M7 with strong positive correlation, show the role RAISA can play in enhancing the efficiency and output of the hotel operations within the Indian tourism and hospitality sector.
- The models M8, M9, M10, and M13 with weak positive correlations show that having a better understanding of RAISA is very necessary for making it a success and keeping in mind all the economic pros and cons, can be very helpful in incorporation of RAISA in the Indian tourism and hospitality industry.
- The findings show that RAISA can be significantly helpful for the industry. It can positively influence various operations of the industry, including service and delivery, customer satisfaction, workforce management and much more which strongly contribute to the overall viability and competitiveness of incorporating RAISA.
- The positive correlation in the models M2, M6, and M7 shows that RAISA can be a beneficiary factor for customer satisfaction, and visitor experience for any organization. The focus still being on the accepting RAISA by the organization rather than adopting. The influence of RAISA on the employees also becomes an influencing factor in Guest satisfaction.
- The negative correlations in the models M18, M19, M21, M22, M23, and M25 throw a bright light on the negative impact of incorporating RAISA on the employees of the organizations. The fear of potential risks can be seen as job displacement, decreased enthusiasm of the employees and low familiarity with RAISA among the employees, guests, and stakeholders. This may lead to employee resentment and bad faith among the employees.
- Whereas the analysis shoes a positive effect of RAISA in case of the suppliers in the tourism and hospitality industry, relating to the ease of their work, and more systematic shape given to the demand and supply chain by the automation in the industry.
- These findings underscore the reliability of RAISA in tourism and hospitality industry covering the diverse impacts of RAISA implementation on the guests, employees, stakeholders, and the suppliers. These findings can be helpful to formulate strategies to address the needs of all the groups involved.
- The observations of collected data emphasize the need for business in the tourism and hospitality industry to closely monitor and evaluate the economic viability of RAISA and the benefits that RAISA technology can bring along with it.
- The adoption of RAISA offer areas of improvement in efficiency, service quality, and productivity. But the industry must also consider and mend the potential challenges to ensure a successful and fruitful adoption and implementation of RAISA in Indian tourism and hospitality industry.

Overall, in conclusion the reviewed study and analysis of collected data enlightens us with some valuable insights into the implementation of RAISA in the Indian tourism and hospitality sector. By keeping in view, the influence of RAISA on customer satisfaction, employee motivation, and associated business implications, the organizations can become more accurate and make more informed decisions and develop strategies to incorporate and use RAISA more effectively and move towards a more innovative and sustainable growth in the tourism and hospitality industry.

https://doi.org/10.52202/074664-0028

7. **Limitations**

On analysing the accumulated data and forming some strong insights for the topic, some limitations can also be seen. These limitations can be a hinderance to implementation of RAISA in the tourism and hospitality industry in India and the robustness the study findings: -

- The focus of the study was on implementation of RAISA in the industry with correlation to guest, employees, and suppliers, but it did not explore the other factors like, cultural, regulatory environment, and infrastructure which could depict a better understanding of the implementation of RAISA in India and follow up with more certain implications.
- The use of convenience sampling may have led to potential bias which can be reduced by using more objective measures and other sampling techniques.
- The study prominently relied on the quantitative data and a thorough analysis of qualitative data can be more helpful.
- The inputs of the respondents can change in future limiting the longitudinal analysis.
- The study did not explore the views of employees already coping with RAISA in their organizations, focused group interview can be of help in analysing the situation more properly.

https://doi.org/10.52202/074664-0028

8. Future suggestions

Based on the research study on, "RAISA in tourism and hospitality industry: A study on technology implementation and implications in Indian tourism and hospitality sector", the suggestions for future research are mentioned below: -

- The future researcher can conduct more subjective research considering only the employee aspect to explore the factors contributing to the negative impact of RAISA on employees specifically.
- Qualitative research can be carried out to attain more contextual information using focused group interviews or case studies.
- Future researchers can also analyse the viability of RAISA keeping in view the policies and regulatory frameworks of governing bodies for implementation of RAISA in India.
- Ethical considerations related to RAISA implementation can also be taken into consideration in the future research, including data privacy, security, and ethical use of automation technologies.
- In future research, focus can be kept on longitudinal study as the preferences of the respondents are constantly changing and it will help to keep track of the long-term effects of RAISA implementation on all involved groups.

References

1. Ashton, M.: Robotics, artificial intelligence and service automation in hotel service process design: can customer dominant logic provide improved value?, http://eprints.bournemouth.ac.uk/35430/, (2021). https://www.researchgate.net/publication/359040395
2. Belanche, D., Casaló, L.V., Flavián, C.: Frontline robots in tourism and hospitality: service enhancement or cost reduction? EM. 31, 477–492 (2020). https://doi.org/10.1007/s12525-020-00432-5.
3. Camilleri, M.A.: The Tourism Industry: An Overview. In: Tourism, hospitality & event management. pp. 3–27 (2017). https://doi.org/10.1007/978-3-319-49849-2_1.
4. Ghosh, M.: ANALYSING IOT AUTOMATION FACILITIES IN HOTEL ROOMS: ENHANCING GUEST EXPERIENCE AND OPERATIONAL EFFICIENCY. Rajasthali Journal (2023).
5. Иванов, С.: The impact of automation on tourism and hospitality jobs. Information Technology & Tourism. 22, 205–215 (2020). https://doi.org/10.1007/s40558-020-00175-1.
6. Иванов, С., Webster, C.: Adoption of robots, artificial intelligence and service automation by travel, tourism and hospitality companies – a Cost-Benefit analysis. Social Science Research Network. (2017).
7. Joseph, S., Brako, P.: An Exploratory Research into the Hotel Industry as a Vibrant Contributor to India's Economic Reforms. Social Science Research Network. (2017). https://www.researchgate.net/publication/319940461
8. Kashyap, G. (2014). Challenges faced by the Hotel Industry: A review of Indian Scenario (Publication No. IOSR Journal of Business and Management (**Volume 16**) e-ISSN: 2278-487X).
9. Koo, C., Xiang, Z., Gretzel, U., Σιγάλα, M.: Artificial intelligence (AI) and robotics in travel, hospitality and leisure. EM. 31, 473–476 (2021). https://doi.org/10.1007/s12525-021-00494-z.
10. Dhiman, M.C., Kumar, A.: Human resource management in Indian hospitality industry. In: Advances in hospitality, tourism and the services industry (AHTSI) book series. pp. 259–275 (2015). https://doi.org/10.4018/978-1-4666-8268-9.ch016.
11. Lukanova, G., Ilieva, G.: Robots, artificial intelligence, and service automation in hotels. In: Emerald Publishing Limited eBooks. pp. 157–183 (2019). https://doi.org/10.1108/978-1-78756-687-320191009.
12. Pandey, K., Mahadevan, K., Joshi, S.: Indian Tourism Industry and COVID-19: A Sustainable Recovery Framework in a Post-Pandemic era. Vision. 097226292110432 (2021). https://doi.org/10.1177/09722629211043298.
13. Ralston, L.S., Stewart, W.P.: Methodological perspectives of festival research studies. Annals of Tourism Research. 17, 289–292 (1990). https://doi.org/10.1016/0160-7383(90)90090-e.
14. Saravanakumar, S., Tjprc: The Service Automation and Robotics in Hospitality Industry, A Study on Business Implications. International Journal of Mechanical and Production Engineering Research and Development. 8, 91–100 (2018). https://doi.org/10.24247/ijmperddec201810.
15. Sharma, R., Uniyal, M., Sharma, N.: Guest attitude towards introducing automation using service robots in hotels of Delhi. International Journal of Advanced Science and Technology. 29, 2930–2937 (2020).
16. Singh, K., Kapoor, D.S., Sharma, A., Thakur, K.: Understanding robotics in the tourism and hospitality industry. In: Advances in hospitality, tourism and the services industry (AHTSI) book series. pp. 64–94 (2023). https://doi.org/10.4018/979-8-3693-0650-5.ch006.

 https://doi.org/10.52202/074664-0028

17. Sofronov, B.: THE DEVELOPMENT OF THE TRAVEL AND TOURISM INDUSTRY IN THE WORLD. Annals of “Spiru Haret” University. Economic Series. 18, 123–137 (2018). https://doi.org/10.26458/1847.
18. Solnet, D., Subramony, M., Ford, R.C., Golubovskaya, M., Kang, H.-T., Hançer, M.: Leveraging human touch in service interactions: lessons from hospitality. Journal of Service Management. 30, 392–409 (2019). https://doi.org/10.1108/josm-12-2018-0380.
19. Tsai, H., Song, H., Wong, K.: TOURISM AND HOTEL COMPETITIVENESS RESEARCH. Journal of Travel & Tourism Marketing. 26, 522–546 (2009). https://doi.org/10.1080/10548400903163079.
20. Tuomi, A., Tussyadiah, I., Stienmetz, J.L.: Applications and implications of service robots in hospitality. Cornell Hospitality Quarterly. 62, 232–247 (2020). https://doi.org/10.1177/1938965520923961.
21. Webster, C., Иванов, C.: Robots in Travel, Tourism and Hospitality: Key Findings from a Global Study. Social Science Research Network. (2020).
22. Wirtz, J., Kunz, W.H., Paluch, S.: What are the Implications of Service Robots in Hospitality for Consumers? ResearchGate. (2022).
23. Ye, H., Sun, S., Law, R.: A review of Robotic applications in Hospitality and tourism research. Sustainability. 14, 10827 (2022). https://doi.org/10.3390/su141710827.

"Feast of Unity": Exploring the Interplay between Awadhi Regional Food and Community Connections

Arya Mohan[1] and Mukul Sahu[2]

[1&2]Postgraduate Scholars, NCHM-Institute of Hospitality, Noida

Abstract.
Food is a necessity of life. With the increase of Global travel, food has become an important aspect for culinary enthusiasts and local travellers in a region like Awadh, which has culturally rich cuisine long served as a unifying force bringing local regional communities together in a celebration of flavours and tradition. This research paper is studied to dive into the interplay between Awadhi cuisine of the Awadh region of Uttar Pradesh and its impact on community gathering. Explorative in nature, the quantitative surveys, and participant responses, focusto unveil the faces of culinary gathering and the role they play in fostering within the community. Our data will explore how authentic Awadhi recipes, cooking techniques, local ingredients and utensils serve as a fuel for social interaction, cultural preservation, and a monopoly identity formation. By evaluating the culture rituals and the chronicle history surrounding regional community feast and culinary drive, this study seeks to clarify the ways in which food serves as a channel for nourishing interpersonal relationships bridging the generational gap and re- installing community bonds. Moreover, this research will scrutinize the wider part of the Awadhi regional food, including its character in contributing inclusivity, elasticity, and cultural diversity. The impact of Globalisation, interchange in dietetic preferences and dispersion of local communities are centre of attention as well as the ways in which local regional communities innovate in response to these changes. In the long run this research strives to throw some light on progressive power of food as a social and cultural occurrence, highlighting its ability to go beyond the boundaries and cultivate a sense of belonging within the spirited amalgams of Awadhi community life. Through a refined relationship between food and culture, this research will contribute to an intense understanding of culinary tradition in creating unity in modern society.

Keywords: Awadhi Cuisine, Community Gatherings, Social Interactions, Cultural Exchange, Food and Culture.

1. Introduction

In the culturally diverse region of Awadh, situated in the heart of India, Food is more than just nourishment – it is a corner stone of communal life, a symbol of heritage, and a source of profound connection [1]. For the past few centuries, Regional cuisine has served as a unifying thread weaving together the fabrics of regional communities through its rich flavor, recipes and community gathering [2]. In the crowded street of Lucknow, the peaceful villages of Faizabad, and the bustling towns of barabanki, the aroma of flavors spices and the delicacies dilutes the air, inviting neighbors, friends and all the peoples to participate in the culinary delights of the region.

 https://doi.org/10.52202/074664-0029

Against the backdrop of gastronomic scarcity and culture diversity our research aims to explore the refined interplay between regional food and community connection [3]. With the center of attraction on community feast, culinary festivals, and daily dining rituals, we seek to reveal the dynamics that under pin the social thread of Awadhi colonies, evaluating how food acts as a trigger for social interaction, identity formation and cultural preservation.

As we begin our journey of exploration, it is important to acknowledge the significance of Awadhi cuisine in-curating the cultural landscape of the region. With its roots to old culinary traditions influenced by renowned cuisines like Mughal, Persian the Awadhi cuisine is famous for its authentic flavor's aromatic spices and extra ordinary preparation techniques. From the iconic biryanis and the lip-smacking kebabs of Lucknow to the delicious sweets and savory snacks of Faizabad, each plate tells a story – a tale of innovation, community tradition and undying bonds.

Though the Awadhi cuisine has a rich culinary heritage, questions abound. How do community feasts and food festivals contribute to the social solidarity and collective identity within Awadhi people? How do authentic recipes and cooking techniques serve as a channel for intergenerational dialogue and cultural transferal?

Awadhi cuisine and the culture of nawabs is the theory of dastarkhwani. Dastarkhwan is a ceremonial dispersion of food – salan (lean sauce), keema (minced meat dish), kebabs, pulao (rice preparation different from biryani, different breads, pudding etc. As per the Awadhi traditions the dinner had to be eaten in a certain way. "Nazakat" and "Tehzeeb" (beauty and conduct) was a important portion of feasting handle.

The deeper understanding of the twisted interchange between food, culture, and community, throwing light on the importance of culinary practices in molding the social cohesion, cultural elasticity, and collective identity. Through the exploration of "feast of unity" we seek to celebrate the richness of Awadhi regional food and its role in growing connection, fostering empathy, and building a much stronger regional community.

2. Literature Review

Food and drink, the most basic culinary goods, are essential to the industry's existence. Traveling is made more enjoyable by the diverse range of regional, international, and fusion culinary options [4]. Culinary advancements have benefited from these occurrences. Travel, frequently referred to as outings seeking exceptional dining experiences, allows individuals to explore the essence of different places through their food [5]. The typical Awadhi cuisine, which combines Mughal, Indian, and indigenous influences in a unique way, is one of the main draws for tourists to Lucknow. It represents the splendour of the Awadhi people and their customs, showcasing the multicultural influences that have shaped the region's culinary identity. Lucknow, with its combination of native and foreign culinary traditions and cuisines, presents an enticing location for those who wish to immerse themselves in the essence of Awadh.

Tourism can play a crucial role in identifying the economic features of heritage and utilizing them to fund conservation efforts by informing the public and raising funds [6]. Promoting tourism not only benefits host communities economically but also motivate them to preserve their environment, cultural customs, and legacy [7]. Furthermore, it fosters community engagement and influences regulatory decisions. Food, beverages, and the cultures surrounding them hold enormous potential to become one of India's main tourist attractions. Initiatives aimed at promoting Indian and Awadhi culinary items to potential overseas tourists contribute significantly to increasing foreign revenue from the sale of food and services abroad.

https://doi.org/10.52202/074664-0029

Scholarly works delve into various aspects of Awadhi cuisine, shedding light on its historical, cultural, and social significance. The cultural and gastronomical heritage of Awadh, tracing its origins and evolution. The historical context of Awadhi cuisine, emphasizing its diverse influences and cultural legacy. The role of communal dining and culinary festivals is significant in fostering social cohesion within cultural contexts, highlighting their role in promoting culinary diversity and community pride [8].

Traditional Awadhi recipes and cooking techniques emphasize the cultural significance, with rituals and symbolism associated with Awadhi culinary traditions revealing deeper cultural meanings. Globalization has impacted Awadhi cuisine, highlighting tensions between preserving traditional culinary heritage and adapting to modern trends, thus addressing the challenges and opportunities for sustainable development in Awadhi cuisine, advocating for both the preservation of culinary heritage and the adoption of sustainable food practices [9].

Through these scholarly works, a holistic understanding of Awadhi cuisine emerges, highlighting its rich cultural tapestry, historical roots, and evolving culinary landscape [10]. The studies contribute to the appreciation and preservation of Awadhi culinary traditions, ensuring their continued relevance in a rapidly changing world.

Awadhi regional food serves as a cultural cornerstone within the vibrant tapestry of Awadh, India, representing a fusion of Mughal, Persian, and indigenous culinary traditions, the cultural and gastronomical heritage of Awadh, tracing its culinary roots to the Nawabi era and highlighting its evolution over time [11]. Author complements this historical perspective by providing insights into the intricate Flavors and cooking techniques that characterize Awadhi cuisine, emphasizing its role as a reflection of the region's rich cultural diversity and historical legacy. Author delve into the social dynamics of regional cuisine, focusing on communal dining practices and culinary festivals as catalysts for social cohesion and community bonding [12]. The study sheds light on the significance of food as a social lubricant and a means of strengthening interpersonal relationships within Awadhi society. Building on this theme, the cultural significance of food festivals in Awadh, underscoring the role in promoting cultural exchange and fostering a sense of community pride [13]. The use of indigenous ingredients and spices that define the unique flavour profile of a regional cuisine [14]. The rituals and symbolism associated with Awadhi culinary traditions, revealing how food rituals serve as expressions of identity and social status within the community. In addressing contemporary challenges. The impact of globalization on Awadhi cuisine, analysing how external influences and changing dietary preferences are reshaping culinary practices in the region. The sustainable development opportunities in Awadhi cuisine, advocating for strategies to preserve culinary heritage while adapting to modern trends. Together, these studies provide a comprehensive overview of the interplay between Awadhi regional food and community connections, highlighting its cultural significance, social dynamics, and evolving culinary landscape.

3. Research Methodology

The research study on feast of unity – exploring the interplay between Awadhi regional food and community connections aims:

- To understand how communal dining events or food do festivals contribute to community bonding in Awadh?
- To find out the cultural significance of Awadhi dishes and culinary practices that are held within the community.

https://doi.org/10.52202/074664-0029

Due to the limited availability of research papers specifically investigating Awadhi regional food and community connections, the study adopts an explorative approach, using quantitative research methodology.

Literature review was conducted to build a foundational understanding about Awadhi regional dishes and community gathering. There lies and extensive amount of literature related to Awadhi cuisine, Awadh region and community gathering. However, very few researchers have studied regional community connections with specific focus on Awadhi regional food. This research paper aims to contribute towards filling these gaps in the current body of knowledge and provide insights about the interplay between the Awadhi regional food and community connections.

To gather firsthand information, a self- administered survey questionnaire was developed. The questionnaire had 15questions out of which 10 sought information regarding the knowledge about Awadhi regional food, satisfaction and rating, local regional gathering, along with 5 questions focused on generating demographic information. With significantly continued migration and settlement from Awadh region across the country Awadhi cuisine has also witnessed continuous increase in recognition and appreciation by the local people of the region where people from Oudh (old name of Awadh) have migrated and settled keeping these in mind the target population was domestic traveler it is always challenging to conduct sampling in tourism research when the study focus is primarily on local residence (Ralston & Stewart, 1990). Therefore, convenient sampling, a non-probability sampling technique was adopted in accordance with the standard sampling methodology.A sample size of 150 respondents was targeted, and in the end, 126 respondents participated in the survey.

Collected survey data was analyzed using MS-excel. MS-excel was used for initial date organization and for conducting statistical analysis and visualizations.

4. Findings

4.1 Data frame summary:

Table 1: Pearson Correlation Coefficient Matrix

	Column 1	Column 2	Column 3	Column 4	Column 5	Column 6	Column 7	Column 8	Column 9	Column 10
Column 1	1									
Column 2	-0.00435	1								
Column 3	-0.05574	-0.03226	1							
Column 4	#DIV/0!	#DIV/0!	#DIV/0!	1						
Column 5	-0.03848	-0.16075	-0.1542	#DIV/0!	1					
Column 6	0.02765	0.026967	-0.01464	#DIV/0!	0.039709	1				
Column 7	0.028135	-0.19607	-0.06985	#DIV/0!	-0.07392	0.127803	1			
Column 8	-0.02141	-0.03891	0.147312	#DIV/0!	0.041399	0.205823	-0.15796	1		
Column 9	-0.09477	0.319701	0.018211	#DIV/0!	0.100952	-0.16092	-0.45522	-0.09217	1	
Column 1C	0.137901	-0.49872	-0.04532	#DIV/0!	0.196061	-0.13402	0.351034	0.017083	-0.5078	1

Demographic: majority of the participants fall in the younger age category with 69.8% participants lying in the age bracket 18-25 and 23% participants falling underage category 26-35 years. There is most males with 70.6% and rest 29.4% are females. 68.3% of the participants are employed and 28.6% of the respondents fall under the category of students. Most of the participants are urban area residents with 92.1% and the rest 6.3% belongs to the sub-urban area.

Community activities and enjoyment: 54% of the participants are rarely engaged in community-based food activities in the Awadh region but on the other hand 43.7% of the participants are occasionally involved in community activity.

Community gatherings and celebration:49.2% of the respondents thinks that social bonding is the outcome of community gathering or celebration in Awadh, whereas 27% of the respondents suggest that cultural identity is the outcome of community gathering or celebration in Awadh. The remaining 13.5% of the respondents think that festive symbolism is the outcome of community gathering or celebration in Awadh.

Dishes with cultural significance: as per the respondents there are a fewdishes that hold cultural significance in Awadh, namely LucknowDum Biryani, mutton nihari, sheermal, galawati kebab, lehsun ki kheer, korma etc.

Culinary events or festival: 96.8% of the respondents agree that there are community – led culinary events or festivals that celebrate Awadhi cuisine in their area. The remaining 3.2% of the respondents do not agree with this statement.

Community engagement and cultural preservation: 86.3% of the respondents thinks that there are various reasons that contribute to community engagement and cultural preservation such as fostering social bonds, social exchange, preserving culinary heritage, passing down tradition, revitalizing culinary heritage and promoting cultural pride. 11.3% of the respondents suggest that social exchange is the outcome of cultural preservation.

Traditional culinary practices vs modern culinary practices: most of the respondents 85.7% think that there is a balance between preserving culinary practices and incorporating modern influences in Awadhi region. The remaining 14.3% of the respondents think that the balance is not maintained.

Awadhi recipes modernization: when we asked the participants how modernization influences traditional Awadhi recipes and cooking methods, 46% of the respondents suggested that culinary fusion influences the traditional recipes and cooking method. 35% of the respondents think that preservation challenge has influenced the traditional recipes and method. 11.1% of the respondents think that there is an impact of urbanization on traditional recipes and methods. The remaining 7.1% of the respondents think that commercialization of cuisine has influenced traditional Awadhi recipes and cooking methods.

Social media influence: Most of the respondents 67.5% are neutral about the influence of social media in spreading the awareness and engagement of Awadhi regional food. The remaining 30.2% of the respondents think that social media has no role in influencing their awareness and engagement with Awadhi regional food.

Culinary discussion or events: majority of the participants 54% have participated or followed community – driven culinary discussion or events. Rest 46% of the respondents have never participated in any such events or discussion.

Recommendation for enhancing community engagement: the participants of the survey have suggested a few recommendations through which we can enhance community engagement through Awadhi regional food.

https://doi.org/10.52202/074664-0029

1. Organize cooking classes where community members can learn about Awadhi cuisine and practices.
2. Organize farm tours, community supported agriculture initiative to connect consumer with producers and build community relationship.
3. Organize and host cooking competition and cultural festival or fairs focused on Awadhi culinary practices.
4. Creating a recipe book with authentic Awadhi recipes, dining experiences and farm to table initiative.

5. Conclusion:

The result of the research throws light on the demographic characteristics, community- led activities, culinary and dietary preferences, and perceptions of community engagement among the participants in the Awadh region of Uttar Pradesh. Most of the participants are youngsters, majorly male, urban residents and employed individual. While there is a high interest in community led culinary events celebrating Awadhi culture and cuisine, a minor portion of participants are occasionally involved in community based culinary activities.

Community gatherings and celebration plays a vital role for maintaining the social bonds and preserving cultural identity, with authentic and traditional dishes like Lucknow Dum Biryani, mutton Nihari and Sheermal serving as a traditional thread within the culinary arena of Awadh. Despite the generality of community driven culinary fairs, the research unveils a desire for more interactive experience as denoted by recommendation for farm tours, cooking competition, cultural festivals mainly focused on traditional Awadhi practices.

Finally, it is very important to address the recommendation as per the respondents that will enhance the community engagement through Awadhi regional food soon. Organizing cooking fairs and classes where people from different parts of the world can learn about traditional Awadhi cooking practices. This data also suggests that the people should organize cultural festival, farm tours in the near feature to build community relationship between consumer and the producer. Finally, documentation of the Awadhi regional food and its technique and practices is necessary for the upcoming generation to learn about Awadhi cuisine and its culture in future.

6. Limitations and Future Scope

Sample of the study consist of younger generation from urban areas, which is not fully representing the broader demographic diversity of the Awadh region the sampling bias limits the finding to other demographics group such as older generation, rural residents, or people from different occupational background. Future researchers could benefit from taking more vast sampling to ensure more accurate data. The data collected rely on the responses from the participants which are self-reported, which are subject to various biases. To mitigate this limitation future researchers could take objective majors from various sources to validate the data collected.

The study focuses on Awadh region that restricts the viability of the findings to other demographic areas with similar cultural and culinary traditional practices. The scope of the study is limited to exploring community engagement or culinary tradition within the Awadh region. While the findings provide valuable information about specific aspects that may not be covered in community gathering dynamics and traditional culinary practices within the Awadh region of

Uttar Pradesh. Future researchers can explore other factors influencing community engagement such as historical events and political dynamics to provide better understanding of the topic.

This research study is limited due to time constraints which have affected the data collection and analysis. As a result, certain fields and problems within the data have not been fully explored or properly addressed. Future researchers could allocate more time and resources to the data collection to provide better understanding of the research topic. By conveying this limitation, future researchers can enhance and ensure a more comprehensive understanding of the Awadhi regional food and community connections.

References

1. Smith, S.L., Xiao, H.: Culinary Tourism Supply Chains: A Preliminary Examination. Journal of Travel Research. 46, 289–299 (2008). https://doi.org/10.1177/0047287506303981.
2. Bhartwal, U., Monika, Tiwari, R.K., Sharma, N.K., Mathur, D., Saxena, S.: Dimensional role of the latest trends in sustainability at religious places in rural Rajasthan. In: Advances in logistics,

https://doi.org/10.52202/074664-0029

operations, and management science book series. pp. 135–145 (2023). https://doi.org/10.4018/979-8-3693-2065-5.ch010.
3. McDaniel, T., Más, F.S., Sussman, A.L.: Growing connections: local food systems and community resilience. Society & Natural Resources. 34, 1375–1393 (2021). https://doi.org/10.1080/08941920.2021.1958965.
4. Stone, M.J., Soulard, J., Migacz, S., Wolf, E.: Elements of memorable food, drink, and culinary tourism experiences. Journal of Travel Research. 57, 1121–1132 (2017). https://doi.org/10.1177/0047287517729758.
5. Hall, C.M.: The experience of consumption or the consumption of experiences? Challenges and issues in food tourism. In: Elsevier eBooks. pp. 314–335 (2003). https://doi.org/10.1016/b978-0-7506-5503-3.50021-x.
6. Björk, P., Kauppinen-Räisänen, H.: Local food: a source for destination attraction. International Journal of Contemporary Hospitality Management. 28, 177–194 (2016). https://doi.org/10.1108/ijchm-05-2014-0214.
7. Green, G.P., Dougherty, M.L.: Localizing Linkages for Food and tourism: Culinary Tourism as a Community Development Strategy. Community Development. 39, 148–158 (2008). https://doi.org/10.1080/15575330809489674.
8. Canter, D.D., Moorachian, M.E., Boyce, J.: The growing importance of food and culinary knowledge and skills in dietetics practice. Topics in Clinical Nutrition. 22, 313–322 (2007). https://doi.org/10.1097/01.tin.0000308468.56942.b8.
9. Sexena, A.N.: The Law of Geographical Indications as a Tool for Economic Development: An application in the Case of Awadhi Cuisine of Uttar Pradesh, India. Social Science Research Network. (2010). https://doi.org/10.2139/ssrn.1682690.
10. Trivedi, M.: Cultural history of the kingdom of awadh, http://ir.amu.ac.in/5952/, (1977).
11. Oldenburg, V.T.: Shaam-E-Awadh: Writings on LuckNow. (2007).
12. Mohapatra, P., Biswas, S.N.: Gastronomy and its impact on Tourism: A Case study on Regional Cuisine of Coastal Odisha, India. International Journal of Research in Social Sciences. 7, 154–168 (2017).
13. Singh, P., Najar, A.H.: Regional food as the catalyst for cultural tourism in India. In: Advances in hospitality, tourism and the services industry (AHTSI) book series. pp. 84–96 (2020). https://doi.org/10.4018/978-1-7998-4330-6.ch006.
14. Gheorghe, G., Bulin, D.: Cuisine – a regional tourism differentiation tool. Knowledge Horizons - Economics. 6, 194–199 (2014).

Exploring Sustainable Gastronomy: Trends, Challenges, and Opportunities in Culinary Innovation with Plant-Based Alternatives

Khilesh Patel[1] and Debarshi Bharati[2]
[1]Research Scholar, School of Hotel Management, Siksha 'O'Anusandhan, Bhubaneswar, India
[2]Assistant Lecturer, Food Production, State Institute of Hotel Management, Dimapur, Nagaland, India

Abstract.
In recent years, the culinary scene has shifted significantly toward sustainability, owing to increased environmental consciousness and a demand for healthy eating habits. This inquiry digs into sustainable cuisine, looking at trends, problems, and prospects for culinary innovation using plant-based resources.The trend toward sustainable gastronomy reflects a global shift toward plant-centric dining experiences in a variety of contexts, including restaurants, foodservice establishments, and homes. Plant-based diets are gaining popularity as an effective way to reduce environmental impact and boost wellness. Consumers are increasingly choosing plant-based alternatives due to ethical issues, health concerns, and a desire to lessen their environmental impact. This transformation is reflected in the rise of plant-based meat substitutes, dairy alternatives, and imaginative plant-based gourmet options targeted to diverse palates.However, embracing sustainable gastronomy has challenges. Overcoming entrenched consumer preferences based on conventional culinary traditions is a huge problem. Furthermore, assuring the affordability and accessibility of plant-based alternatives remains a challenge, particularly in economically disadvantaged areas and nations with limited agricultural resources.
Sustainable gastronomy promotes culinary innovation by encouraging chefs to experiment with locally sourced, seasonal ingredients and traditional cooking methods. This not only improves the flavor profile of plant-based dishes, but also promotes cultural diversity and culinary history. Furthermore, incorporating sustainable practices into foodservice operations can lead to cost savings, resource optimization, and improved brand reputation, attracting environmentally concerned customers and investors.
This study takes a qualitative method, combining content analysis and in-depth interviews to investigate consumption trends, difficulties, and opportunities in culinary innovation for plant-based alternatives. Its goal is to provide deep insights into the issue by bringing together specialists from several disciplines.In essence, the pursuit of sustainable cuisine represents a transformative path toward a more resilient, equitable, and environmentally sensitive food system. Embracing plant-based alternatives and cultivating culinary inventiveness can help to conserve biodiversity, reduce carbon emissions, and promote better eating habits. To create the future of food in a sustainable manner, it is necessary to address the associated difficulties through joint efforts, legislative measures, and investments in research and development.

Keywords: sustainable, gastronomy, plant-based, culinary, innovation.

https://doi.org/10.52202/074664-0030

1. Introduction

A culinary concept known as "sustainable gastronomy" places equal emphasis on the enjoyment of food and its effects on the environment, society, and human health. Fundamentally, sustainable gastronomy aims to reduce the environmental impact of food production and consumption while advancing economic viability and social equality. Sustainable gastronomy methods have seen a noticeable movement in favor of plant-based food substitutes in recent years. These replacements, which include plant-based meats, dairy alternatives, and vegetable-based meals, provide a number of benefits. In comparison to their equivalents derived from animals, they usually need fewer resources to manufacture, emit fewer greenhouse gases, and have a smaller environmental effect. Furthermore, sustainable gastronomy tackles health issues related to excessive meat consumption, promotes biodiversity, and lessens animal suffering by increasing the number of plant-based choices on menus. Accepting plant-based food substitutes promotes a more peaceful coexistence of humans, food, and the environment in line with the principles of sustainable cuisine.

Gastronomy is a collection of standards for good cooking that vary by social structure and country [1]. The cultural and emotional aspects of food extend beyond nutrition. Creativity and the capacity to make new dishes from properly selected materials are important aspects of cuisine. Sustainable gastronomy aims to benefit both the environment and society [2]. Consumers are more concerned about the health and environmental impact of their food choices, resulting in shifts in eating patterns. PBFs are increasingly being consumed to promote health, conservation, and environmental sustainability [3]. The Food and Agriculture Organization (FAO) defines a sustainable diet as safe, healthy, nutrient-dense, and environmentally friendly. Consumers are increasingly adopting plant-based foods (PBFs) due to ethical concerns, animal rights/welfare campaigns, and increased greenhouse gas emissions from animal-based food production [4].

Figure 1: Plant Based diets

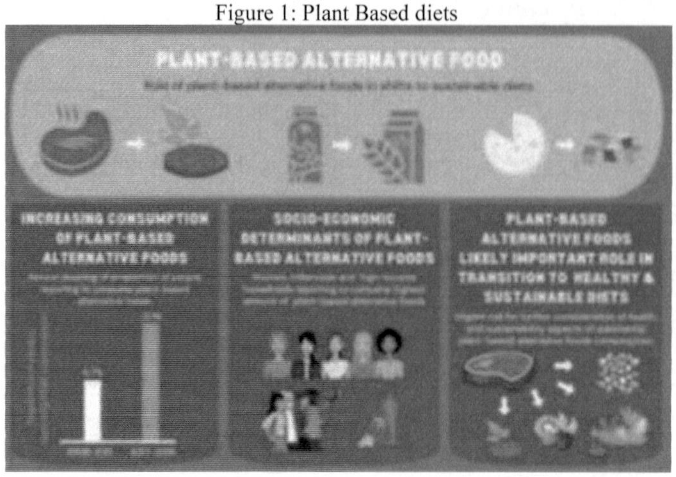

Plant-based diets, including vegetarianism and veganism, are becoming increasingly popular in Western countries. Plant-based diets differ from vegetarian diets in terms of their eating patterns. The term "plant-based" is broader as it focuses on plant-based foods including fruits, vegetables, nuts, oil, whole grains, and legumes, with occasional ingestion of animal products such milk, eggs, meat, and fish. Plant-based diets may allow for the substitution of animal products with vegetables, without the need for permanent restrictions on animal foods [5].Plant-based diets are often chosen for health, environmental, animal welfare, meat-free, or religious reasons [6]. Flexitarian diets should not only focus on processed meat alternatives, but also include raw proteins from both animals and plants. However, flexitarians eat a wide range of foods. PBMAs are primarily consumed by their target demographic. Flexitarians, who abstain from meat on a regular basis, seek out meat alternatives to maintain their decreased meat diet. Flexitarians are logically identified as potential launch customers. According to PBMAs, the current market strategy is to closely replicate traditional meat products. The term "PBMAs" refers to plant-based meat analogs and can also be interpreted as "plant-based meat analogs" based on factors such as flavor, texture, appearance, nutritional content, and product names [7]. The food industry's goal of creating meat-like plant-based foods makes it easier for flexitarians, vegetarians, and vegans to avoid PBMAs, which are often associated with meat [8]. Although the adoption and acceptance of PBMAs by food consumers may not be obvious, this review will demonstrate that they facilitate the consumers in high-income countries require good, inexpensive, and accessible alternative protein products to meet their cravings for meat-free meals [9].

2. Review of Literature

Plant-based food or feed innovation is mostly focused on developing new goods, such as meat alternatives [10], plant-based beverages, and new food ingredients. Research suggests that PBFs can delay, lessen, or possibly prevent certain conditions have all reported on disease conditions. PBFs contain various carbohydrate types and micronutrients, including vitamins, minerals, and bioactive compounds, in addition to macronutrients like protein [11]. Soy products contain calcium and is flavones, which can promote bone health and reduce the risk of osteoporosis. PBFs are considered incomplete proteins because they lack some essential amino acids. Research also suggests that consuming PBFs can reduce the risk of cardiovascular disease, improve cardiovascular health, control glycemic levels, and lower blood cholesterol, obesity, and blood pressure [12]. PBFs in soy products can improve insulin sensitivity and modulate glycemic index [13].

Researchers have improved the nutritional and nutraceutical value of plant-based foods using novel ways. Food innovation and recipe reformulation have the ability to contribute to a variety of sustainable development goals, including reducing food waste and changing dietary habits [14]. The food pyramid is not without innovation. Proposed recommendations for modern Mediterranean people based on diet, socio-culture, environment, health, and lifestyle problems. The main recommendations are to consume plant-based meals, practice economic prudence, consider qualitative lifestyle factors and cultural elements, such as sociability, culinary activities, physical activity, rest and balance, and follow recommended food consumption frequency [15]. The authors developed a new diet pyramid that takes into account food production, seasonality, biodiversity, traditional, local, and eco-friendly items. The food industry is increasingly focusing on sustainability challenges. However, the policy and market framework are positive only for the short term. To expand the business opportunity, consumers'

https://doi.org/10.52202/074664-0030

beliefs, perceptions, and understanding of novel alternatives, such as cultured meat and highly processed plant-based meat alternatives, insects, and novel protein sources (such as algae) must change [16]. Plant-based and flexitarian diets are gaining popularity, and the food supply industry has responded with convenience goods. However, consumer opinions remain unclear [17].

Various research finds that ruminant meat has a higher environmental impact than grains, fruits, and vegetables [18]. Young adults and women are more likely to adopt plant-based diets .Transitioning to plant-based diets is crucial for enhancing human health and animal welfare, as well as reducing the negative environmental impact of the food system. However, meat consumption continues to rise and is expected to increase more in the next decade.To reduce the expected global surge in meat consumption, it is crucial to significantly reduce average meat consumption levels, beginning in rich cultures. Meat reduction can be implemented in a variety of ways. To reduce meat consumption, consider reducing portion sizes, replacing meat with plant-based alternatives (hybrid meats), following the "less but better" principle (less quantity, more quality), and removing meat from dishes without a replacement. Substitute animal-based foods like eggs or cheese with plant-based alternatives like legumes, mushrooms, or tofu. Other protein sources with limited market share, such as insects and seaweed, or cultured meat, should also be considered plant-based meat alternatives (PBMAs) [19].

Figure 2: Plant-based protein sources

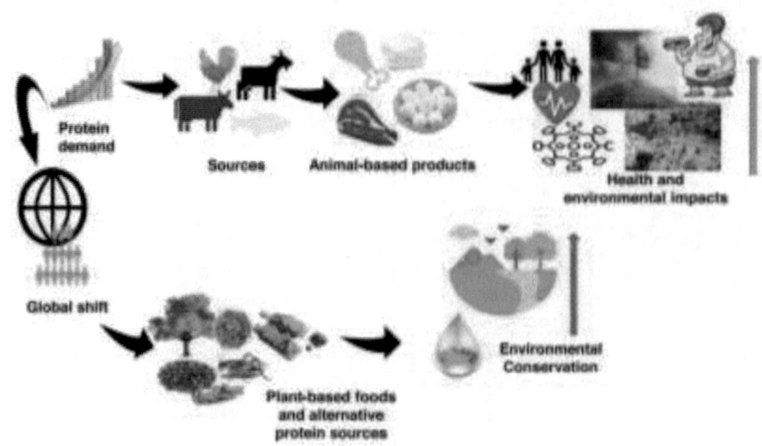

Plant-based protein sources such as legumes, seeds, and nuts, as well as byproducts from food production, have been explored and used to mimic meat and other analogues. Plant and alternative protein sources have been used to create meat, egg, dairy, and pastry products, as well as beverages and dietary supplements [20]. Traditional sources of items, such as cheese, have specific flavors and qualities. To optimize alternative cheese production, it's important to consider the relationships between ingredients, processing methods, proportions, and formulations. Consumers want plant-based meat products to have intense flavor, firm texture, and nutritional value [21].

Consumers want food innovation to address sustainability indicators. Customers prioritize sustainable, healthful, and animal-friendly products. However, individuals must overcome obstacles to modifying their diets, cooking talents, taste preferences, and familial support [22].

Plant-based solutions are gaining traction as a sustainable alternative to animal-derived proteins. Oat proteins have potential as performance components to customize taste and texture. Enzymatic modification can provide customized plant-protein components with specific functional characteristics [23].

Figure 3: Plant-based Alternative Meat

Producers aim to provide consumers with innovative and appealing goods, including classic fermented meat and sausage aromas have not been realized.

Raw plant materials sometimes lack important qualities like umami and sweetness, necessitating careful attention to mouthfeel. A condiment "taste rack" is a collection of spices that can transform vegetables into delicious dishes by "umamification". It was designed to be effective in a flexitarian framework.Acceptance of alternative proteins is driven by factors such as taste, health, familiarity, attitudes, food neophobia, disgust, and social norms. Several techniques have been attempted to enhance plant-based protein functioning, such as enhancing solubility and hydrolysis. However, there is a possibility of negative nutritional consequences. Combining plant and animal proteins can produce healthful and appealing diets. Soy milk, an

https://doi.org/10.52202/074664-0030

alternative to cow milk, has health and nutritional benefits. New technology and research are being developed to produce high-quality soy milk [24].Incorporating plant-based proteins into milk can result in high-protein, multisource, functional foods. Low solubility of plant proteins, such pea protein, makes them problematic to use in food systems with unpleasant taste. Low-temperature homogenization creates stable colloidal dispersions. Plant-based proteins in the dairy industry can encourage product innovation and match current processing conditions and consumer needs [25].

Sustainable gastronomy, which includes culinary practices that promote environmental conservation, social fairness, and economic sustainability, is gaining popularity globally. One of the most significant trends in this field is the use of plant-based alternatives into culinary innovation [26].

Trends:

1. Rise of Plant-Based Diets: Plant-based diets are becoming increasingly popular because of concerns about health, animal welfare, and environmental sustainability. This trend is evident in the growing popularity of vegetarian and vegan offerings in restaurants and other foodservice businesses.

2. Innovative Plant-Based Products: Food manufacturers are always creating novel plant-based alternatives to classic animal-derived ingredients. These include plant-based meats, dairy-free cheeses, egg alternatives, and more, giving consumers a broader range of choices.

3. Endorsements from Celebrities and Chefs: Influential figures in the culinary world, such as chefs, celebrities, and food bloggers, are promoting plant-based diets and developing novel plant-based recipes. Their impact contributes to the widespread popularity of plant-based cuisine.

4. Integration in Fine Dining: Plant-based cuisine is no longer just for health food restaurants; it is also being adopted by fine dining places. Michelin-starred chefs are ingeniously using plant-based ingredients, highlighting their gastronomic potential.

Challenges:

1. Taste and texture: One of the most difficult issues in plant-based cuisine is reproducing the taste, texture, and mouthfeel of animal-derived goods. To achieve this, substantial research and development are required to generate plant-based alternatives that appeal to customers' sensory preferences.

2. Cost and Accessibility: Plant-based components can be more expensive and less available than animal-derived alternatives, providing issues for both consumers and businesses, especially in areas with limited access to various plant-based products.

3. Cultural Acceptance: In some civilizations, meat and dairy products are culturally and culinarily important. To encourage the acceptance of plant-based diets, it may be necessary to overcome cultural obstacles and promote the cultural diversity and richness of plant-based cuisines.

4. Supply Chain Issues: Ensuring a consistent supply chain of high-quality plant-based components is critical to the success of plant-based cuisine. Crop failures, transportation constraints, and supply chain disruptions all have an impact on plant-based product availability and affordability.

Opportunities:

1. Health & Wellness: Plant-based diets have various health benefits, including a lower risk of chronic diseases including heart disease, diabetes, and certain malignancies. Leveraging the health benefits of plant-based gastronomy allows businesses to target health-conscious customers.
2. Environmental Sustainability: Plant-based diets have a substantially lower environmental footprint than animal-based diets, as they use fewer natural resources and emit less greenhouse gases. Plant-based gastronomy can attract ecologically concerned consumers while also contributing to sustainability aims.
3. Market Growth: The growing demand for plant-based products opens up attractive potential for firms to enter and expand into the plant-based market segment. Investing in research and development, marketing, and distribution of plant-based alternatives can result in significant growth and market distinctiveness.
4. Collaboration and Innovation: Chefs, food scientists, nutritionists, and food manufacturers can work together to foster innovation in plant-based gastronomy. By pooling their knowledge and resources, stakeholders can create new plant-based goods and culinary skills that meet customer demand for delicious, nutritious, and sustainable food options.
To summarize, plant-based gastronomy is a vibrant and evolving subject within sustainable gastronomy, with exciting prospects for culinary innovation. Despite hurdles such as taste, cost, and cultural acceptance, growing awareness of health, environmental, and ethical concerns is propelling the global adoption of plant-based diets. By tackling these problems and capitalizing on possibilities, stakeholders may help plant-based cuisine flourish and gain general recognition, ultimately supporting a more sustainable and equitable food system.

3. Materials & Methodology

The literature evaluation for this study included a thorough examination of current studies on sustainable gastronomy, culinary trends, plant-based diets, food innovation, and the environmental impact of food production. Based on this foundation, the study design was developed to use qualitative methodologies such as interviews, focus groups, and case studies to delve into the subject matter in depth and nuance. Open-ended questions were used in interviews and focus groups with key stakeholders, such as chefs, restaurateurs, and food industry experts, to elicit thorough replies and capture diverse viewpoints on sustainable cuisine and plant-based alternatives. The qualitative data obtained was thematically analyzed, allowing for the identification of recurring themes, patterns, and insights that contribute to a better understanding of the topic under inquiry.

https://doi.org/10.52202/074664-0030

4. Conclusion

The study digs into the developing landscape of gastronomy, with a special emphasis on the growing importance of sustainable practices and the introduction of plant-based alternatives into culinary innovation. Through rigorous study, it becomes clear that there is a substantial shift in people's attitudes toward food consumption, driven by concerns about environmental sustainability, health consciousness, and ethical considerations. This study's findings show an increasing consumer demand for plant-based alternatives, reflecting a desire to minimize carbon footprints and ameliorate the negative consequences of animal agriculture on the environment. It also emphasizes the difficulties faced by chefs and food business professionals in developing plant-based foods that not only meet nutritional criteria but also appeal to the different palates of consumers accustomed to traditional meat-centric diets. Despite these limitations, there are several potential for culinary innovation and commercial expansion in the field of sustainable gastronomy. Chefs can take plant-based food to new heights by using innovative techniques, flavor profiles, and ingredient combinations, attracting even the most hesitant guests. Furthermore, this study emphasizes the need of collaboration among stakeholders throughout the food supply chain, including farmers, producers, chefs, and consumers, in driving positive change and promoting the adoption of sustainable culinary practices on a worldwide scale. To summarize, the study of sustainable gastronomy offers culinary experts not just obstacles but also numerous chances to embrace innovation, stimulate creativity, and contribute to a more sustainable future for both the earth and its inhabitants.

References

1. Hegarty, J.A., O'Mahony, G.B.: Gastronomy: a phenomenon of cultural expressionism and an aesthetic for living. International Journal of Hospitality Management. 20, 3–13 (2001). https://doi.org/10.1016/s0278-4319(00)00028-1.
2. Rinaldi, C.: Food and Gastronomy for Sustainable place Development: A multidisciplinary analysis of different theoretical approaches. Sustainability. 9, 1748 (2017). https://doi.org/10.3390/su9101748.
3. Gálvez, J.C.P., Granda, M.J., López-Guzmán, T., Coronel, J.R.: Local gastronomy, culture and tourism sustainable cities: The behavior of the American tourist. Sustainable Cities and Society. 32, 604–612 (2017). https://doi.org/10.1016/j.scs.2017.04.021.
4. Pimentel, D., Pimentel, M.: Sustainability of meat-based and plant-based diets and the environment. ˜ the œAmerican Journal of Clinical Nutrition. 78, 660S-663S (2003). https://doi.org/10.1093/ajcn/78.3.660s.
5. Key, T.J., Appleby, P.N., Rosell, M.: Health effects of vegetarian and vegan diets. Proceedings of the Nutrition Society. 65, 35–41 (2006). https://doi.org/10.1079/pns2005481.
6. Ploll, U., Stern, T.: From diet to behaviour: exploring environmental- and animal-conscious behaviour among Austrian vegetarians and vegans. British Food Journal. 122, 3249–3265 (2020). https://doi.org/10.1108/bfj-06-2019-0418.
7. Rosenfeld, D.L.: The psychology of vegetarianism: Recent advances and future directions. Appetite. 131, 125–138 (2018). https://doi.org/10.1016/j.appet.2018.09.011.
8. Sabaté, J., Soret, S.: Sustainability of plant-based diets: back to the future. ˜ the œAmerican Journal of Clinical Nutrition. 100, 476S-482S (2014). https://doi.org/10.3945/ajcn.113.071522.
9. Giacalone, D., Clausen, M.P., Jaeger, S.R.: Understanding barriers to consumption of plant-based foods and beverages: insights from sensory and consumer science. Current Opinion in Food Science. 48, 100919 (2022). https://doi.org/10.1016/j.cofs.2022.100919.
10. Pérez-Cueto, F.J.A.: Sustainability, health and consumer insights for plant-based food innovation. International Journal of Food Design. 5, 139–148 (2020). https://doi.org/10.1386/ijfd_00017_3.

11. Patel, H., Chandra, S., Alexander, S., Soble, J.S., Williams, K.A.: Plant-Based nutrition: an essential component of cardiovascular disease prevention and management. Current Cardiology Reports. 19, (2017). https://doi.org/10.1007/s11886-017-0909-z.
12. Hu, F.B.: Plant-based foods and prevention of cardiovascular disease: an overview. ˜ the œAmerican Journal of Clinical Nutrition. 78, 544S-551S (2003). https://doi.org/10.1093/ajcn/78.3.544s.
13. Barnard, N.D., Scialli, A.R., Turner-McGrievy, G., Lanou, A.J., Glass, J.: The effects of a low-fat, plant-based dietary intervention on body weight, metabolism, and insulin sensitivity. ˜ the œAmerican Journal of Medicine. 118, 991–997 (2005). https://doi.org/10.1016/j.amjmed.2005.03.039.
14. Gibbs, J., Cappuccio, F.P.: Plant-Based Dietary Patterns for human and planetary health. Nutrients. 14, 1614 (2022). https://doi.org/10.3390/nu14081614.
15. Haddad, E., Sabaté, J., Whitten, C.: Vegetarian food guide pyramid: a conceptual framework. ˜ the œAmerican Journal of Clinical Nutrition. 70, 615S-619S (1999). https://doi.org/10.1093/ajcn/70.3.615s.
16. Fehér, A., Gazdecki, M., Véha, M., Szakály, M., Szakály, Z.: A comprehensive review of the benefits of and the barriers to the switch to a Plant-Based diet. Sustainability. 12, 4136 (2020). https://doi.org/10.3390/su12104136.
17. Dakin, B.C., Ching, A.E., Teperman, E., Klebl, C., Moshel, M.L., Bastian, B.: Prescribing vegetarian or flexitarian diets leads to sustained reduction in meat intake. Appetite. 164, 105285 (2021). https://doi.org/10.1016/j.appet.2021.105285.
18. Resconi, V.C., Escudero, A., Campo, M.M.: The development of aromas in ruminant meat. Molecules/Molecules Online/Molecules Annual. 18, 6748–6781 (2013). https://doi.org/10.3390/molecules18066748.
19. Rubio, N.R., Xiang, N., Kaplan, D.L.: Plant-based and cell-based approaches to meat production. Nature Communications. 11, (2020). https://doi.org/10.1038/s41467-020-20061-y.
20. Santo, R., Kim, B.F., Goldman, S.E., Dutkiewicz, J., Biehl, E., Bloem, M.W., Neff, R.A., Nachman, K.E.: Considering Plant-Based Meat Substitutes and Cell-Based Meats: A Public Health and Food Systems perspective. Frontiers in Sustainable Food Systems. 4, (2020). https://doi.org/10.3389/fsufs.2020.00134.
21. Van Loo, E.J., Caputo, V., Lusk, J.L.: Consumer preferences for farm-raised meat, lab-grown meat, and plant-based meat alternatives: Does information or brand matter? Food Policy. 95, 101931 (2020). https://doi.org/10.1016/j.foodpol.2020.101931.
22. McCauley, D.: Market Drivers and Barriers for Plant-Based Protein Foods. In: Springer eBooks. pp. 485–501 (2022). https://doi.org/10.1007/978-3-030-91206-2_17.
23. Martinelli, E., De Canio, F.: Non-vegan consumers buying vegan food: the moderating role of conformity. British Food Journal. 124, 14–30 (2021). https://doi.org/10.1108/bfj-01-2021-0023.
24. Phillips, F.: Vegetarian nutrition. Nutrition Bulletin. 30, 132–167 (2005). https://doi.org/10.1111/j.1467-3010.2005.00467.x.
25. Bhatia, V., Gopi, G., Behera, P.: Plant-based diet: A solution to the sustainability of life and environment. Indian Journal of Community and Family Medicine. 7, 19 (2021). https://doi.org/10.4103/ijcfm.ijcfm_123_20.
26. Pérez-Cueto, F.J.A.: Sustainability, health and consumer insights for plant-based food innovation. International Journal of Food Design. 5, 139–148 (2020). https://doi.org/10.1386/ijfd_00017_3.

Exploring the Impact of Technology on Services: Insights from the Hospitality and Tourism Industries

Priya Saroj[1] and ATAM Sadiqur Rahman[2]
[1,2] University Institute of Legal Studies, Chandigarh University, Mohali, Punjab, India

Abstract.
Intelligent environments are emerging as a result of technological upheavals like the Internet of Things (IoT), autonomous devices, artificial intelligence (AI), and rich media (VR/AR), which arealtering the structures, procedures, and practices of several industries. To shed light on service breakthroughs affecting ecosystems, this article examines significant technical advances via valueco-creation. Tourism and hospitality are used as examples in the article. In this study, new technicaladvances are brought together with established theories of co-creation, service ecosystems, networks, and technological disruption. By definition, guests' experiences in the hospitality and tourist industries occur in unfriendly settings that are foreign to their usual routine and comfort zone. Customers' experience navigation will be rethought with the advent of intelligent surroundings. To stay competitive, stakeholders must reimagine how they connect with technology, reengineer services and use these technologies at a conceptual level. This article showshow technology is changing business models and encourages people to work together to create value locally and globally. Societal. Technology disruptions affect every part of our lives. To betterprepare for forthcoming change, policymakers might benefit from a holistic view of developments in this area, which offers nuanced viewpoints. The hospitality and tourist industries thrive despite various institutional, cultural, and physical limitations. Without delving into specific destination kinds and conditions, this article outlines possible interruptions and developments. This calls for more research into how stakeholders may adjust in different settings.
Keywords: Technology disruptions; tourism and hospitality; artificial intelligence, virtual reality, augmented reality, co-creation

1. Introduction:

Information and communication technology (ICT) are having a profound impact when it comesto creating new goods and services. Intelligent technologies are reshaping industry standards and driving service providers towards dynamic performance optimization; these systems range from simple assembly lines to complex multi-stakeholder ones that integrate sensors, data storage, microprocessors, software, and connectivity [1]. With the help of processing power and pervasive network connectivity, innovative connected devices are reshaping markets, upending value chains,and reengineering corporate processes and economies Life, work, and travel are all affected by this. They are considering that it brings about dynamic structures in all areas. A variety of shapesand forms are emerging from this transformation. Economies are increasingly taking the shape ofdecentralized networks of owners, suppliers, intermediaries, and stakeholders that engage in dynamic demand-supply interactions via decentralized platforms [2]. There used to be just two options for getting around in the days before the sharing economy: having a vehicle or paying individuals or organizations to do it for you. Recent technological advancements and intelligencehave unavoidably had an impact on service management. When service providers began using newplatforms, the availability and

https://doi.org/10.52202/074664-0031

accessibility of their offerings skyrocketed, allowing users all overthe globe to plug and play instantly. Furthermore, thanks to the sharing economy, consumers can connect with others adept at using new platforms to provide their services. Businesses like Uber Eats and Deliveroo expanded quickly to help service providers satisfy customer demand. The emergence of several innovative business models made the market's expansion possible. Micro and macro economies are being reengineered by the growth of Internet connectivity, big data, and the Internet of Everything, revolutionizing production and consumption. Businesses in the tourism and hospitality industries provide services to customers in unfamiliar and often dangerous places,far from the comforts of home and the linguistic, cultural, and physical resources typical of those locations. Consumers require more information to operate in challenging circumstances and makethe most of their time and money. As tourism and other forms of information and communicationtechnology have become more widely used, this type of consumption has also been changing. Since the early 1960s, technology in the form of global distribution systems (GDS), booking and reservation systems, social media and mobile applications, recommender systems, context-drivensearch, search engines, and Web data mining, has helped consumers find relevant information andservice providers to facilitate transactions. As a frontline service industry incorporating new knowledge about technological advancements into strategic planning processes, tourism can teachus a lot about dealing with the ever-changing wave of technological advancements and managingand marketing our services disruptions in service management and the evolution of value co- creation processes brought about by technology are examined in this article. Based on the premisethat information and communication technologies promote value co-creation, this article delves into how technology progress permits value co-creation within the ecosystem of tourism services.The study shows how new technological trends and projects might encourage micro-foundationallevel value co-creation. Also covered is the newly created value in regular tourist services and howit could shake things up. It describes how revolutionary innovation changes the market structures on a societal level. The article concludes by extrapolating marketing and service management principles from the tourist and hospitality industries. This study explores the implications, both at the macro and micro levels, of technological advances and disruptions in developing new ecosystems for service management [3].

1.1 Advances in Service Management Due to Technology

Service providers, consumer interaction techniques, and their expectations have been influenced by recent technology breakthroughs. Figure 1 is a summary based on a survey of recent research into emerging and new technologies that have direct consequences for service sectors. The synthesizing innovations that have been identified are backed by various enabling technologies, which have the potential to drive disruptive change. Provide the groundwork for implementing technology innovations in the firm/customer sphere. There has been a deluge of new service innovations brought about by these enabling technologies, which have had previously unseen effects on both the macro and micro levels of society. Their effects can be better assessed within the theoretical framework of value creation [4]. The four forms of service innovation have emerged due to technological advancements: service experiences, service systems, resource integration, andthe last many years. With the customer's point of view as its starting point, the service experiencearchetype reflects the statement by Vargo and Lusch that "value is always uniquely and phenomenological Nine determined by the beneficiary". It suggests that value co-creation can include something other than direct service. Crucially, technology is merely a tool for customer service; it is not the service itself. The consumer plays a vital role in the service systems typology,which emphasizes social connectivity and the dynamic interaction of resources. This provides a more dynamic and comprehensive perspective on co-creating value, which utilizes S-D logic for synthesis given this, we take a look at the

https://doi.org/10.52202/074664-0031

innovations in technology and their possible effects onthe innovation and management of services. Technologies that allow for Modern service innovations are based on seven crucial technical breakthroughs that affect how firms and customers interact, affecting marketing and service management. These include 5G mobile networks, A.I., RFID, mobile devices, applications and application programming interfaces, cryptocurrency, and blockchain technology. We begin with a brief description of the technology and its possibilities for service innovation before discussing the consequences for service marketing and tourist management. The next generation of mobile networks is a wireless telecommunications system known as 5G. Mobile networks will be able to transfer massive amounts of data much more quickly after 5G is implemented laying the groundwork for the IoT, interconnection, and ambient intelligence solutions. Upgraded consumer involvement is on the horizon thanks to this, and developments in data-rich streaming services, like games and films. Applications facilitate interactions between machines and customers, and they are fundamental tothe development of intelligent, autonomous devices and agents, mission-critical systems, and infrastructure. Natural settings. Scalable application of the technology is anticipated, and it is presently in the early phases of implementation around the globe. Companies in the telecommunications industry, including Ericsson, Telenor, E.E., China Unicom, and others, are considered stakeholders [5]. 5G comprises several different Advancements that include edge computing services that connect data from remotely located sensors to wireless devices, reducing latency in cloud-based applications and · new types of video delivery services that are likely to compete against existing channels, such as Services like Netflix and Amazon contribute significantly. Immediate effects of 5G on service innovation include faster customer-facing onlinecontent delivery and enhanced Internet of Things connectivity. Services leveraging urban automated networks, such as the Internet of Things and autonomous devices, would surge in acceptance once 5G telecoms are widely used. Future automotive, healthcare, energy, media, andentertainment industries will all feel the effects of 5G. The development of service ecosystems, which will be co-created by various actors, is expected to positively impact rural/urban integration,decentralization of work, reduced mobility needs, energy efficiencies, increased security, and overall improved life expectancy [6]. The perceived reliability of equipment has already been affected by recently brought light international security concerns with one prominent 5G provider, prompting demands for future network evolution that require further investigation. As a result of developments in technology, four types of service innovation have emerged in the last many years.With the customer's point of view as its starting point, the service experience archetype reflects the statement by Vargo that "value is always uniquely and phenomenological nine determined by the beneficiary". It suggests that value co-creation can include something other than direct service. Crucially, technology is merely a tool for customer service; it is not the service itself. Theconsumer plays a vital role in the service systems typology, which emphasizes social connectivityand the dynamic interaction of resources. This provides a more dynamic and comprehensive perspective on co-creating value, which utilizes S-D logic for synthesis. Given this, we take a lookat the innovations in technology and their possible effects on the innovation and management of services. Local storage on tiny (microchip-sized) devices enabled with near-field communicationsis the basis of Radio Frequency Identification technology. Environmental data can be sensed, stored, and transmitted by these devices. It is an already popular technology that is integrated intoa variety of products, including credit cards. Products include hotel keys, airport baggage and transportation systems, tolls and security systems, animals, and inanimate objects. Along with reading devices like cell phones and nearby RFID allows the ever-expanding use of data stored onchips in access machines, especially when combined with AI technology. Identifying closeness with a geographical global positioning system, time and location-tracking apps can make various services dependent on place possible. RFID technology, when integrated with new types of sensors, allows intelligent settings built on the principles of optimizing resources and operations, which also facilitate the maintenance of real-time approaches to engaging customers. RFID enables the Internet of Things (IoT), an emerging set of applications and services. Sensor-based technologies make service delivery advances possible. Modern technology has made it possible for businesses and their customers to work together to build an experience driven by data in the context of a cyber-physical system setting. They have spread over. As a result of services capes, dataflow and data volume have increased, laying the groundwork for socio-technical service. Mobile telephony and micro computing capabilities are integrated into smartphones and other mobile devices. They surpassed all other internet connection methods in 2018 and are used by more than 40% of the global population. Consumers have quickly embraced smartphones as theirpreferred tool, even if the debut of the first smartphone occurred in 2007[7]. In doing so, they are essentially underserved and geographically isolated populations to acquire internet connectivity service capabilities. Smartphones are assemblages of technology that are becoming more and morecomplex. With built-in or externally hosted APIs, sensors, recorders, GPS, cameras, and wirelesstechnologies apps. The capacity to reply to specific users through face, voice, and other biometricdata will be a future feature. As they evolve into more physically flexible designs, technologies for movement recognition, projection, and holographic displays will combine with wearables. The widespread use of augmented and virtual reality is supported by technology, with 5G allowing forthe transmission of quicker and richer material that incorporates real-time telepresence Research and the creation of novel kinds of energy and charging accessibility are made possible by environmentally friendly battery materials. Changes to resources are also expected to significantly impact how they are used and adopted in the future. Typically using an unconventional platform interface design, wearables are a subset of intelligent mobile devices. Things like bracelets, tattoos,or watches. Based on their capacity to enhance the individual, the so-called notion of the extended self. More and more, cell phones will respond to their surroundings by utilizing data both proactively, so they can experience the owners' personalized environment, and reactively, so they can deal with unforeseen events in the scene. Soon, smartphones will serve as virtual personal assistants, collaborating in real-time with all available resources to enhance customer satisfaction.Apps, in conjunction with APIs, make wearables and smartphones functional. These interfaces developed by outside parties lie atop platform technology, such as operating systems and hardware.Despite being a well-known and widely-used set of technologies, the continuous improvement ofapp accelerated pages, particularly those impacted by Google's open-access Accelerated Mobile Pages initiative (AMP, see. It is expected that https://www.ampproject.org/ will significantly improve their performance. Therefore, it paves the way for the utilization of augmented and virtualreality content in the future, especially in the realm of service management. The military,healthcare, and other industries will likely see new mobile-enabled engagement techniques. Domains, from engineering to education and services. More than four hundred and fifty travel appsare available in the Apple App Store & tourist-related applications like maps, personal trip guides,and information about attractions at destinations. Travel agencies, translators (both spoken and written), currency converters, and specialized hotel service providers. Quite a few of these apps have minimal levels of interaction, but new types of apps made possible by technological advancements make them far more useful. When people buy and sell digital currency with one another online, they are engaging in what is known as cryptocurrency. Includes procedures to prevent counterfeit goods. The word describes a combination of secure communication protocolsand monetary systems. It offers a safe and verifiable record of transactions. Digital currency algorithms dictate the creation and control of the system, the recording and processing oftransactions, and the three is the creation and transfer of new tokens. A unique identification is assigned to each coin during its creation by encryption, which is enhanced during transmission andexchange. The peers play an essential role in logging and validating all transactions involving eachcoin and serving as a shared financial institution. A 'blockchain' encompasses the recording process. Digital decentralized ledger technology is what blockchain is all about. The ledger is

https://doi.org/10.52202/074664-0031

an ever-expanding collection of records called blocks connected by cryptography. Blocks cannot be changed once written; this is the principle. Blockchain technology is open. It is Computer-readableand dispersed across a network, so it lacks a centralized supreme power. The community that holds the blocks collectively ensures the chain's legitimacy, successfully partnering to create shared value. Cryptos, ever since their 2009 debut are currently traded on stock markets and serve as a common currency. On every continent [8]. They are marketed as a digital commodity and exchanged similarly to gold and silver because of their rarity and high value. Several advancementsin cryptocurrency allow novel kinds of to occur without the usual market-based intermediaries, such as banks and government entities, along with the related professional and financial services.That being said, they are profoundly altering the commercial landscape, particularly regarding the evolution of contracts and value-based transactions between businesses and their clients, particularly in cases when minimal trust exists between them (Tillier,2018 was the year. As an advantage of using it, you will not have to worry about scams or transactions, as the sole option isto utilize cryptocurrency, which, by extension, verifies ownership by the immutability of block chain data. As a result, purchasing services becomes more accessible, efficient and uncomplicated. Providers of services stand to gain significantly from this development as they evolve progressively depending on the interaction between service systems and their consumers. Since the introduction of these enabling technologies, new synthesizing service developments haveexploded. The synthesis process exemplifies a paradigm shift towards valuing co-creation across disciplines and dimensions. Three groundbreaking innovations and the far-reaching consequences of each. Next, we will review the role of intelligent environments, cybersecurity, and gamificationfor service industries in shaking up service management and value co-creation in the marketplace.In 2015, Harwood and Garry worked together. Even though they provide unique service advancements, they utilize the technology described earlier. See, for example, which synthesizes innovations that rely on mid-range theories of value co-creation and which may pave the way for empirical studies to investigate and expand upon the innovation's theory and operationalization. Following this introduction, the study delves more into these topics. An intelligent environment uses everyday technology to fulfil various needs. Lasting improvements in resource utilization and novel understandings of business processes derived from intricate data to companies and their party involved. The IoT, or Internet of Things, represents a technological change. It linkseverything and everyone, everywhere, at any time, and sparks creative new uses together with services. The medical and hospitality industries and urban areas have innovative service environments. Ambient technologies are being used to simulate spaces like hospital rooms, whichcan help patients and medical professionals by using integrated sensor technology. By utilizing these medical spaces individuals have agency over their surroundings and engage with the healthcare institution's infrastructure and staff. To control the televisions, lights, window coverings, and other electronic devices in each patient's room, the hospital medical equipment, beds, etc., are all linked to a system to ensure smooth operation. Inns are Journal more and more people are researching and experimenting with intelligent room setups to help manage their surroundings and collaborate on developing new service offerings. Both Hilton and Marriott are testing the waters [9]. When integrating ambient technology with sensors on gadgets and equipment. For service experiences that guests control. The Singapore Tourism Board spearheadsthe brilliant tourism initiative. 'STAN,' a tourism analytics network, can be used to collect and evaluate data from mobile devices used by tourists. Cellphones, transit, and gridlock and safety systems. The aim of one that a strategy that is based on data targeting specific visitor offers basedon their time, location, and budget. To safeguard and defend against cyber threats, a confluence of technology and processes has emerged. Data, software, and hardware from malicious, unapproved,or fraudulent use and implement data protection by design. Exploits in cyberspace that use Malicious software, ransomware, spyware, etc., are prevalent in today's market, and A significantrisk to the ongoing success of any organization is computer vulnerabilities. However, this is only one viewpoint. In light of the anticipated exponential expansion of linked things, gadgets, and humans, numerous new technological developments have resulted. An increasingly crucial function of AIs is expected to monitor security threats in real-time and respond accordingly, evenwith the same technology [10].

Attackers also have access to advancements, which suggests the vulnerability is significant. Especially vulnerable are systems, mission, and safety-critical systems, including healthcare, financial, and identity systems. Only recently, a breach at Marriott affected the accounts of almost500 million guests. Database for Starwood reservations. Many people are upset that it took PaneraBread eight months to contain the data breach that affected their online delivery system. Strong Many companies choose not to implement cryptographic measures due to concerns about the potential impact on their resources. To overcome these obstacles, researchers are looking into cryptocurrencies and block chain technologies. For a long time now, people have been urging us to reevaluate cybersecurity from every angle. Concerns arise from engineering, systems, networks,and human behavior viewpoints. The massive success of the video game industry in ideologicallyengaging consumers all over the globe has given rise to a widespread movement known as gamification. G45amification involves incorporating game mechanics into non-game contexts, such as websites, employee engagement processes, marketing, branding, customer retentionstrategies, and achievement badges and goals. New devices, AIs, VR, AR, etc., are all part of the highlighted technical breakthroughs that this method takes advantage of. The tourism and hospitality industries stand to benefit significantly from gamifications many possible uses and ancillary activities. Club Med's Opie Holiday Village in Provence, France, used gamification in their JADE app to encourage users to learn about the village's history, participate in sustainability efforts, and get to know the surrounding landscape through augmented reality (Atelier Nature, 2016). Loyal customers of Starbucks, a restaurant chain that has integrated game mechanics into its rewards program, can earn stars, badges, and other perks like menu challenges, grocery store rewards, and freebies that help them build their custom. It can enhance satisfaction levels and positively impact interactions, boosting brand recognition and loyalty for organizations and locations related to tourism. Future applications of AR/VR and LBS are also exciting for people. Utilized to pique users' interest in new service offerings, made possible by AI-powered gamified experiences tailored to each user's unique tastes.

1.2 As a Technologically Enabled Co-Creation Ecosystem, Tourism

Companies in the travel industry would do well to reframe their brands as technology-driven platforms that improve the customer experience and foster value co-creation rather than attractingaudiences with superficial technological trends. Development. Online travel agencies, lodging, transportation, and sightseeing are all parts of a larger tourism ecosystem that includes these platforms. Technological progress can help bring people together in the tourist industry's sometimes-hostile service ecology to foster co-creation. Fear of the unknown and anxiousness arecommon emotions experienced by tourists when visiting new places. Tourist locations are not always idyllic escapes; sometimes, they are unpleasant and strange compared to the tourists' homelives. A traveler without travelers may experience anxiety when confronted with strangers, particularly those speaking a language they are not acquainted with, or it may lead them to avoid trying local cuisine. Equally challenging for international tourists is getting to their destination, which involves navigating unfamiliar social mores and cultural norms, exchanging money, goingthrough security checks, and making contingency plans. Tourists usually take an active role in their own experiences but when they encounter animosity, it hinders their natural inclination to co-create. The lack of familiarity with local environments and resources makes tourists especially susceptible and ill-prepared to respond during terrorist attacks or natural disasters. The worldwidetourism industry has been rocked by terrorist incidents, which have led

https://doi.org/10.52202/074664-0031

several visitors to the concerned with whether they ought to cease traveling or how to be ready as a result, we need to know how tourists can presume and co-create experiences during the service exchange process indangerous and unfamiliar places, mainly when there are unforeseen incidents and turbulence. It iscrucial to offer cutting-edge resources to facilitate experience co-creation and provide emergencysupport in case of necessity [11]. Traditional tourism research has viewed the industry as a web ofinterdependent service providers whose efforts, when put together, provide visitors with memorable adventures at the locale. Service quality in the tourism industry greatly depends on thecommunication and mutual understanding of all parties involved. Service ecosystems offer a perspective on the interplay between institutions, individuals, and technology in the co-creation ofservice value, which aligns with the system theory perspective on tourism. There has been a changein emphasis from macro processes to micro-foundations in value co-creation across stakeholders in service ecosystems states that co-creation on a micro level involves the everyday service exchange procedures of the organizations and individuals that comprise a service ecosystem. Whileco-creation techniques have historically been seen as interactions between businesses and their customers, the focus is now shifting to interactions between customers themselves, as customer- to-customer processes have a more significant impact on the final product. Thanks to ICTs, humans can interact with other humans, such as visitors, service providers, suppliers, intermediaries, and locals. Tourists can co-create their experiences and enjoy value propositions by interacting with various stakeholders in tourism ecosystems, including transportation, lodging, attractions, retail, and food service sectors. Information and communication technologies also enable the co-creation of value between customers and continuously change the face of service management. As impersonal agents, ICTs facilitate the development of co-creation ecosystems and alter the character of human-mediated social interaction. Incorporating both the micro and macro perspectives on service innovations, ICTs provide opportunities for the joint production ofvalue, which may lessen travel's perceived dangers while increasing familiarity with the location and the ease with which information can be proactively provided. Using ICTs may also make people feel more welcome. It promotes accessible tourism, which benefits individuals withdisabilities and other special needs who travel. Many people dissuade people with mobility, visual,auditory, or cognitive disabilities from going on vacation because they are obstacles at destinationsin terms of both physical and service quality. By combining and supplementing human labor, the technology eliminates or dramatically reduces the need for customer service problems. People withtrouble moving around can use virtual reality to see how easily accessible a place is and how the trip will go without leaving their house [12]. Similarly, people on the autistic spectrum could use virtual reality to familiarize themselves with their destinations before the journey. Virtual reality platforms for destinations allow autistic tourists to co-create their experience and practice it until they find a routine, which may reduce anxiety and discomfort during the actual trip. Introducing new competitors and changing market conditions can cause established industries to undergo a period of disruptive innovation. These have altered market structures in the tourism industry by impacting stakeholder relationships, one example of a tech-led disruption is Airbnb, a platform that facilitates connections between hosts and guests, allowing them to co-create value. Ultimately,Airbnb became a formidable competitor to traditional hotel chains by facilitating direct connections between tourists and locals at various locations. Customers searched for cheaper, more authentic, locally sourced experiences directly from residents, upsetting travel and industryhierarchies. The Internet of Things (IoT) opens the door to intelligent ecosystems in the tourist industry, thanks to the interconnection of systems and gadgets visitors may personalize. Using thevarious sources, we will review three significant technological developments that might soon shake up the service management industry. These include augmented and virtual reality, autonomous devices or agents, and location-based services encompassing social context through social media. VR and AR Using a combination of screen-based technologies, haptic devices, and exoskeletons, virtual reality (VR) creates a simulated environment that completely engrosses the user, altering their perceptions of sight, hearing, touch, and somatosensory input noted that virtualreality (VR) uses these gadgets to fool the human mind into perceiving external signals as embodied sensations of controlling a body in a virtual setting. Virtual reality is changing the faceof service industry marketing and micro-level management, not just the gaming industry. The ideaof physical travel is challenged by virtual reality, which offers new ways to imagine one's body ina service setting, regardless of where the service is located. Its use in service design enables full- stack customer journeys and high-fidelity prototyping. Direct providers, like healthcare workers, are already trained to empathize with patients and manage emergency rooms through service walkthroughs. Similarly, hotels employ them for marketing purposes. Organizations in the tourism and hospitality industries are embracing virtual reality (VR) to give clients a taste of faraway placesvia virtual tours and pre-arrival facility experiences. Electronic commerce giants like Alibaba, Amazon, and Walmart are preparing to launch virtual reality markets, which will revolutionize how customers purchase [13]. Lowe's Hoolboom's How to VR Experience and other virtual reality (VR) experiences put clients at the center of stories and allow them to try out services before theybuy them. Using virtual reality, cultural heritage sites can make visitors feel like they have traveledacross time and space, fulfilling the fantasies of those who bring a sense of wonder and adventureto their archaeological digs. Similarly, augmented reality (AR) uses mobile screen devices (such as glasses, smartphones, and wearables) to exhibit consumers layered data. To enhance the user's sensory experience, layers often combine real-time location data with graphical and informative material. Pokémon GO, created by Niantic, is a computer game in which users capture charactersthat are situated in real-world locations. This game has made the technology more popular. The way service businesses convey their identity and promote their products is transformed byaugmented reality (AR), which is made feasible by faster processing rates and allows for rich real-time experiences. Brands in the service industry leverage these kinds of interactive platforms to increase engagement by bringing in-home service experiences to consumers. More content is available through augmented reality than other display formats, which could improve travelers' on-trip experience. Museums, galleries, and significant events use augmented reality to create more immersive visitor experiences. Service management may reengineer tourism experiences, provide interactive customer experiences, facilitate interpretation, and interact with avatars using augmented reality. By reducing the time, it takes to retrieve visual information, AR makes it possible for complex service procedures to have added value. Augmented reality (AR) can hyper-personalize content for each individual by engaging customers who desire to utilize and learn knowledge in keeping with the trend of user-centered design. While it is too soon to tell which companies will be able to implement such layers, it is probable that those with expertise in game-based technologies, where augmented reality is created through design-led interventions, will be the ones to disrupt the integration of sensor-based technologies that allow customers to access locallevels of detail. Robots, cars, or aero planes- both the macro and micro levels of an organization'sstructure and operations- are susceptible to technological disruption. Virtual assistants (VAs), robots, drones, and connected autonomous vehicles (CAVs) are just a few examples of the new autonomous devices and agents that quickly alter service delivery. AI and robots in the restaurantindustry are already causing operational disruption. You can find robots doing everything from serving to cooking at fast food and quick casual restaurants. Spice, a robot-driven fast-casual concept in Boston, USA, and Reis and Ivy, an automated frozen yogurt concept restaurant with a franchising structure, has attracted substantial funds for expansion. Several other low-level operational projects standardize a service environment that has historically relied on human interaction by using robots to do tasks such as flipping burgers and making cocktails [14]. However, the tourism industry that depends on locals to create one-of-a-kind experiences in the framework of local service management could suffer due to the ensuing upheaval. Artificial intelligence can potentially displace human workers in many occupations and expand the reach of the service

https://doi.org/10.52202/074664-0031

industry. The initial focus is on modeling repetitive operations, with machine learningenabling the creation of tailored solutions. When external characteristics and empathy are crucial to the experience's co-creation, human connection is necessary in service management. Numerous organizations utilize virtual assistants (VAs), often called chatbots, to assist with customer servicesand personalize experiences based on individual needs. VAs are now the most advanced type of artificial intelligence. They help businesses save money on labor and provide instantaneous assistance to clients by answering common questions in any language. The creation of accurate anthropomorphic representations by robot developers to simulate natural interactions is on the rise.Like the velociraptor receptionist at Japan's Hotel Okura, they may belong to the uncanny valley of creepiness that customers perceive as just marketing gimmicks. Robots can do mundane, repetitive jobs, assist with physical labor, and even serve consumers at inconvenient times. The supervisor cites several smart environments used for drones and CAVs, including services cape monitoring, data collection, visitor guidance, remote service delivery, and integrated mobility service systems. They can be utilized for product delivery in locations with optional infrastructure.These advancements can reduce the likelihood of human mistakes while allowing service workersmore time to engage with consumers on a more personal level by eliminating monotonous activities. Autonomous devices have the potential to revolutionize service industries by enabling a plethora of on-demand services, such as on-demand laundry and ATMs. These services could ease ground traffic congestion, improve road safety, and enhance the quality of life in heavily populated tourist destinations. Customers' service expectations are being raised by the widespread use of deep learning technologies in several sectors. The highlighted technologies lead to a need for more user-friendly services that provide better advantages, including optimized relationship management through personalized real-time recommendations, reasoning, and decision support. Autonomous technologies, on the other hand, could dehumanize co-created experiences and endanger traditional low-skilled or low-paid service sector professions. Considerably altering thelandscape of the travel market. Using chatbots to substitute human hotel receptionists is a prime illustration of this. The possibility that autonomous vehicles may shake up established service businesses and due to economies of scope, agents will be impacted by the performance of the services embedded in the devices and agents. Location-based services (LBS) use data collected from sensors and autonomous agents to offer products and services through mobile devices and screens [15]. Using beacons equipped with RFID technologies to transmit signals to customers' smart devices as they approach the waypoints and record data, LBS technologies can be used in sensors that measure or detect physical property. Billboards integrated into the services cape can identify the users' profiles and provide personalized recommendations based on their actions. Withthe increasing accuracy of signal processing to determine location along with facial, voice, and movement recognition powered by AI, these technological advancements can enhance the serviceexperience and evolve the placement of advertisements and sponsored content. Language-based services allow for the instantaneous, location-dependent personalization of customer service experiences. Location-based services (LBS) track customers' actions, habits, and business interactions in real-world settings, providing massive datasets on consumer behavior. In the hospitality industry, similar apps are used to monitor guests' whereabouts; this information is thenutilized to enhance the reliability of service delivery and recovery times. To provide users with socially interactive, individually tailored experiences, context-based services take into account theusers' actual physical surroundings and combine this data with their social networks. Tourism marketers are increasingly seeing the value of these tools in enhancing social encounters. In general, incumbents in specific contexts face disruption from gadgets and technology providing contextualized services. To be relevant, local businesses must change how they deliver informationto clients. For instance, when Google Maps becomes the gateway by which tourists locate a restaurant near their hotel, they will need to adapt. Similarly, when it comes to social media, it instantaneously provides suggestions and guidance, thereby obstructing the impact of

https://doi.org/10.52202/074664-0031

hotelconcierges and local advertisements. In such instances, the co-creation process's personal, social, and geographical aspects are brought to the forefront by networked intelligent gadgets, which change the setting to the point where even the term "local" is changed. As a reflection on these three domains of technological progress, we provide a literature review that examines the functionsof these technologies, considering how the results bring attention to potential future disturbances to service experiences and research implications. These stand for three significant domains of study in the hospitality and tourist industry. Extra-sensory experiences, reflecting the enhanced sensory experiences possible with virtual and augmented technologies · Hyper-personalized experiences, reflecting the merger of location and social context in service experiences · Beyond-automated experiences, reflecting the nature of experiences beyond a process of standardization through automation of services [16].

1.3 Final Thoughts: New Disruptive Insights in Service Management a Broadcast:

Research in service management that investigates service actors' co-creation and market distribution of value is urgently needed in light of recent, dramatic technological advances. Governments worldwide are putting money into "smart cities" programmed to boost sustainability,efficiency, and quality of life via new technology, group decision-making, and economic cooperation. To achieve and ensure a high quality of life for stakeholders and people, smart cities must strike a balance between software and hardware, technology, and human resources. Althoughmost new frameworks for innovative sustainable development have been developed for urban contexts, some can be applied to other regions. To create a customer-dominant logic-based ecosystem that offers hyper-personalized and beyond-automated experiences, attain a sustained competitive advantage, and improve the quality of life for both locals and visitors in smart tourismdestinations, it is necessary to have technology, leadership, innovation, social capital, and human capital. Technological improvements are expediting the continuous trend toward the co-creation of value between firms and customers in several sectors. Companies throughout the classical manufacturing period created almost all of the worth in the goods they offered for sale. Nowadays,services, not manufacturing, are the main engines of economic growth, and in most cases, customers and businesses work hand in hand to provide value for consumers. The technical developments discussed in this paper—including location-based services, robots, virtual reality, and augmented reality—accelerated that trend and affected businesses and customers. Moving closer to the customer, consumer-facing technologies like social media, mobile devices, and the Internet of Things (IoT) are repositioned as potent operant tools. New legal frameworks and a shift in the power dynamic between businesses and consumers are necessary to realize the full potentialof these resources' leverage. Companies, particularly those operating on platforms, may feel pressured to take advantage of their dominant positions, but this strategy could be unsustainable. Companies cannot just sit on their human capital and expect to reap the benefits. Customers engage with businesses that provide services or items on their preferred platforms and bring theirpreferences and networks into these engagements. Ultimately, this benefits everyone involved because the ecosystem constantly adapts to new information. The intersection of consumer behavior with macro-level processes offers a wealth of potential study subjects. Even smarter citieswill continue to attract the same kind of people who have always visited cities: residents and tourists. How will people in these cities rate the business services they get? How will they respondwhen faced with expanded consumer data lifecycle management and "smart" city components? When foretelling those responses, which theories could be the most helpful? As previously mentioned, value co-creation offers a fascinating framework for comprehending customer desires, motives, and responses. Will customers be interested in and supportive of the mutually beneficialpartnership with technology if we undertake all that co-creation? When they engage in episodes ofco-creation exchange, what drives them to do so? Given the increased involvement required for co-creation,

https://doi.org/10.52202/074664-0031

what are the marketing consequences? Where does this leave us if customers opt out?Given recent technology developments, how may new regulatory frameworks influence consumer desires, motives, and responses? On a micro level, a firm's ecological position is becoming increasingly important in determining the relative importance of exploration and exploitation tactics. Platform businesses, including Google and Amazon, employ both approaches. Regarding exploitation, they use beta and alpha testing to increase efficiency by leveraging their size and distinctive corporate culture. Additionally, they make use of economies of scale and network effects. Their innovations have inherent benefits in exploration because their brand recognition isstrong, and millions of users are already familiar with and adapted to their products. Several multinational corporations have comparable impacts, including Tencent, Alibaba, WeChat, Apple,Facebook, and Microsoft. These companies can only use the highlighted technologies if they adaptto meet the growing demand for innovative services among consumers. Future service improvements may originate from non-traditional sources, such as start-ups, mergers, or competitors, and these developments in foundational technology are predicted to shake up established market structures. For instance, the need for P2P sharing led to the rise of YouTube inthe last several years. Despite being restricted and controlled by the original service providers at first, it improved social presence in the digital world. Think of it as a cross between Netflix, Noton the high street, or eBay, but for mainstream television. The consumer economy, including non-profit companies, is essentially reduced to exploitative tactics. As illustrated by technological advancements, the telecoms sector is driving ecological breakthroughs by leveraging 5G connection and ambient intelligence. Everybody else has to hone their craft inside the ecosystem of service supply, which these platform corporations define. Companies can only afford to lose ground in the market if their systems are not easily integrated. Numerous companies will try to increase their market share by introducing small-scale socio-technical service innovations that usestakeholder co-creation. Amazon Alexa, Google Assistant, and Samsung SmartThings are just a few examples of home appliance manufacturers who are now compelled to commit to and executea strategy within an IoT ecosystem. In each scenario, the other businesses in that ecosystem feel the effects of a platform company's illness. Numerous study opportunities have arisen due to technological advancements: How can platform-level companies, such as telecoms and others mentioned before, realistically strike a balance between exploration and exploitation? How can they find a middle ground using novel models or approaches? When established platforms becometoo dependent on exploiting their users, what new competitors will emerge to shake things up? Numerous micro-level research inquiries concern consumers' desires, motives, and responses. Companies must focus more on capitalizing on the present to avoid missing out on opportunities to meet customers' evolving needs. Investigating these desires in their current or future forms in aworld where autonomous devices and agents, virtual reality and augmented reality, and services tailored to specific locations and social systems are all pervasive could yield some interesting insights. Could customers' socialization and decision-making abilities be negatively impacted by the overwhelming nature of hyper-personalized and extra-sensory experiences? How do customers feel about being involuntarily used as test subjects in an experiment? Once upon a time, Amazon customers discovered that the company was experimenting with the price-demand curve by changing prices randomly. Amazon did not do well, but more studies could identify moderating elements that could improve the taste of the trial. Companies conducting consumer research are engaging in a transaction, and our theories of fair and reciprocal exchange should be helpful in thiscontext.

References

1. Al-Fuqaha, A., Guizani, M., Mohammadi, M., Aledhari, M., Ayyash, M.: Internet of Things: A survey on enabling technologies, protocols, and applications. IEEE Communications Surveys and

Tutorials/IEEE Communications Surveys and Tutorials. 17, 2347–2376 (2015). https://doi.org/10.1109/comst.2015.2444095.
2. Alyasiri, O.M., Selvaraj, K., Younis, H.A., Sahib, T.M., Almasoodi, M.F., Hayder, I.M.: A survey on the potential of artificial intelligence tools in tourism information services. Babylonian Journal of Artificial Intelligence. 2024, 1–8 (2024). https://doi.org/10.58496/bjai/2024/001.
3. Armstrong, J.S., Morwitz, V.G., Kumar, V.: Sales forecasts for existing consumer products and services: Do purchase intentions contribute to accuracy? International Journal of Forecasting. 16, 383–397 (2000). https://doi.org/10.1016/s0169-2070(00)00058-3.
4. Bilgihan, A., Ricci, P.: The new era of hotel marketing: integrating cutting-edge technologies with core marketing principles. Journal of Hospitality and Tourism Technology. 15, 123–137 (2023). https://doi.org/10.1108/jhtt-04-2023-0095.
5. Buhalis, D., Harwood, T., Bogicevic, V., Viglia, G., Beldona, S., & Hofacker, C. F. (2019). Technological disruptions in services: lessons from tourism and hospitality. *Journal of Service Management*, *30*(4), 484–506. https://doi.org/10.1108/josm-12-2018-0398
6. Grobelna, A. (2015). Extraversion and its Importance in the Hospitality Workplace. Analysing the Selected Job Outcomes. *Ekonomiczne Problemy Turystyki*, *31*, 105–124. https://doi.org/10.18276/ept.2015.3.31-06
7. Kalyar, M.N., Ali, F., Shafique, I.: Green mindfulness and green creativity nexus in hospitality industry: examining the effects of green process engagement and CSR. International Journal of Contemporary Hospitality Management. 33, 2653–2675 (2021). https://doi.org/10.1108/ijchm-09-2020-1079.
8. Kasem, M. S., Hamada, M., & Taj-Eddin, I. a. T. F. (2023b). Customer profiling, segmentation, and sales prediction using AI in direct marketing. *Neural Computing & Applications*, *36*(9), 4995–5005. https://doi.org/10.1007/s00521-023-09339-6
9. Ogbeide, G.-C., Fu, Y., Cecil, A.: Are hospitality/tourism curricula ready for big data? Journal of Hospitality and Tourism Technology. 12, 112–123 (2020). https://doi.org/10.1108/jhtt-09-2017-0081.
10. Önder, I., Açikgöz, F.: Blockchain for tourism and hospitality industries. (2023). https://doi.org/10.4324/9781003351917.
11. Schmidtke, H.R.: Location-aware systems or location-based services: a survey with applications to CoViD-19 contact tracking. Journal of Reliable Intelligent Environments. 6, 191–214 (2020). https://doi.org/10.1007/s40860-020-00111-4.
12. Shamim, S., Cang, S., Yu, H., & Li, Y. (2017b). Examining the Feasibilities of Industry 4.0 for the Hospitality Sector with the Lens of Management Practice. *Energies*, *10*(4), 499. https://doi.org/10.3390/en10040499
13. Stylos, N., Zwiegelaar, J., Buhalis, D.: Big data empowered agility for dynamic, volatile, and time-sensitive service industries: the case of tourism sector. International Journal of Contemporary Hospitality Management. 33, 1015–1036 (2021). https://doi.org/10.1108/ijchm-07-2020-0644.
14. Torres, E.N., Singh, D., Robertson-Ring, A.: Consumer reviews and the creation of booking transaction value: Lessons from the hotel industry. International Journal of Hospitality Management. 50, 77–83 (2015). https://doi.org/10.1016/j.ijhm.2015.07.012.
15. Truong, N., Dang-Pham, D., McClelland, R., Nkhoma, M.: Exploring the impact of innovativeness of hospitality service operation on customer satisfaction. Operations and Supply Chain Management : An International Journal. 307–319 (2020). https://doi.org/10.31387/oscm0420272.
16. Vicario, R.B., Ruíz-Palomo, D., Fernández-Miguélez, S.M., Gutiérrez-Ruiz, A.M.: Examining the effects of hotel reputation in the relationship between environmental performance and hotel financial performance. Journal of Hospitality and Tourism Management. 53, 10–20 (2022). https://doi.org/10.1016/j.jhtm.2022.08.012.

 https://doi.org/10.52202/074664-0031

Empowering Women in Hospitality: Exploring HR Practices for Career Advancement in Hotels

Garima Malik[1], Ashish Raina[2], and Bhupender Dighliya[3]

[1]Research Scholar, CT University, Ludhiana, Punjab, India
[2]Associate Professor, CT University, Ludhiana, Punjab, India
[3]Assistant Professor, Starex University, Gurugram, Haryna, India

Abstract.
This study delves into the pivotal role of Human Resource (HR) practices in facilitating the career advancement of women in the hospitality industry, particularly within hotels. Recognizing the importance of gender diversity and empowerment in the workplace, the research aims to identify and analyze HR strategies geared towards supporting female employees' professional growth. Focusing on five-star and four-star hotels in Pune, India, the study solicits insights directly from HR managers through a meticulously designed questionnaire. By examining the responses garnered from HR professionals, this research illuminates the specific practices and policies implemented within hotel settings to empower women in their careers. The analysis employs the Statistical Package for the Social Sciences (SPSS) to discern patterns and trends in the data, providing a quantitative assessment of the effectiveness of various HR interventions. Through this investigation, the study seeks to contribute to the existing body of knowledge on gender equality and workplace empowerment, particularly within the context of the hospitality industry. Ultimately, the findings aim to inform stakeholders, policymakers, and industry practitioners about best practices and strategies for fostering inclusive work environments conducive to the advancement of women in hospitality careers.
Keywords: Empowering women, Hospitality industry, Human resource practices, Career advancement, Hotels

1. Introduction

The hospitality industry serves as a vital sector in the global economy, offering employment opportunities and career advancement prospects for individuals from diverse backgrounds. Within this realm, the role of women has been increasingly recognized, yet challenges persist in achieving gender equality and fostering their professional growth. The present research endeavors to shed light on this critical aspect by delving into Human Resource (HR) practices geared towards empowering women in hospitality, with a specific focus on five-star and four-star hotels situated in Pune, India.

In recent decades, the hospitality industry has witnessed significant transformations, marked by globalization, technological advancements, and evolving consumer preferences. Amidst these changes, gender diversity and inclusion have emerged as focal points for organizations striving to enhance their competitiveness and sustainability. However, despite concerted efforts to promote gender equality, women continue to encounter barriers hindering their career advancement within the hospitality sector.

The rationale behind this research stems from the recognition of the persistent gender disparities prevalent in the hospitality industry, particularly concerning leadership roles and career progression

https://doi.org/10.52202/074664-0032

opportunities for women. While numerous studies have explored gender-related issues in various organizational contexts, there exists a dearth of comprehensive research specifically examining HR strategies aimed at supporting female employees' professional growth in the hospitality sector, particularly in the context of Pune, India.

2. Literature Review

The hospitality industry, characterized by its dynamic and customer-centric nature, has historically been marked by gender imbalances, particularly at higher levels of leadership and management. Studies examining gender diversity in hospitality have highlighted persistent disparities in representation, with women often underrepresented in executive positions [1]. Despite the industry's labor-intensive nature, which traditionally attracted a significant female workforce, women continue to face barriers to career advancement, including glass ceilings, gender biases, and limited access to developmental opportunities [2-5].

1.1 HR Practices and Women's Career Advancement

The role of HR practices in facilitating women's career advancement within the hospitality sector has garnered increasing attention from scholars and practitioners alike. Research suggests that supportive HR policies and practices, such as mentoring programs, leadership development initiatives, and work-life balance policies, play a crucial role in mitigating gender disparities and promoting women's professional growth [6, 7]. For instance, organizations with gender-sensitive HR practices, such as flexible work arrangements and diversity training, exhibited higher levels of gender diversity in leadership roles [8].

Moreover, studies have emphasized the importance of organizational culture and climate in shaping women's experiences in the workplace. A positive organizational culture that values diversity, fosters inclusivity, and promotes meritocracy has been linked to higher levels of job satisfaction, commitment, and career advancement opportunities for female employees [8]. Calinaud emphasizes the need for family-friendly policies, gender equality measures, support networks, and personal development plans [9]. Zhong highlights the role of hospitality education in preparing women for leadership roles and reducing industry barriers [10]. Walsh suggests that organizations should focus on long-term career development, provide autonomy, and offer support during mid-career stages [11]. Finally, Liu underscores the importance of women's career expectations, work-family balance, and organizational commitment in their career advancement [12]. These findings collectively underscore the need for a holistic approach that encompasses both organizational and individual factors.

While the literature offers valuable insights into HR practices and women's career advancement in hospitality, it is essential to consider the influence of contextual factors, including organizational culture, national culture, and industry dynamics. In the context of India, cultural norms and societal expectations regarding gender roles may impact HR practices and women's participation in the workforce [13]. Moreover, the unique challenges faced by women in emerging economies, such as access to education, childcare facilities, and social support networks, may shape their career trajectories and opportunities for advancement within the hospitality industry [14].

Additionally, the competitive landscape and market dynamics of the hospitality sector in Pune, India, may have implications for HR strategies aimed at promoting gender diversity and inclusion [15]. As a burgeoning hub for tourism and hospitality, Pune presents both opportunities and

 https://doi.org/10.52202/074664-0032

challenges for organizations seeking to attract and retain talent, particularly women, in a highly competitive market [16].

3. Research Methodology

The questionnaire, meticulously designed following an in-depth review of pertinent literature. The questionnaire consists of two sections. Section A is dedicated to acquiring essential basic information from the respondent about the hotel. This consistsof questions regardingnumber of employees, composition of the employees, and number of female employees. This information gives a brief overview of the hotel employees. Section B developed to assess the information about the HR practices that are adopted by the hotelto assist female employees in their "career advancement." It includes variables that measure HR practices, which are rated "yes" or "no"

The questionnaire was distributed through various channels, including email, WhatsApp, Facebook Messenger, and other social media platforms to the HR managers of five and four star hotels situated in Pune. The online survey results were obtained in Microsoft Excel spreadsheet format and after the coding the data was imported in SPSS 26 (Statistical Program for the Social Sciences) for analyzing the data. Descriptive analysis was used to analyzing the data.

Table 1: List of Hotel Included in the Study

S. No.	5-star Hotel	S. No.	4-star Hotel
1.	Conrad Hotel	1.	The Pride hotel
2.	Blue Diamond	2.	Royal Orchid Golden Suites
3.	Hyatt Pune	3.	Oakwood Residence
4.	Double Tree Hilton	4.	E-Square, The Fern Hotel
5.	Ritz Carlton	5.	Royal Orchid Central
6.	The Corinthains Resorts & Club	6.	Lemon Tree Premier
7.	The O Hotel	7.	Sayaji Hotels Ltd.
8.	Fort Jadhavgadh	8.	Ramada Plaza
9.	Courtyard By Marriott – Hinjewadi	9.	Fairfield by Marriott
10.	The Westin Pune	10.	Courtyard Marriott Pune Chakan
11.	Marriott Suites (Former Oakwood Premier)	11.	Amanora Fern Hotel & Club
12.	Radisson Blu Pune		
13.	JW Marriott		
14.	Crowne Plaza - Pune City Centre		
15.	Novotel Hotels & Resorts		
16.	Hyatt Regency – Pune		
17.	Sheraton Grand Pune Bund Garden Hotel		
18.	Vivanta Pune Hinjawadi		
Total (5 star and 4 star)			**29 Hotels**

4. Results

4.1 Basic Details of Hotel

Representation of Women in the Workforce:

The majority of hotels surveyed have a total workforce of 100 to 300 employees. Approximately half of the hotels surveyed have less than 20 female employees, indicating a potentially low representation of women in the workforce.

Female Employee Distribution:

The distribution of female employees across different ranges is relatively even, with a significant portion falling within the 20-40 range. However, there is a decline in the number of female employees as the range increases, with very few hotels having more than 60 female employees.

Policy Regarding Female-Male Employee Ratio:

The majority of hotels surveyed (approximately 93%) have a policy regarding the female-male employee ratio, which suggests a conscious effort towards gender diversity.

Table 2: Basic Details of Hotel

Variable		Number	Percentage
Total number of employees working in your hotel	Less than 50	0	0
	50-100	7	24.14
	100-300	18	62.07
	300-500	3	10.34
	Above 500	1	3.45
Number of Female Employee	Below 20	13	44.83
	20-40	13	44.83
	40-60	3	10.34
Total number of female employees working in hotel	Less than 20	12	41.38
	20-40	12	41.38
	40-60	03	10.34
	60-80	1	3.45
	Above 80	1	3.45
Organization have any policy regarding female-male employee ratio	Yes	27	93.10
	No	2	6.90
At which level women employees are preferred by management in your hotel?	Managerial	0	0.00
	Supervisory	1	3.45
	Operational	2	6.90
	Above all positions	26	89.66
Your organization give preference to female who's is?	Married	0	0.00
	Unmarried	8	27.59
	Equal preference to both	21	72.41

Preference for Female Employees by Management:

Most hotels surveyed give preference to female employees across all positions, indicating a commitment to gender equality in hiring practices. The preference for female employees is consistent regardless of marital status, with a slight preference towards unmarried females.

https://doi.org/10.52202/074664-0032

Level of Preference for Female Employees:
The data suggests that management in most hotels surveyed prefers to hire women across all levels, including supervisory, operational, and managerial positions.
With a preference for female employees across all positions, it can be inferred that hotels may provide opportunities for career advancement for women. However, the lack of female representation in managerial positions suggests potential barriers to advancement that need to be explored further.

4.2 HR's Policies/strategies employed by the Hotel

Flexible Working Hours:
5 hotels (17.24%) have implemented flexible working hour policies, which allow employees to adjust their work schedules to better suit their personal needs or commitments. 24 hotels (82.76%) do not offer flexible working hour policies, indicating a lack of flexibility in work schedules for employees.
Gender Equality Policies:
All 29 hotels (100.00%) surveyed have implemented gender equality policies. These policies likely promote equal opportunities and treatment for all employees regardless of gender, aiming to create an inclusive and fair workplace environment. None of the surveyed hotels reported not having gender equality policies in place.

Equal Employment Opportunities:
All 29 hotels (100.00%) provide equal employment opportunities. This suggests that these hotels are committed to fair recruitment and employment practices, ensuring that hiring decisions are based on merit rather than discriminatory factors such as gender, race, or ethnicity. None of the surveyed hotels indicated a lack of equal employment opportunities.

Table 3: Statistics of HR's Policies/strategies employed by the Hotel

HR's Policies/strategies		N	Percentage
Flexible Working Hours	Yes	5	17.24
	No	24	82.76
Gender equality policies	Yes	29	100
	No	0	0
Equal Employment Opportunities	Yes	29	100
	No	0	0
Maternity or Parental leave	Yes	29	100
	No	0	0
Special leave for female employees	Yes	13	44.83
	No	16	55.17
Creche Facility	Yes	2	6.90
	No	27	93.1
Special quota for women employees at the recruitment level	Yes	6	20.69
	No	23	79.31
No allocation of night shifts for women employees	Yes	25	86.21

	No	4	13.79
	Yes	25	86.21
	No	4	13.79
Pick up and drop facility	Yes	25	86.21
	No	4	13.79
Women Grievance cell for handling complaints	Yes	3	10.34
	No	26	89.66
Comeback policy for women employees after career break	Yes	16	55.17
	No	13	44.83
No job transfers for women employees	Yes	8	27.59
	No	21	72.41
Special perks /incentives for women employees	Yes	3	10.34
	No	26	89.66
Women's networking within the company	Yes	9	31.03
	No	20	68.97
Formal mentoring program for females	Yes	9	31.03
	No	20	68.97
Identification & fast-tracking of high potential female	Yes	5	17.24
	No	24	82.76
Job Rotation Program	Yes	9	31.03
	No	20	68.97
Provision of Career related executive development programs	Yes	4	13.79
	No	25	86.21
Periodic career development reviews	Yes	4	13.79
	No	25	86.21

Maternity or Parental Leave:
All 29 hotels (100.00%) offer maternity or parental leave benefits to their employees. These benefits likely include paid time off for new parents to care for their newborns or adopted children, reflecting a commitment to supporting work-life balance and family responsibilities. None of the surveyed hotels reported not offering maternity or parental leave benefits.

Special leave for female employees:
13 hotels (44.83%) provide special leave for female employees, which could include additional time off for medical reasons, family emergencies, or other specific needs. 16 hotels (55.17%) do not offer special leave specifically designated for female employees.

Creche Facility:
Only 2 hotels (6.90%) offer a creche facility, indicating a relatively low provision of on-site childcare services for employees. 27 hotels (93.10%) do not have a creche facility available for their employees. It suggests that there may be room for improvement in supporting employees with family responsibilities.

Special quota for women employees at recruitment level:

https://doi.org/10.52202/074664-0032

6 hotels (20.69%) have implemented a special quota for recruiting women employees; aiming to increase gender diversity in their workforce. 23 hotels (79.31%) do not have such a quota in place. It suggests that there may be opportunities to enhance efforts to improve gender diversity in recruitment.

No allocation of night shifts for women employees:
25 hotels (86.21%) do not allocate night shifts to female employees, potentially due to safety concerns or other considerations. Only 4 hotels (13.79%) allocate night shifts to female employees.

Provision of transportation for women employees in night shift:
Similar to the previous point, 25 hotels (86.21%) provide transportation for female employees working night shifts. Only 4 hotels (13.79%) do not offer transportation for female employees during night shifts.

Pick up and drop facility:
25 hotels (86.21%) offer pick-up and drop facilities for female employees, particularly beneficial for those working during odd hours or night shifts. Only 4 hotels (13.79%) do not provide this service.

Women Grievance cell for handling complaints:
3 hotels (10.34%) have a dedicated women's grievance cell for handling complaints related to harassment, discrimination, or other issues specific to female employees. 26 hotels (89.66%) do not have such a dedicated cell. It potentially indicates a gap in formal support mechanisms for addressing issues specific to female employees.

Comeback policy for women employees after career break:
16 hotels (55.17%) have a comeback policy in place, facilitating the return of female employees after a career break, possibly through retraining, flexible work arrangements, or other support mechanisms. 13 hotels (44.83%) do not offer such a policy.

No job transfers for women employees:
8 hotels (27.59%) do not transfer female employees between job roles or departments. 21 hotels (72.41%) allow job transfers for female employees. It potentially indicates limitations in career mobility within these establishments.

Special perks/incentives for women employees:
3 hotels (10.34%) offer special perks or incentives specifically tailored for female employees. 26 hotels (89.66%) do not provide such benefits. It indicates that there may be a lack of tailored recognition or support for female employees in these establishments.

5. Discussion

The hospitality industry is known for its dynamic and diverse workforce, yet it has historically faced challenges in achieving gender equality and supporting the professional growth of female employees. In the context of Pune, India, where the hospitality sector is thriving with a mix of five-star and four-star hotels, understanding the HR practices geared towards empowering women is crucial for fostering a more inclusive and equitable work environment.

It is encouraging to note that all surveyed hotels have implemented gender equality policies, indicating a strong commitment to promoting fairness and equal treatment regardless of gender. These policies likely serve as the foundation for creating an inclusive workplace culture where female employees feel valued and respected. The provision of maternity or parental leave by all surveyed hotels underscores the recognition of the importance of supporting employees during significant life events. However, the relatively low implementation of special leave for female employees and creche facilities suggests that there may be opportunities for hotels to further enhance family-friendly policies to better support employees with family responsibilities.

While some hotels offer specialized support such as job rotation programs and formal mentoring programs for female employees, there is a need for more tailored initiatives to address the unique challenges and opportunities faced by women in the workplace. This could include providing special perks or incentives for female employees and establishing dedicated grievance mechanisms to address issues specific to female employees. While there are initiatives in place such as job rotation programs and formal mentoring programs, there is room for improvement in providing career development and advancement opportunities for female employees. Enhancing these programs and implementing strategies to identify and fast-track high-potential female employees could contribute to narrowing the gender gap in leadership positions within the hospitality sector. The majority of hotels surveyed do not allocate night shifts to female employees, indicating a focus on safety and security for female employees working during non-standard hours. However, providing transportation and ensuring safe working environments for female employees during night shifts remain critical considerations for fostering a supportive workplace environment.

In conclusion, our study highlights the importance of HR practices in empowering women and promoting gender equality within the hospitality industry, particularly in five-star and four-star hotels in Pune, India. While there are commendable efforts in implementing gender equality policies and family-friendly initiatives, there are also areas for improvement in providing specific support for female employees and enhancing career development opportunities.

Moving forward, it is imperative for hotels to continuously evaluate and strengthen their HR strategies to create a more inclusive and equitable workplace environment. This may involve implementing targeted initiatives to support the professional growth of female employees, fostering a culture of diversity and inclusion, and addressing any barriers to career advancement faced by women in the hospitality sector. By prioritizing gender equality and empowering women, hotels can not only attract and retain top talent but also contribute to building a more sustainable and resilient workforce for the future.

https://doi.org/10.52202/074664-0032

6. Limitations and Future Direction

The study on "Empowering Women in Hospitality: Exploring HR Practices for Career Advancement in Hotels" sheds light on HR strategies supporting female employees in Pune's five-star and four-star hotels. However, limitations exist. Firstly, the research's focus on Pune may limit generalizability to other regions. Additionally, the study primarily examines HR policies without delving into their implementation or effectiveness. Future research could address these limitations by conducting comparative studies across diverse geographical locations and exploring the practical implications of HR strategies on female employees' career advancement. Furthermore, qualitative research could provide deeper insights into the experiences and perceptions of female employees, allowing for a more comprehensive understanding of the challenges and opportunities they encounter in the hospitality industry. Such endeavors would contribute significantly to fostering gender equality and empowerment in the hospitality sector.

References:

1. Choi, D.-H.: Impact of organizational culture on the organizational commitment in hotel employees. Indian Journal of Public Health Research and Development. 9, 841 (2018). https://doi.org/10.5958/0976-5506.2018.00838.0.
2. Barreto, M., Ryan, M.K., Schmitt, M.: The glass ceiling in the 21st century: Understanding barriers to gender equality. In: American Psychological Association eBooks (2009). https://doi.org/10.1037/11863-000.
3. Bertrand, M., Goldin, C., Katz, L.F.: Dynamics of the gender gap for young professionals in the financial and corporate sectors. American Economic Journal. Applied Economics. 2, 228–255 (2010). https://doi.org/10.1257/app.2.3.228.
4. Calinaud, V., Kokkranikal, J., Gebbels, M.: Career advancement for women in the British hospitality industry: the enabling factors. Work, Employment and Society. 35, 677–695 (2020). https://doi.org/10.1177/0950017020967208.
5. Crites, S.N., Dickson, K.E., Lorenz, A.: Nurturing Gender Stereotypes in the face of Experience: A study of leader gender, leadership style, and satisfaction. Nurturing Gender Stereotypes in the Face of Experience: 19, 1 (2015).
6. Crompton, R., Lyonette, C.: Women's career success and work-life adaptations in the accountancy and medical professions in Britain. Gender, Work and Organization. 18, 231–254 (2011). https://doi.org/10.1111/j.1468-0432.2009.00511.x.
7. Da Silva, H.B., Yukongdi, V., Kusakabe, K., Zimmermann, W.: Career advancement in the hotel industry in Timor-Leste: Study of individual, organizational, and gender differences. SAGE Open. 11, 215824402110326 (2021). https://doi.org/10.1177/21582440211032683.
8. Dashper, K.: Mentoring for gender equality: Supporting female leaders in the hospitality industry. International Journal of Hospitality Management. 88, 102397 (2020). https://doi.org/10.1016/j.ijhm.2019.102397.
9. Eagly, A.H., Karau, S.J.: Role congruity theory of prejudice toward female leaders. Psychological Review. 109, 573–598 (2002). https://doi.org/10.1037/0033-295x.109.3.573.
10. Elhoushy, S., El-Said, O.A., Smith, M.F., Dar, H.: Gender equality: caught between policy reforms and manager beliefs. Journal of Sustainable Tourism. 1–24 (2023). https://doi.org/10.1080/09669582.2023.2218067.
11. Elomami, A.: Job satisfaction and employees' turnover in Libyan oil companies: the application of the job characteristics model in a different cultural setting, https://ethos.bl.uk/OrderDetails.do?uin=uk.bl.ethos.678804, (2015).
12. Gill, A., Fitzgerald, S.P., Bhutani, S., Mand, H.S., Sharma, S.: The relationship between transformational leadership and employee desire for empowerment. International Journal of Contemporary Hospitality Management. 22, 263–273 (2010). https://doi.org/10.1108/09596111011018223.

13. Gupta, V.: Talent management dimensions and their relationship with retention of Generation-Y employees in the hospitality industry. International Journal of Contemporary Hospitality Management. 31, 4150–4169 (2019). https://doi.org/10.1108/ijchm-10-2018-0859.
14. Li, L., Leung, R.W.: Female managers in Asian hotels: profile and career challenges. International Journal of Contemporary Hospitality Management. 13, 189–196 (2001). https://doi.org/10.1108/09596110110389511.
15. Sigler, T.H., Pearson, C.M.: Creating an empowering culture: examining the relationship between organizational culture and perceptions of empowerment. Journal of Quality Management. 5, 27–52 (2000). https://doi.org/10.1016/s1084-8568(00)00011-0.
16. Yan, Z., Couch, S., Blum, S.C.: Factors affecting women's career advancement in the hospitality industry: perceptions of students, educators and industry recruiters. Journal of Hospitality & Tourism Education. 23, 5–13 (2011). https://doi.org/10.1080/10963758.2011.10697015.

 https://doi.org/10.52202/074664-0032

Regional Food and its Impact on the Community Engagement

Vidushi Gupta[1], Divya Bose[2]

[1]Ph.D. Scholar, Department of Tourism and Hotel Management, Bundelkhand University, Jhansi, Uttar Pradesh, India.
[2]Senior Lecturer, Institute of Hotel Management, Catering and Nutrition – PUSA, New Delhi, India.

Abstract

Today's global environment has seen the food industry play a key role in tourism. The cuisine of the area draws tourists in addition to the natural beauty of the locations. Given the volume of visitors over time, it has become increasingly important for generating income and promoting tourism globally. It also encourages people to pursue opportunities for entrepreneurship and looks for promising careers in fields that will influence and improve sustainable practices and local development. In light of the aforementioned, it is suggested to conduct the study in Lucknow, which is renowned as the "City of Nawabs and Kebabs," due to its popularity and level of tourism. Because there are many devoted food tourists, our demographic is profoundly matched to the people's interests. According to earlier researchers, local cuisine has a significant influence on tourists' inclinations to visit any given location and can even serve as a catalyst for return visits. This study focuses on a significant portion of culinary tourism that revolves around local cuisine found in the city's center. Additionally, the attraction of local cuisine to visitors due to its genuineness and appeal has been noted as a lucrative aspect of generating income. Thus, this study meets a pressing need by presenting the findings of the beneficial effects of local cuisine and community involvement on both visitors and residents. According to the study's findings, exploration has significantly improved local cuisine while expanding the variety of lost and recovered items in the area. The study's findings indicate that people are starting to prefer local cuisine over quick food while dining out. Additionally, as previously indicated, the city's niche market has strongly urged community involvement, supporting preferences as a successful criterion of choice and the sincerity of food enthusiasts to eat meals from the actual location. Considering the study, regional food is preferred and suggested.

Keywords: Community engagement, Entrepreneurship, Regional food, Revenue generation, Tourism.

1 Introduction

Food has been an integral part of our lives. It has not only provided growth and nourishment but also, has provided all sources of livelihood. With lifestyles of today, the intake of food has changed a lot. In timings, habits and the resources that has built our body to somewhere save on time to spend time over our meals. Similarly, the need of fast food has come over and taken on to the market, which has not only affected the food markets but also the regional food of the world. The travelers across the country have a major affection towards regional food or native food for which they revisit places [1].

Regional food, which is specifically belonging from a region represents the cultural and heritage value of the regions. It showcases the richness and use of staple ingredients and meals prepared from the same. Traditional foods reflect cultural inheritance and have left their imprints on contemporary dietary patterns. They are key elements for the dietary patterns in different countries and consequently are important to accurately estimate population dietary intakes. Regional food has always been in demand in the world and the need of it has been always created by the native public. Keeping the tourist attractions in mind, the need of local food has always been observed. The food not only attracts the vibe but also the feeling of attachment and nativity to one's belonging to the place.

Spreading light on the food, it is an emotion to a lot of people. Lucknow, the city itself is miraculously having so much heritage and culture within **[2].** The regional and traditional food of Lucknow has an outstanding impact over the world with its melting Tunday ke Kebab and Namish along with Raheem ki Nalli Nihari with the best short crusted kulcha's and topping the list with the famous Idris ki biryani. Every year the tourists come in huge numbers to see the heritage of the city with its finest monuments, famous chikankari work and mouth-watering delicacies **[3]**As per the survey, it has been observed that the travellers always prefer regional food of Lucknow than any other thing to eat. The famous attractions include the tunday's kebabs for which the city is known for. The engagement of the entire community is inclined towards the richness of the flavours and taste which is always in demand due to the hype and the economic value of the food. As per the travellers and tourists, the need of the hour, includes the food which is highly addictive as being the first preference in the list provided. Not only the tourists but also the local people, prefer the regional food over any fancy dining in Lucknow.

The capital of the Awadh region then and being capital of Uttar Pradesh now which is Lucknow, that was ruled by the Delhi Sultanate and then the Mughal Empire. The Nawabs of Lucknow, who used to be the Nawabs of Awadh, got their name after Lucknow, became their capital during the reign of the third Nawab[4].The city rose to prominence as the artistic capital of North India, and its nawabs—who are best remembered for their lavish and cultured lifestyles, were art admirers as well. Inculcating the same values in their foods and creating magnificent delicacies, with every inch of richness in them and every single bite to taste like a heavenly bliss [5].

Considering the facts and the popularity of the region, diversity in the food can be seen in both categories of vegetarian food and non- vegetarian food. The region happily serves the exquisiteness in affordable ranges, which initiates the travellers to even invest more in the cuisine. The community engagement is fully determined to satisfy their people being it local or an outsider in every possible way. The hospitality includes wide range of use of 'Hindi language' and 'Urdu language' which showcases the respect or 'Tehzeeb' or 'Ahetram' which comprises to comfort the guest every possible way [6].

https://doi.org/10.52202/074664-0033

Bringing the Lucknow's food in light and comprehending it towards a heritage trail, the food itself describes it as comfort food which is always in demand for everyone.

Considering the aim of the study, few objectives were determined, such as:

- To find the preference of people to select regional food over any other food to eat when visiting and staying in Lucknow.
- To identify the need of selection on food as per people's choice based on their likeability and affection.
- To identify the presence and impact of regional food of Old Lucknow.
- To acknowledge the establishment of the community engagement within the communal zones of Lucknow.

2 Materials and methods

Lucknow, capital of Uttar Pradesh has a longitude of 80.949997 and a latitude of 26.850000 and with the city place category with the gps coordinates of 26° 51' 0.0000" N and 80° 56' 59.9892" E. The population as per the census count is 3,945,000 of 2023 and the yearly travellers are approximately 5755323 as per the annual tourists visit stats. The collection of data was done in three ways: survey methods, feedbacks and personal interviews. The demography in the study was Old Lucknow, areas like Aminabad, Chowk, Nakhhas, Saadatganj, Thakurganj, Alambagh, Alam Nagar, Charbagh and surrounding areas, which were used for the data collection as these areas are in the city centre or downtown [7].
Considering the tourist locations in these areas and the local markets present around, the fact was to find maximum number of tourists in these areas only. Also, in findings, majority of famous food joints selling regional food as well as the fast food were located here only. So, to conduct the research, this area was perfectly apt.

The method of data collection was divided into three parts:

1. General survey
2. Feedbacks
3. Personal interview

The study was focussing on major findings where the review of the tourists and the local people was very important. So, the data was collected in one week, starting from the general surveys. The general survey was conducted with local people asking about the preference of food and choice of food in Lucknow. The collection started with targeting the famous monuments and the crowd first, where lot of people were from outside Lucknow including foreigners were present.

In conversations with the local rickshaw drivers and auto drivers, the information was collected about what are the general questions of people related to food tourism. Also, major finding was that the tanga drivers (horse carriage) were the major source of travel to seek opportunity to drop travellers and tourists to famous eateries of Lucknow in Old Lucknow [8]. The tanga drivers had a specific route which was in a stretch of 2 kms from the famous heritage monument – BadaImambada to chowk where the chikankari factories were present, in between covering maximum areas like 'Clock Tower' and 'Teele Wali Masjid' [9].

The general survey included visiting all the places and collecting the data. From visiting TundayKebabi to exploring the entire Old Lucknow. The questions included about numbers of visitors daily, history of places, economic values of items priced and choice of people for food. The major data was collected through the local drivers of that area. Also, the aim of the study was to find the community engagement and so, the survey was conducted with travellers in group and pairs. In fact, areas like Akbari Gate were very successful in gathering information as the entire market was flooded with loads of regional food and travellers.

The feedbacks were conducted within 3 days at the end of the days, the gathering of data was done from the travellers and local people about the food and their experience in Lucknow. The feedbacks was collected from the shop workers about the taste of dishes, pricing and daily sales. Area such as Nakkhas was found highly populated on Sundays, where the collection was much easier as it is fleeing market.

Adding to the Old Lucknow, the entire place was densely populated with travellers and Tourists and could be spotted easily in the city found exploring the heritage and essence of the cultures of the place. Feedbacks were also collected from the railway stations from the travellers returning back to their native places. Also, the feedbacks gave very crucial information of types of guests coming from all parts of the world. The data was collected upon few parameters such as the taste preference, daily sales, number of customers visiting on an average and their choices of food and eating joint. Collection of genuine reviews were observed as it matches both responses of service taker and service provider [10].

Even counting the children and youth, their preference was found to be regional food of Lucknow because of its economic value, portion sizes and taste including the varieties. The zeal to purchase fresh malaiMakhan and fresh baked kulcha's was on another level and having an early morning breakfast at Raheem's NalliNihari was cherry on top. Also, the breath-taking short crust khasta's of Lucknow in morning breakfast was found to be in habit of lot of local people as well [11].

The final method used to collect data was personal interviews with the people of Old Lucknow. Measuring the count to 23 people, the data collected was in favour of regional food rather than local fast food because of reasons of economic value, easy to find, varieties, flavour and the heritage carried with.

The personal interviews included questions related to the preference of food, favourite dishes, selection of eateries and the value of money. Concluded and summarised the findings in a very positive aspect.

3 Results and discussion

The study was focussed on finding the popularity of regional food amongst the travellers and tourists and their community engagement with local food. The regional food of Lucknow is very famous over the world. From across the nations, people come to eat the food. In this entire time, the data has been collected in all possible ways from the Old Lucknow. In the surveys, the travellers and tourists both were very happy to answer and they were happy to share the feedback. The hospitality of Lucknow, irresistible of its pleasure carrying the legacy within its culture was generalised within the parameters of sustaining the guests with comfort [12].

The general survey was also conducted at all the famous food joints to gather data as per their daily customers visit. The surveys included a detailed discussions with the shop workers and about their best-selling dishes, which included the dishes which were truly famous and true to the value as far as finances are concerned. The questions included a lot of details about the history of the places, starting from the very first outlet of 'TundayKebabi' based in Chowk, in the heart of the market [13]. Where the dishes ranged from Rs.5 to Rs. 150 only, where the world-famous kebabs were only starting from Rs.5 (beef) and mutton kebab's for Rs. 20. Where the average daily customers are in thousands, and as per the survey we could find that in huge

 https://doi.org/10.52202/074664-0033

numbers tourists and travellers visit it daily. The entire menu was very short and the service was very fast.

Moving towards the Raheemsnihari and kulcha's which was only costing Rs.70 for half (beef) and Rs.140 for mutton followed by kulcha's for Rs.15 only per piece. Situated in the heart to city and near to TundayKebabi, this place serves very limited menu and is very famous for its NihariKulcha's. Opening the shop at 4 am and selling the nihari since 4:30 am onwards, majority people come for breakfast here. During Ramzan, this place opens 24*7 and is always in demand due to the food served during sehri time. The personal interviews were conducted here and we found maximum no. of people were from outside Lucknow came to taste and eat the food. They discussed about the quality of food and how cheap it was and daily it was fresh as prepared same day and they also told they have waited here to eat as the Nihari sometimes gets finished due to its high popularity. This place only accepts cash.

Discovering the delicacies in the same area, Akbari gate is filled with all the Mughlai and Awadhi Street food and very famous Namish which is also known as MalaiMakhan in Hindi is served here only in winter season which is very light weighted just like foam but is very flavourful and little expensive. Rs. 800 per kg. Followed by the streets food where Haleem was found only for Rs. 30 per plate and this man makes atleast 15 kg haleem per day yet sometimes it was finished. As per the general survey, people only prefer it as a starter and it is full of protein as it is made with dal and meat of choice. Topped with fried onions and lemon, this makes a really good starter as per the people. Both local and outsider all come here to eat, usual crowd found in evening only near Clock Tower.

In conversation with the travellers and tourists they also told us about a spot selling beef and chicken skewers only for Rs.5 and Rs.20 specifically by roasting the marinate over the coal. So, data collected here was through a general survey and good amount of people come here. People gave feedbacks about the easy to go meal it was and obviously the range is so low [14].

Final place to collect the data was the most famous the Idris Biryani and Rs.150 per plate, only mutton biryani and place opens at 8am and finishes biryani at 11am, second shift starts as second batch is prepared. This place is very famous amongst all, from all people throughout the survey they mentioned the popularity of this place. And undoubtedly could find the people from across the world coming here. The visitors visiting areas are really fond of eating and these people are offered the best hospitality here.

In the entire survey, data was collected from local shops also selling the local food and fast food like burgers, parathas and other foods such as wraps etc. The areas situated near the KGMC hospital, which is in the vicinity of Old Lucknow, had numerous options of eateries of fast food and major shops were pure vegetarian. Survey included questions based on the general food preference, costs and varieties present.

Use of all listed tools were done to collect the information and yet to find that majority of people preferred eating regional food as it was cheap, fresh and had taste. Also, the availability of the food was present throughout the city coming to the eateries serving both vegetarian and non-vegetarian food. Throughout the entire data collection, the inclination of people was found to be towards the regional food as the survey, feedbacks and interview displayed. Considering the entire data collected and verifying the data, majority people visited Lucknow to explore the heritage of the city and cultures and taste the legacy which enriches the values of its history [15]

A lot of the information regarding how food stores might already be involved in this effort is unknown because previous research on the food environment has mostly focused on community engagement as an intervention component. This study set out to investigate the community engagement strategies used by neighbourhood food retailers in lower-class neighbourhoods with clear health goals in order to comprehend how these businesses interact and collaborate with their local communities [16].

Fig. 1. Menu from Raheem Hotel, Chowk

Fig. 2. Nihari Kulcha

Fig. 3. Idris Biryani, Nakhas

Fig.4. Tunday Kebabi, Chowk

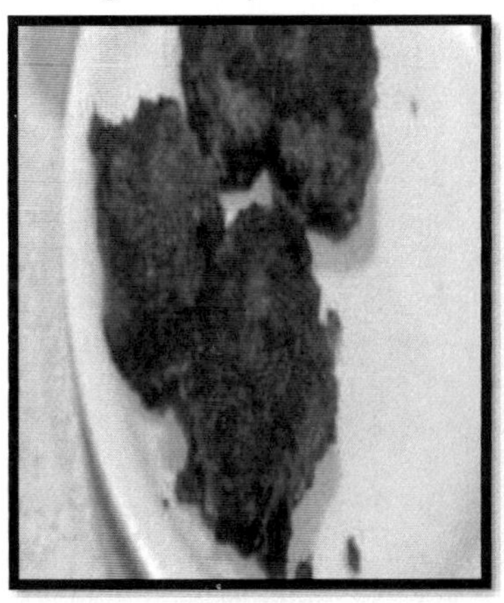

https://doi.org/10.52202/074664-0033

Fig. 5. Namish (Malai Makkhan)

4 Conclusion

Comparing the study, the entrepreneurship is raising in the areas of Lucknow with its community engagement. Keeping the data in mind, the inclination of the people can be seen towards the regional food of Lucknow. In the entire discussion, the findings were in favour of preference of regional food over the fast food or other food present around. Choosing the heritage of the city and following the legacy to inherit the taste of the culture not only represented sense of satisfaction but also sense of responsibility and longingness towards a native society. Producing revenue for the capital, Lucknow with its tourists' engagements is a reflection of sustaining the heritage of the place and the famous hospitality.

Analysing the study and the objectives, the entire study was focussing on regional food and its impact on community engagement. The analysis not only provided relevant data but also gave details about the people's choice when it comes to select the region as well as regional food. Throughout the observations received from all travellers and tourists, a sigh of happiness was seen. Talking about cultures, heritage and ofcourse food, gave immense pleasure to people. Seeing their enthusiasm in this aspect was just cherry on the top. And as far as we could consider, the entire love towards regional food could have been seen.

Ranging from the breakfast to supper and including the tea's within, the generosity of the people was at its highest. It was very considerate to listen to people from outside Lucknow talking about their favourite food and how frequent visits they made to Lucknow just for having their favourite meals. In the entire visit and conversation with people over the food, one could find the momentarily happiness on faces and still can be enlightened by facts and figures of Lucknow. The results also reflected the zeal of fascination on people's behaviour with utmost hospitality and gestures. One could easily feel the comfort with food that people have and they really stress on the fact to actually visit Lucknow and eat.

It not only represents the belonginess but also the affection towards the community and place. It determines moral values and support they represent to any new comer. Talking of comfort, the people were definitely inclined over the regional food and they also mentioned about waiting to have their favourite meals throughout the day.

Summarising the entire study in a positive direction, the travellers are not only attracted towards the 'City of Nawabs' but also the 'City of Kebabs'. The place well defines itself suitable enough to eat here and stay and enjoy the traditional atmosphere of the city with its heritage monuments and culture. Generating people to revisit the place in all forms to always find love towards the food and affection to the place and representing the entire world by using clothes made of Chikankari (embroidery) to showcase the value of such communities where the entire work is done by hands. Concluding with the aspect of community engagement, this demography was well suited to conduct the study, yet proofing it to completely apt for choosing regional food over any other.

References

1. Lin, L., Mao, P.-C.: Food for memories and culture – A content analysis study of food specialties and souvenirs. Journal of Hospitality and Tourism Management. 22, 19–29 (2015). https://doi.org/10.1016/j.jhtm.2014.12.001.
2. Ansari, G.: The City of Palace Culture. Urban Symbolism, 329-335, (1993)
3. Singh, M.M. and Srivastava, A., 2022. Heritage Arch of Uttar Pradesh: The Cities of Living Heritage (Agra, Lucknow & Varanasi) with Special Reference to Lucknow. *International journal of economic perspectives*, *16*(4), pp.52-57.
4. Singh, S.: A study in Cultural Expressions of the Nawabs of Avadh. Tourism Recreation Research/Tourism Recreation Research. 14, 75–79 (1989). https://doi.org/10.1080/02508281.1989.11014557.
5. Shastri, S.: Cultural/Heritage tourism in Lucknow: An Impact on environment. International Journal of Research in Social Sciences. 5, 180–186 (2015).
6. Siddiqi, W.H.: Lucknow - the historic city. (2000).
7. Menon, S., Bhatt, S., Sharma, S.: A study on envisioning Indian tourism – Through cultural tourism and sustainable digitalization. Cogent Social Sciences. 7, (2021). https://doi.org/10.1080/23311886.2021.1903149.
8. PS, Pratheep.: The impact of tourism on Indian culture. KnE Social Sciences. 1, 429 (2017). https://doi.org/10.18502/kss.v1i3.765.
9. Haider, S.H.S., Tjprc: The architecture of the Imamabadas in Lucknow; ImambadaSibtainabad. International Journal of History and Research. 8, 1–10 (2018). https://doi.org/10.24247/ijhrdec20181.
10. Sharma P. Lucknow: The Synonym of Culture. SIT Journal of Management, 3(2):134-146, 2013.
11. Inglis, J.S.: The siege of Lucknow. A diary. (1892).
12. O'Neill, M.: Increasing community engagement in collective impact approaches to advance social change. Community Development. 51, 17–35 (2020). https://doi.org/10.1080/15575330.2020.1714684.
13. Sears, L.B.: The Public Voiceand Sustainable Food Systems: Community engagement in food action plans, https://kuscholarworks.ku.edu/handle/1808/26899, (2017).
14. Agrahari, R., Brar, K.S.: Traditional potpourri of Chikankari flat stitches: glory of Lucknow. Asian Journal of Home Science/Asian Journal of Home Science. 12, 489–495 (2017). https://doi.org/10.15740/has/ajhs/12.2/489-495.
15. Singh, S.: Urban Development and Tourism. Tourism Recreation Research/Tourism Recreation Research. 17, 71–78 (1992). https://doi.org/10.1080/02508281.1992.11014651.
16. Kaur, R., Winkler, M.R., John, S., DeAngelo, J., Dombrowski, R.D., Hickson, A., Sundermeir, S.M., Kasprzak, C., Bode, B., Hill, A.B., Lewis, E., Colón-Ramos, U., Munch, J., Witting, L.L., Odoms-Young, A., Gittelsohn, J., Leone, L.A.: Forms of Community engagement in Neighborhood Food Retail: Healthy Community Stores Case Study project. International Journal of Environmental Research and Public Health/International Journal of Environmental Research and Public Health. 19, 6986 (2022). https://doi.org/10.3390/ijerph19126986.

 https://doi.org/10.52202/074664-0033

The Impact of Technology on Guest Experience in the Hospitality Industry: An Analysis of Enhancements and Detraction

Mo. Maharaj[1], Rohit Sharma[2] and Dr. Amit Kumar[3]

[1]Research Scholar CT University, Ludiana, Punjab
[2]Professor, CT University, Ludiana, Punjab, India
[3]Assistant professor, Maharishi Markandeshwar University, Ambala, Haryana

Abstract.
The hospitality industry is rapidly experimenting with technology and it is important to understand the small impact, this technology has on the overall guest experience. This study explores how various technological innovations improve and impact the guest experience by delving deeper into the various relationships between technology and guest satisfaction. The research will use a mixed method combining quantitative and qualitative research to collect feedback from guests, hotel employees and business professionals. This series will focus on collecting information about visitor preferences and interests in business areas with different levels of technology integration. Key topics examined will include the impact of mobile apps, keyless entry systems, in-room automation and other technology solutions that increase efficiency, streamline and personalize the guest experience. The study will also investigate potential issues such as privacy issues, poor technology and human intervention in the hotel environment. This research, which brings together the good and the bad, is designed to provide a better understanding of hotels and resorts. Business partners strive to develop integrated technology to increase guest satisfaction. These findings will provide an understanding of the role of technology in creating a modern guest experience and provide an insight into the use of technology in the hotel industry.

Keywords – Hospitality industry, Rapidly, Innovations, keyless entry, efficiency, satisfaction, streamline, technology, Guest

1. Introduction

In India, hotel sector is one of the fastest growing sectors. The hotel industry in India comprises a variety of businesses from major international brands or chains and large independent hotels to medium and small hotels. HVS (2016) also shows the similar trends increase in new branded or organized supply and increase in demand over previous year. The report also showed that domestic demand in India is on rise as compared to inbound demand due to increased spending capacity and lower cost carriers in India.

Mainly, the hotels operating in the developing market are highly benefitted by adopting new and advanced technologies and contributing a lot to the overall success [1]. This phenomenon also considered by researchers from the Indian perspective [2] and assessed that the use of technology has shifted to 4.0 in the age of the internet and adoptions of smart technologies tremendously adopted by different hospitality firms in order to ensure innovating and efficient customer services particularly room division operations of a hotel. The dynamic technology also indicated rising numbers of hotels

https://doi.org/10.52202/074664-0034

adopting different business operations shortly. Whereas from different sides, when employees of the firm are unable to adopt the latest tools quickly, it is the most challenging task for the firm to carry out operations smoothly, and it acts as one of the biggest hurdles in front of the management sustain in the market. Technology readiness level is considered the approach through which understanding the technology's technical maturity level during the acquisition phase can know easily [3]. Further, the most significant part understands the maturity level of the technology, and then only its proper implementation or acceptance is possible within the workplace. It can be stated that various trends are present that linked with the new type of digital and smart technologies [4].

1.1 Meaning and Concept of Technologies in Hospitality

The term "smart technology" refers to the incorporation of computing and communications technology into other types of technology that did not before have these capabilities. In order to be considered "smart," a piece of technology must be able to interact with and learn from other similar devices through a network, allowing for features like automatic or adaptive functioning, as well as remote access or control. It's a term used to describe the combination of hardware and software features that create a unified connection in the online community. Both smart gadgets and smart systems are possible manifestations of the technology being discussed here [5]. Technologies that are both smart and enabling may be broken down into two categories: applied and enabling. Video analytics, robotics, the Internet of Things (IoT), data analytics, the cloud, AI, and machine learning are all examples of enabling technologies that form the basis upon which applied technologies are constructed. One type of applied technology is known as facial recognition, which was developed as a result of the use of video analytics. Using technology from the internet of things and robotics, new developments such as robot concierges, robot vacuum cleaners, robots that give in-room service, and other similar inventions are now being developed [6]. Baggage tags in closets, light switches in guest rooms, minibars with self-checkout capabilities, and so on are all examples of the Internet of Things in its many technological guises. IoT-connected gadgets are what make up a smart hotel room, which are then leveraged to deliver a seamless experience for the hotel's guests [7]. The Internet of Things (IoT) has allowed for the social space to expand beyond its physical boundaries and into the realm of the virtual [8].

1.2 Growth and Development of Technology

The decade of the 1990s marked the beginning of the implementation of several technologies and applications. Electronic commerce and Enterprise Resource Planning are only two examples of the types of technological applications that have recently become crucial to a company's survival and growth. The intense competitiveness among organizations has compelled them to investigate new avenues of expansion and strengthen their hold on the market. Because of the novelty of the connection between the production of commercial value and conducting business over the internet, there is a pressing need for more research [9].

https://doi.org/10.52202/074664-0034

2. Review of Literature

Research into the potential benefits of technology has been ongoing since the advent of digital tools, which has allowed for widespread adoption of technological innovations. Perceived usefulness, perceived ease of use, and perceived joy of usage are all factors that have been studied as potential predictors of technology adoption [10].

Many developed countries have found that adopting smart technologies like artificial intelligence, chatbost, virtual reality, robotics, mobile apps, and wifi-5 networks improves organizational performance, as argued by Mouelhi [11].

Sharma (2016) examined the various difficulties encountered by various types of hotels throughout the process of implementing technological change and presented a comparative study. From the viewpoint of a practitioner, it also investigates the part that technical advances have in improving consumer satisfaction with respect to service quality [12].

The research conducted by the author to gain an understanding of the degree to which hotels in India are participating in digitization [13]. The introduction of new technologies and the proliferation of online marketplaces have fundamentally altered the ways in which consumers may purchase goods and services. The automation of guest services made possible by digitalization allows hotels to lower their overall operating costs. The adoption of digital technology is becoming increasingly common in hotels. In order to determine the capacity of Indian hotels to integrate digital technology into their business practices, it is essential to have a solid understanding of the role that digitalization plays in hotels.

3. Objectives of the Research

The following are objectives for this study

1. To examine the influence of technology on overall guest satisfaction in hospitality.
2. To examine specific preference and experience related to technology in guest services.
3. To assess the impact of tech-assisted customer service on overall guest contentment and loyalty.

4. Research Methodology

Sample

Total 104 participated in the survey out of which 82 were male and 22 were female, The questionnaire was filled by the participants for the research paper on "The Impact of Technology on Guest Experience in the Hospitality Industry: An Analysis of Enhancements and Detraction"

Primary Data

A structured questionnaire used with 18 questions which consists of MCQ to record the perception of participants towards The Impact of Technology on Guest Experience in the Hospitality Industry: An Analysis of Enhancements and Detraction.

Secondary Data
Was gathered from the literature previously written through research papers, government reports, news articles etc.

Data interpretation
Demographic profile of respondents is elaborated by pie diagram.

Figure 1. Age

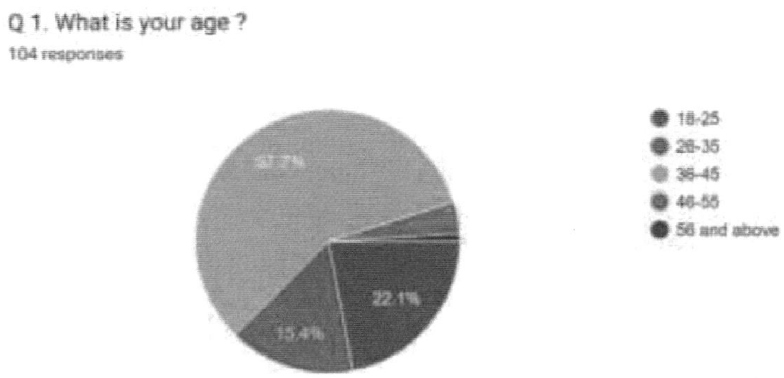

As shown in the figure 1 total 104 participated in the survey out of which maximum numbers of respondents were 60 between in the age group of 36 years to 40, 16 respondents falls between in the age group of 26 to 35, 23respondents falls between in the age group of 18 to 25, 4 respondents falls between in the age group of 46 to 55. 1 respondents falls between in the age group of 56 to above.

Figure 2. Gender

 https://doi.org/10.52202/074664-0034

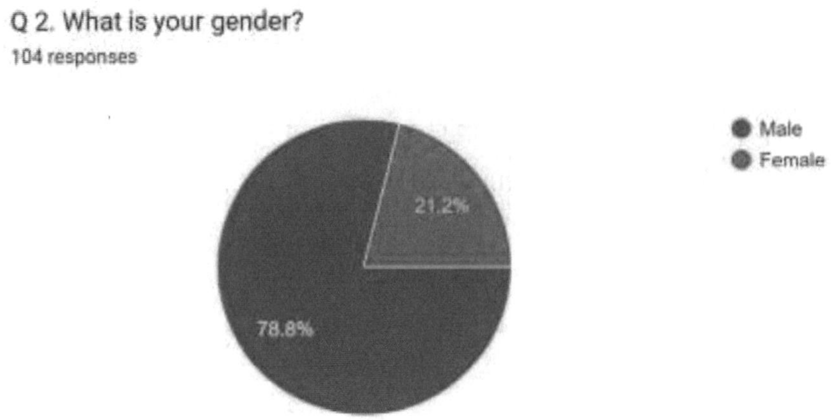

Data given in the figure 2 indicate that 82 (78.8%) of respondents were male, and 22 (21.2%)) were female.

Figure 3. Rate the overall satisfaction with recent hospitality experiences.

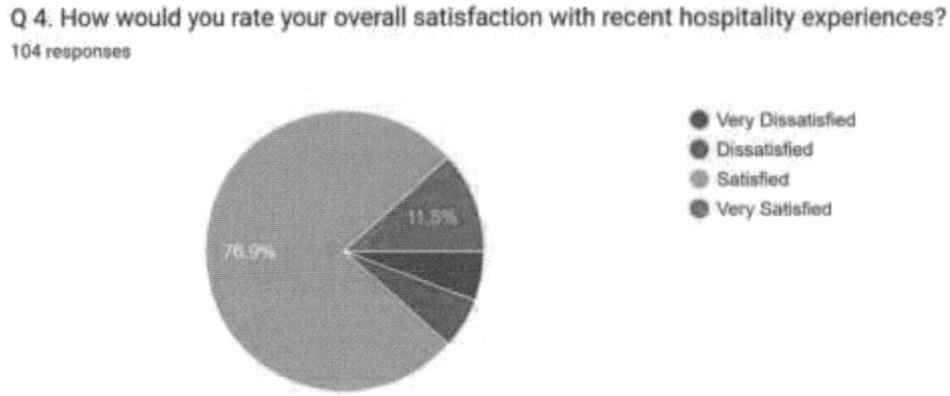

Data given in figure 3 indicate that out of 104 respondents, 80 respondents that is 76.9% were mentioned satisfied on rate to overall satisfaction with recent hospitality experiences , 12 respondents mentioned very satisfied on rate to overall satisfaction with recent hospitality experiences which is 11.5%, 06 respondents mentioned dissatisfied on rate to overall satisfaction with recent hospitality experiences that is 5.8%, 06 respondents mentioned very dissatisfied on rate to overall satisfaction with recent hospitality experiences that is 5.8% which is very less % as per the above research.

Figure 4. Encountered any issues or inconveniences due to technological failures during the stay.

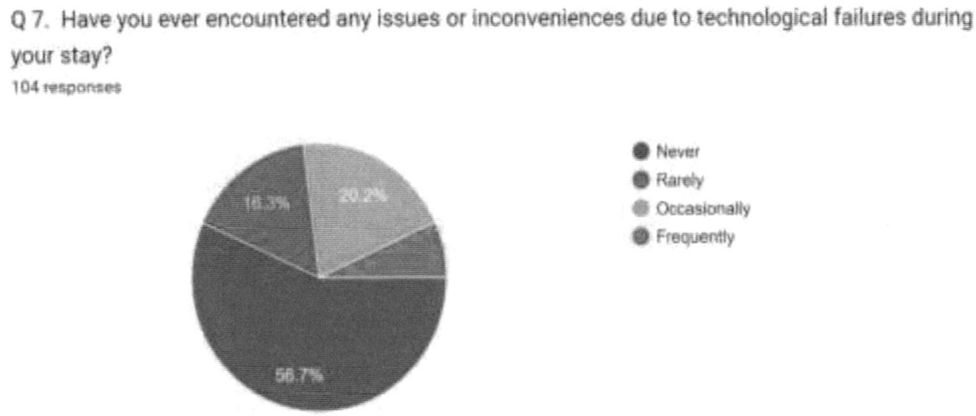

Data given in the figure 4 indicate that out of 104 respondents, 59 respondents that is 56.7% were mentioned, Never encountered any issues or inconveniences due to technological failures during their stay, 17 respondents that is 16.3% were mentioned, Rarely encountered any issues or inconveniences due to technological failures during their stay, 21 respondents that is 20.2% were mentioned, occasionally encountered any issues or inconveniences due to technological failures during their stay, 7 respondents that is 6.7% were mentioned, Frequently encountered any issues or inconveniences due to technological failures during their stay.

Figure 5. Personalized services facilitated by technology during the stay

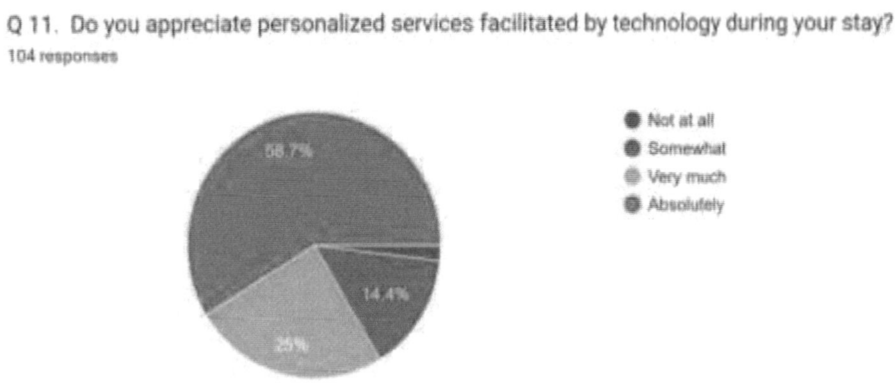

As mentioned in the figure number 5, out of 104 respondents, 61 respondents that is 58.7% were mentioned, absolutely appreciate personalized services facilitated by technology during their stay. 26 respondents that is 25% were mentioned, Very much appreciate personalized services facilitated by technology during their stay. 15 respondents that is 14.4% were mentioned, somewhat appreciate personalized services facilitated by technology during their stay. 2 respondents that is 1.9% were mentioned, Not at all appreciate personalized services facilitated by technology during their stay.

https://doi.org/10.52202/074664-0034

Figure 6. Advancements in technology

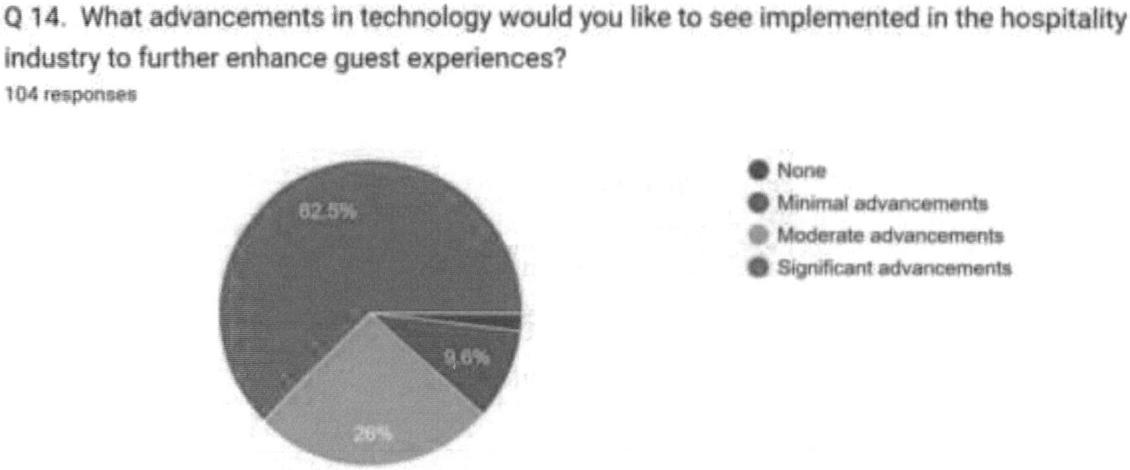

As mentioned in the figure number 6 out of 104 respondents, 65 respondents that is 62.5% were mentioned, significantly advancements in technology would like to see implemented in the hospitality industry to further enhance guest experiences. 27 respondents that is 26% were mentioned, Moderate advancements in technology would like to see implemented in the hospitality industry to further enhance guest experiences. 10 respondents that is 9.6% were mentioned, Minimal advancements in technology would like to see implemented in the hospitality industry to further enhance guest experiences. 2 respondents that is 1.9% were mentioned, None advancements in technology would like to see implemented in the hospitality industry to further enhance guest experiences.

Figure 7. Customer service interactions that involve technology, such as chat bots or virtual concierge services

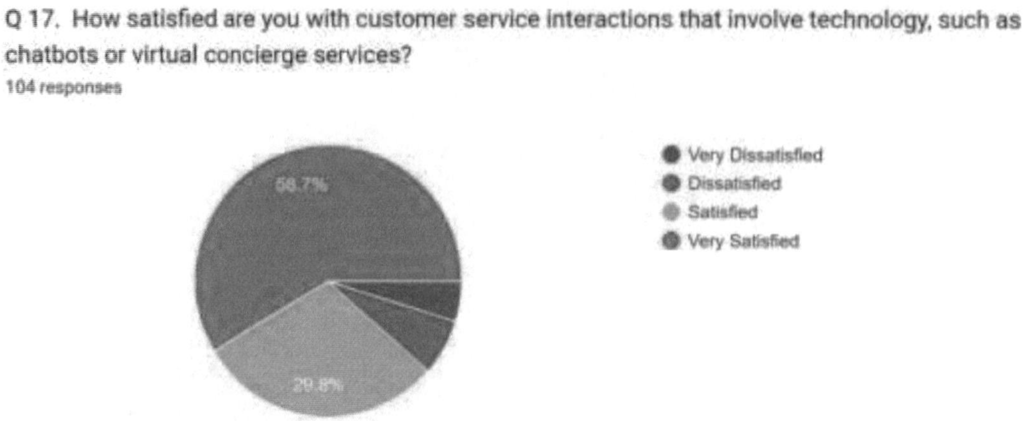

Data given in the figure 7 indicate that out of 104 respondents, 61 respondents that is 58.7% were mentioned, Very satisfied with customer service interactions that involve technology, such as chat

bots or virtual concierge services. 31 respondents that is 29.8% were mentioned, Satisfied with customer service interactions that involve technology, such as chat bots or virtual concierge services. 7 respondents that is 6.7% were mentioned, Dissatisfied with customer service interactions that involve technology, such as chat bots or virtual concierge services. 5 respondents that is 4.8% were mentioned, Very dissatisfied with customer service interactions that involve technology, such as chat bots or virtual concierge services.

Figure 8. Impact of technology on quality of hospitality experience.

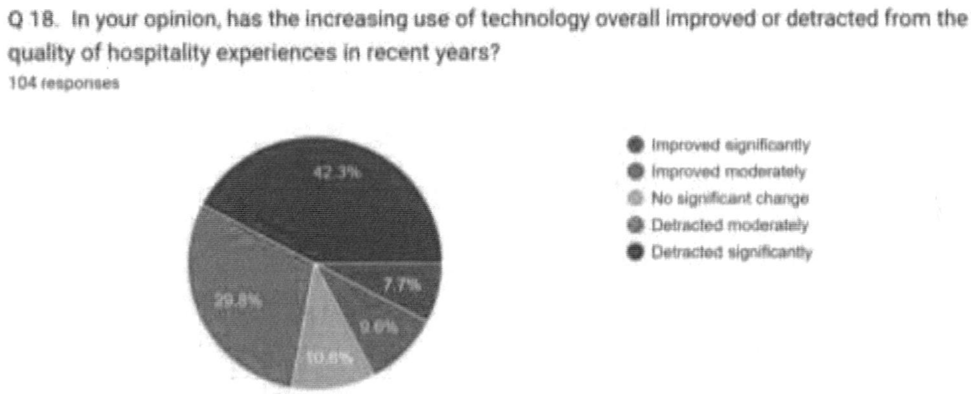

Data given in the figure 8.Indicate that out of 104 respondents, 44 respondents that is 42.3% were mentioned, Detracted significantly has the increasing use of technology overall improved or detracted from the quality of hospitality experiences in recent year. 31 respondents that is 29.8% were mentioned, Detracted moderately has the increasing use of technology overall improved or detracted from the quality of hospitality experiences in recent year. 11 respondents that is 10.6% were mentioned, No significantly has the increasing use of technology overall improved or detracted from the quality of hospitality experiences in recent year. 10 respondents that is 9.5% were mentioned, Improved moderately has the increasing use of technology overall improved or detracted from the quality of hospitality experiences in recent year. 8 respondents that is 7.7% were mentioned, Improved significantly has the increasing use of technology overall improved or detracted from the quality of hospitality experiences in recent year.

5. Limitations of the Study

Confined to player views:

The look at often relies at the perspectives of guests, doubtlessly overlooking the inner operational factors and demanding situations faced by hospitality establishments in imposing and managing generation.

Move-Sectional Nature:

The study captures a selected moment in time, restricting its potential to assess the longitudinal consequences of technology on guest enjoy or potential modifications in perceptions over the years.

Small pattern length:

https://doi.org/10.52202/074664-0034

A limited pattern size may also effect the generalizability of the findings, and the observe may not capture the entire spectrum of visitor reports in the hospitality industry.

Bias in online Responses:
Online survey responses may be prompted via the digital literacy of members, potentially aside from the views of these much less relaxed without or with get right of entry to to online systems.

Outside factors:
Unforeseen external factors, which includes financial downturns or global crises, aren't explicitly taken into consideration, and their ability effect on visitor experiences might not be properly addressed.

Recognizing and addressing these barriers is critical for decoding the study's consequences as it should be and offering a comprehensive know-how of the effect of technology on guest experiences inside the hospitality industry.

6. Future Scope of the Study

Longitudinal study: To gain a more thorough knowledge of the changing impact of technology, future research could perform a longitudinal study to monitor changes in visitor preferences and experiences over an extended period of time.

Comparative Research Across Regions: Examining and contrasting how technology affects visitor experiences in various geographic and cultural contexts would increase the study's relevance and provide light on regional differences.

Integration of Operational Views: To provide a comprehensive picture that takes into account the opinions of both guests and the industry, future study might examine the operational difficulties and triumphs encountered by hospitality facilities when implementing and managing technology.

Qualitative Exploration: To acquire deeper insights into the complex experiences and preferences of visitors, qualitative research techniques like focus groups and in-depth interviews could be used.

7. Conclusion

Hereby in the conclusion made from this study by analyzing the primary and secondary sources of data. The conclusion derived with the three objectives, these are as follows-

Objective 1: To examine the influence of technology on overall guest satisfaction in hospitality.

In summary, research on the impact of technology on overall guest satisfaction in the hotel industry reveals a strong relationship and multiple relationships. It is undeniable that technology has become a part of today's guests and affects satisfaction. Research shows that technological interventions such as seamless booking processes, ease of mobile check-in and state-of-the-art amenities result in excellent service for all guests. Guests appreciate the efficiency, convenience and enhancement technology that allows them to better understand their stay. However, it must be acknowledged that the effect of technology on satisfaction is not the same. Visitors have different preferences, and while some accept and value technological advancements, others may prefer a more social, human-centered approach. It is very important to strike a balance between technological development and human resources management to meet the different needs of guests.

Considering the situation like the COVID-19 pandemic that has forced hotels to adopt modern technologies for delivering services in a human contactless manner to focus on customer service and efficient business operations in a smooth manner [14]

Objective 2: To examine specific preference and experience related to technology in guest services.

Exploring unique preferences and experiences with guest technology provides insight into how guests interact with and perceive technology interventions in the hospitality industry. Survey results show a wide range of preferences among guests regarding the use of technology across a range of services. While most are pleased with the convenience and efficiency achieved through the use of technologies such as online booking and mobile check-in, others express a preference for different methods. This duality underscores the importance of providing a flexible and customizable technology experience that suits the individual preferences of different groups of guests.

Depending on the use of technology in the room, the research found differences in interest in devices such as smart phones. Lighting, climate control and entertainment, It is clear that the success of these features in improving visitors depends on features such as ease of use, reliability, and the ability to change the individual's preferences. Additionally, the analysis shows interest in contactless services such as mobile payments and digital room keys, as well as the social quality of their use and the understanding of safety and hygiene. This reflects the changing expectations of guests and highlights the role of technology in providing a safe and seamless service. In summary, analysis of customer service technology's unique interests and experiences highlights the need to be personalized and focused on people. Use technology. Restaurant establishments should use technology to increase convenience and safety while being mindful of personal preferences. Insights from this brand provide a better understanding of how technology can improve guest experience and satisfaction in a hospitality environment.

In room division operations, computers can be termed as among the most common tools underneath smart technologies used in hotels. Apart from this, digital cameras, projectors, tablets, etc., are some other common tools that are being considered to conduct out a smooth flow of information while functioning specifically for room division [13]

Objective 3: To assess the impact of tech-assisted customer service on overall guest contentment and loyalty

Research on technology-enabled customer service in the hospitality industry provides insight into its impact on guest satisfaction and loyalty. The findings reveal a relationship that highlights the benefits and challenges of using technology through customer interactions. On the positive side, guests are generally satisfied with the technology's customer service and the available and efficient service. Services like chatbots and virtual concierge systems have proven effective at quickly responding to questions, resolving issues, and offering advice. This performance results in great customer satisfaction, especially when immediate benefits are provided. However, it must be acknowledged that sometimes guests still prefer human interaction with certain services. Although technology improves efficiency, the human touch is still important to meet guests' more challenging or unique

https://doi.org/10.52202/074664-0034

needs. Striking the right balance between technology-enabled and human customer service is crucial to increasing overall customer satisfaction and loyalty. The study also reveals the importance of integrating the customer's different channels. Whether it's interacting with a chatbot, virtual concierge, or human staff, integration and collaboration ensure a great experience with guests. Conflicts or disruptions in this integration can cause dissatisfaction and affect customer loyalty. In summary, technology-supported customer services play an important role in shaping overall customer satisfaction and loyalty. Hospitality establishments need to use technology to increase efficiency by recognizing the inefficiency of human interaction. Harmony between these two elements is essential to improving guest satisfaction, trust and the overall feel of the hotel experience.

These results compliment the results that suggested that the adoption and application of smart technologies in the hospitality industry change faster, which depends on the market trend and the rising customer preferences in the market. Further, they suggested that in the era of modernization, every industry and company are operating with the help of new technology. The biggest benefit is that it allows serving customers efficiently in the market and raising their financial and operational efficiency.

References

1. Buhalis, D., Leung, R.: Smart hospitality—Interconnectivity and interoperability towards an ecosystem. International Journal of Hospitality Management. 71, 41–50 (2018). https://doi.org/10.1016/j.ijhm.2017.11.011.
2. Prentice, C., Dominique-Ferreira, S., Wang, X.: The impact of artificial intelligence and employee service quality on customer satisfaction and loyalty. Journal of Hospitality Marketing & Management. 29, 739–756 (2020). https://doi.org/10.1080/19368623.2020.1722304.
3. Chahal, N. and Kumar, M., 2014. The impact of information communication technology and it's Application's usage in lodging Industry: An exploratory study. *International Journal of Techno-Management Research*, *2*(1), pp.1-17.

a. Kinley, N., Ben-Hur, S.: Changing Employee Behavior: A Practical guide for Managers. (2015).

4. Chawla, D., Joshi, H.: Consumer attitude and intention to adopt mobile wallet in India – An empirical study. International Journal of Bank Marketing. 37, 1590–1618 (2019). https://doi.org/10.1108/ijbm-09-2018-0256.
5. Chen, S.H., Tzeng, S.Y., Tham, A., Chu, P.X.: Hospitality services in the post COVID-19 era: Are we ready for high-tech and no touch service delivery in smart hotels? Journal of Hospitality Marketing & Management. 30, 905–928 (2021). https://doi.org/10.1080/19368623.2021.1916669.
6. Cheng, Y.: Towards an understanding of the factors affecting m-learning acceptance: Roles of technological characteristics and compatibility. Yà-tài Guǎnlǐ Pínglùn/Asia Pacific Management Review. 20, 109–119 (2015). https://doi.org/10.1016/j.apmrv.2014.12.011.
7. Chittipaka, V., Kumar, S., Sivarajah, U., Bowden, J., Baral, M.M.: Blockchain Technology for Supply Chains operating in emerging markets: an empirical examination of technology-organization-environment (TOE) framework. Annals of Operation Research/Annals of Operations Research. 327, 465–492 (2022). https://doi.org/10.1007/s10479-022-04801-5.
8. Anderson, C., 2012. The impact of social media on lodging performance. *Cornell Center for Hospitality Research Report,* 12(15).
9. Anderson, R.E.: Consumer dissatisfaction: the effect of disconfirmed expectancy on perceived product performance. Journal of Marketing Research. 10, 38–44 (1973). https://doi.org/10.1177/002224377301000106.
10. Andreassen, T.W., Lindestad, B.: The effect of corporate image in the formation of customer loyalty. Journal of Service Research. 1, 82–92 (1998). https://doi.org/10.1177/109467059800100107.

11. Angelova, B., Zekiri, J.: Measuring Customer Satisfaction with Service Quality Using American Customer Satisfaction Model (ACSI Model). International Journal of Academic Research in Business & Social Sciences. 1, 27 (2011). https://doi.org/10.6007/ijarbss.v1i2.35.
12. Barsky, J., Labagh, R.: A strategy for customer satisfaction. Cornell Hotel and Restaurant Administration Quarterly/˜ the œCornell Hotel and Restaurant Administration Quarterly. 33, 32–40 (1992). https://doi.org/10.1177/001088049203300524.
13. Bradić, M., Kosar, L., Калењук, Б.: Business guests satisfaction in the hotel industry: A case study of North American hotel chains. Turizam. 17, 60–70 (2013). https://doi.org/10.5937/turizam1302060b.
14. Butcher, K.J.: Differential impact of social influence in the hospitality encounter. International Journal of Contemporary Hospitality Management. 17, 125–135 (2005). https://doi.org/10.1108/09596110510582323.

 https://doi.org/10.52202/074664-0034

A Study on Madhya Pradesh as a Culinary Tourism Destination

Vishal Valuskar[1], Anant Kumar Verma[2], and Parikshit Sharma[3]

[1&2]Assistant Professor, School of Hospitality and Tourism, Jagran Lakecity University, Bhopal

[3]Associate Professor, School of Hospitality and Tourism, Jagran Lakecity University, Bhopal

Abstract:

Foods are inescapable in any discussion of human existence. Being a combination of biological, sociocultural, and touristic aspects, they form an essential element of the tourism offering in popular places. Foods are unique to tourism because they are a biological necessity for every traveller, just as they are for every human being. However, the mets by which this requirement is met vary greatly depending on the sociocultural characteristics of the cuisine and hospitality of the travel destination, as well as the specific ways in which the food is prepared and served to the travellers. Madhya Pradesh has been a rich in culinary tourism because of its cultural food and beverages opportunities. This paper aimed to review past literatures based culinary tourism destination to attract tourist in Madhya Pradesh.

Keywords: Culinary tourism, food tourism, Madhya Pradesh, tourism.

1. Introduction

Culinary tourism, sometimes known as "food tourism," is a growing subset of the travel industry in which visitors seek out local specialties as a means of thoroughly immersing themselves in a new culture. A lot of the time they also include regional specialties like drinks, cheeses, chocolates, etc. [1]

Dedicated Food Tours and Experiences

As a result of its increasing appeal among chefs, oenophiles, and foodies, culinary toulim has exploded in popularity in recent decades. Consequently, towns with a stellar reputation for their culinary and alcoholic offerings have started providing guided tours of their finest establishments. [2]
Sightseeing and other optional activities could be a part of these trips. They show tour groups sights that the average traveller wouldn't know about. If you want to avoid typical tourist fare and dine like a native, a cuisine tour is a fantastic option. In addition, joining a culinary tour is a great way to meet new people and share in the experience of trying the local cuisine. [3]

Culinary Tourism Coupled with Cooking Lessons

Indulging their passion for food to a new level, many travellers take advantage of local culinary lessons while on vacation. These may be provided by restaurants, resorts, and even private cooks who have mastered the local cuisine. [4], [5]

Countries Known for their Fantastic Offerings to Food Tourists

While the cuisines of every civilization serve as a distinctive marker of that culture, there are a few that have gained international renown for their exceptional ingredients and preparation methods. Countries like France, Italy, Thailand, Mexico, Greece, and Japan are prominent examples.
Each of these locations is home to a storied culinary tradition and time-tested recipes. In addition, they take into account the most up-to-date trends and fusion cuisine, which contribute to the enjoyment of going to a new place only to dine.
Tourists come from far and wide to sample the finest alcoholic beverages, but the food is what draws the most attention. Delicate French wines , whisky from the Scottish Highlands, Japanese sake, and dramatic Port wine all deserve to be savoured in the heart of their vineyards. [6], [7]

For people with a passion for both travel and culinary experiences, culinary tourism is the perfect option. It skilfully mixes culinary treats with cultural significance, mythology, and breath-taking landscapes. More and more vacationers are acquiring a taste for it, and it's no surprise that they're travelling the globe in search of culinary perfection. [8]

1.1. Culinary tourism in India

A total of US$19,127,000,000 would be spent on culinary tourism in India by 2022. From 2022 to 2032, market sales are predicted to increase at a healthy CAGR of 20.4%. 6-9% of worldwide culinary tourism revenue is generated in the Indian sector.

"2017-2021 India Culinary Tourism Demand Outlook Compared to 2022-2032 Forecast"

India's diverse landscape, kind people, and rich history make it an ideal destination for those interested in trying new cuisines. Culinary, or the art of preparing, presenting, and enjoying delicious food, is only one of many factors that impact the travel industry.

So the value of food-related tourism in India has skyrocketed in recent years. Travelers are venturing to all corners of the United States in search of the best local cuisine and drinks.

The local food tourism industry in India is predicted to grow throughout the assessment period as an increasing number of food enthusiasts travel to learn about culinary cultures and beverages. Growth would also be aided by things like a rising middle class and accelerated globalisation.

The Five Most Important Factors Driving the Growth of India's Food Tourism

Dine-in Practices in a Native's House

It is common for visitors visiting India to experience home cooking and family bonding firsthand by participating in a culinary tour or eating at a local restaurant. They do this to disseminate knowledge about their country's culinary traditions based on first-hand experience.

Increasing Shift towards Localism

The rising desire for genuine cultural immersion is likely to increase the number of tourists eating in people' homes. People in rural regions are turning away from globalisation and toward localism, which bodes well for the demand of regional foods and drinks. It'sdisigned to help them be true to themselves culturally rather than succumbing to a worldwide trend.

Rising Popularity of Cooking Lessons

With the rise of "culinary tourism," visitors are showing a growing interest in taking cooking classes. While eating out at a restaurant is a great way for tourists to sample regional specialties, taking in a cooking demonstration or enrolling in a cooking class is a great way to really immerse yourself in the local culinary culture.

Nowadays, cooking classes are becoming more popular with young people, families with small children, and solo travellers as a pleasant way to interact with locals and people from other countries. In addition, by participating in the first steps of the culinary process, tourists may learn about the numerous influences that went into producing a cuisine.

Availability of Traditional Ingredients

The abundance of traditional grocery shops throughout India that provide ingredients for traditional cookery is expected to contribute to the country's progress. Tourists from other countries typically see the use of spices, curry, and other Indian culinary techniques as an art form. Furthermore, the industry would be propelled by the rising interest in exploring India's rich cultural heritage and contemporary practises.

Government Initiatives to Boost Tourism

A number of programmes are being implemented by the Indian government to encourage more tourists to visit the nation. New, low-cost accommodations for tourists from other countries are also a priority.

The Food Safety and Standards Authority of India (FSSAI) has implemented numerous programmes to guarantee the nation's citizens' access to safe and nutritiousiod. This is in response to frequent concerns voiced by foreign visitors about the safety, cleanliness, and provenance of ingredients used in street food. With its first phase kicking out in 2016, Project Clean Street Food aims to educate the over 20,000 street vendors in Delhi on the significance of health and hygiene in their line of work.

1.2. Culinary tourism in Madhya Pradesh

https://doi.org/10.52202/074664-0035

Traveling around Madhya Pradesh, you'll see places you've never seen before, have encounters that will become tales, and most importantly, go on a culinary adventure. Due to its rich cultural and historical background, the cuisines draw inspiration from many different sources. The food of Bundelkhand reflects the region's history, culture, and environment. The famous warmth of the local Bhundeli people, however, is something that can only be found in this part of India. Murar Ke Kebabs, Kodu Til Ka Baat (rice with black sesame), and Bhundeli Ghost are just a few examples of the Mughal, Rajput, and Tribal influences on the cuisine here. (9)

Like many other cuisines Mahalkhoshal food has come a long way from its humble beginnigs. It is inspired by Rajasthani, Gujarati, Maharashtrian, and tribal cuisines, and is located in the upper eastern valley of the Narmada. But it has changed through time to accommodate contemporary tastes while still keeping a substantial amount of its original integrity. Just try out some old standbys like the classic Tikkis. Eat a steaming plate of Gulab Jamun. And if it doesn't do it, the Meva Bati should.

Meat eaters, however, have to try out Bhopal's cuisine. The Sheek Kebab, a dish of minced lamb, fresh mint, curry paste, green chilies, and other spices, the Keema Pulav, and the Chicken or Mutton pilaf live up to the name "Meat Lovers Paradise." The "Bhopaali Paan," however, is certain to steal the show. Islam and Hinduism both contribute significantly to the cuisine. After years of working together, each dish is at its peak. [10], [11]

Gwalior was governed by the Guptas, the Sindhias, and the Mughals throughout the course of many centuries. Then the British arrived. Further, the four separate empires each had its own unique culture. This was entirely absorbed by Gwalior over time. Gwalia kebabs are so tasty that just one mouthful will make you want more. Next to it comes the mouthwatering rogan josh cooked in the traditional style of Gwalior. The Gwalior biryani is another "can't miss" item here. Throughout the city, you may find restaurants serving this delicious and satisfying dish.

Madhya Pradesh is home to several delicious regional **specialties,** including the unique Malwa food, the Nimar cuisine, and the Bagelkhand cuisine. Rass Kheer, Lapsi, Kalakand, Lavang Latiki, Pek Puri, Bhutte Ki Khees, or any of the other sweets and savoury dishes from the Punjab (grated corn roasted in ghee and later cooked in milk with spices). A single trip here won't be sufficient to sample all of the delicacies on offer, such as the tamarind-based Amli Ri Kadhi, the savoury Khoprapak, and the sweet Malpua.

The magnificent architecture may give you a taste of what it was like to be a monarch, but it is the cuisine that will really take you back in time. (12] Madhya Pradesh is home to a diverse population and rich cultural heritage, and its cuisine reflects it. If you want to experience authentic Indian cuisine, you must go to the country's very core. In spite of inevitable interference from progress, traditional culinary techniques have been kept remarkably intact. In Madhya Pradesh, residents still regularly practise the hospitality that has made India famous across the world. People are deified when the words "Atithi Devo Bhava" are said. Despite the passage of time, this age-old custom is still strongly associated with migration in the thoughts of the people. The people of this state are known for their warmth and welcome, so you should visit and see for yourself. Their attitude towards you will astound you.

INFLECTIONS ON THE CUISINE

History shows that the state's diverse cultural heritage had an impact on the cuisine of its people. The state's cuisine thoughtfully incorporates flavours and techniques from neighbouring states Rajasthan and Gujarat. Madhya Pradesh is a large state, and as a result, its northern and southern regions have very different cuisines. People from all across India have relocated in Madhya Pradesh, adding to the state's already diverse culinary scene. The ancient Malwa diet eliminated the flavorful core of Gujarati and Rajasthani cooking. Jowar, once used as a staple, has been eclipsed by wheat as a crop of choice. Since the soil is arid and unproductive for the majority of the year, the consumption of vegetables has been strictly regulated by pulses. For a taste of the favourable distinction that Madhya Pradesh cuisine has over other cuisines, you must try some of the state's food.

MULTI DIVERSE CUISINE TO GUSTO YOUR SENSES

If you're a vegetarian, a non-vegetarian, or neither, you should come to Madhya Pradesh to taste the cuisine. Every tourist in Madhya Pradesh will find something to enjoy. In other words, everyone can find what they're looking for. True to their origins and with a distinct flavour profile. Madhya Pradesh dishes are easily distinguished from those of other Indian regions by their liberal use of spices, oil, and other ingredients. Bafla (wheat cakes) dipped in thick ghee and eaten with daal are the most beloved of the many wheat-based delicacies (a pungent lentil broth). If you're trying to eat vegetarian, one option is a dessert called "Bhutte ki kheers," which is made with maize and milk. One of the most well-known Madhya Pradesh delicacies is chak k i ki shaak. Wheat dough is cooked and then combined with curd to make this meal. [13]

Those who previously couldn't find enough meat or fish to sustain themselves now have many more choices. The most well-liked dishes are the fiery rogan josh, the tender korma, and the juicy seekh and shami kebabs. These are all part of the sumptuous banquet that is life. Extremely sweet and delectable sweets from the state are the perfect antidote to the state's tongue-tingling savoury dishes. It's no secret that Madhya Pradesh has carved out a special place for itself in the after-dinner entertainment industry. Thelmawa-bati,"shreekhand; 'khoprapak,' and'malpua; to mention a few, are all dishes you just must taste. In addition to kulfi and jalebi, the cashew-based barfi and kusli are other must-haves. Traditional and mild beverages like as lassi (buttermilk) and sugarcane juice are must-trys. If you're in the mood for something stronger, try the native sulfi or date palm toddy, both of which are distilled from the blooms of the mahua tree. [14]

2. Literature Reviews

This study analysed how tourism and travel affects national economies and how this effect is related to other facets of the economy. Every economy relies on tourism as a means of bringing in money. The most significant developments in the tourist industry as they have occurred recently are outlined in this study. It has been calculated that the culinary tourism submarket accounts for almost one-third of all tourist income. The European Union (EU) is seeing rapid growth in a relatively new kind of tourism: the culinary tourism market. The European Union (EU) recognises the importance of the culinary sector to rural tourism and actively promotes culinary tourism at events such as fairs and festivals. Recent years have seen growth in rural tourism both internationally and in Ukraine. This is largely attributable to the sector's distinguishing characteristics, such as the promotion of sustainable community development, the preservation of local traditions and cultural values, and the hosting of gastronomic events. The purpose of this essay is to examine the theoretical facets of gastronomy tourism's global growth, to examine the current state of gastronomic tourism in the European Union and Ukraine, and to find opportunities and strategies to expand regional gastronomic tourism. The book relies heavily on analytical reports and studies about the growth of event tourism and statistics for its methodology and data. [15]

Recent years have seen a rise in discussions on the significance of food in destination marketing, tourism tourism development, and academic research. Modern tourists are more interested in multisensory, holistic experiences, and a large part of that is gaining a deeper understanding for local cuisine. The overriding topic of ' analysis is the culinary experience. In doing so, it tills a need in our understanding of how image, satisfaction, and behavioural objectives (which have been studied extensively from a destination's viewpoint) relate to food and the travel experience. The goal of this research was to ascertain whether or not tourists would suggest India as a destination for foodies based on their impressions of Indian food and their degree of satisfaction with their dining experiences there. The study also reveals other aspects of the eating habits of international visitors to India, such as the number of times they ate at Indian restaurants, the types of restaurants they frequented, the most important resources they used to learn about Indian cuisine, and their overall impressions of Indian cuisine as a result of their experiences. The survey also included a comparison of Indian and Thai cuisines, two of the most popular cuisines in the world. [16]

 https://doi.org/10.52202/074664-0035

The purpose of this research is to get an understanding of current trends specific to the tourism industry in Madhya Pradesh and to provide potential avenues for improvement. This research tries to evaluate the tourism industry in Madhya Pradesh from a statewide perspective, looking at how it varies from county to county. As a result, a study was conducted, with aspects including tourism traffic volume being taken into account. Travel and tourism in Madhya Pradesh have tremendous potential. Therefore, there is a need for more work to be done on developing appropriate methods for advertising to and informing clients with middle- and lower-incomes. The goal of this research was to determine how the tourism industry in Madhya Pradesh may help to meet the goals of the state's limited economic growth by analysing existing partnerships and providing ongoing support to prospective tourism entrepreneurs. [17]

One in ten people throughout the globe are employed in the tourism industry, and the industry accounts for 10 percent of global GDP. However, there are numerous threats to the environment, and the tourism industry must become more sustainable. The food tourism is increasingly dependent on visitors' ability to buy and consume food and drink, and it is well acknowledged that the production, transportation, and disposal of food is a major source of pollution. As a result, there has been a rise in the food for seasonal and regionally sourced foods and beverages. For the same reason, more and more establishments are sourcing their ingredients locally, and more and more food spots are promoting their local food and drink as part of their packages. Everyone has to eat while on vacation, and studies suggest that visitors are willing to spend more on authentic culinary food. To learn how local food is used in the promotion of UK tourist destinations, we conducted online research and conducted eight in-depth interviews with residents of five locations that provide unique food and/or alcoholic experiences. [18]

The study's goal is to provide a macroeconomic analysis of the role that transaction costs and benefits play in influencing consumers' gourmet tourism decisions. This chapter provides a high-level overview of the culinary tourism industry in India, discussing the costs and advantages associated with various aspects of the industry such as searching for information, negotiating prices, making decisions, and enforcing laws. The suggested structure is inspired by the cuisines that India offers to many tourists. Future research implications and context within the post-COVID-19 situation are highlighted. (19) A growing trend in travel, "culinary tourism" is sampling the local food at each This study discusses the development, present state, and potential future difficulties of culinary tourism. This study explains the history of culinary tourism as well as its contemporary destination. In terms of both gross domestic product and the number of people directly employed, tourism is among the world's top industries. And so, food is a huge contributor to it, and it has a tremendous bearing on the growth of tourism to a location. This is why the concept of culinary tourism has been quietly gaining ground as a means to boost a region's tourism industry. Culinary tourism is mostly undiscovered even in India. The potential for drawing visitors to far-flung parts of the nation based on the quality of their authentic culinary cuisine is enormous. As a matter of pride, each state may lay claim to its own distinctive flavour profile. It could do a lot of good if it was directed correctly. There are few things more compelling than a destination's culinary scene to inspire potential visitors to book a trip there. This study emphasises culinary tourism and its potential in light of these considerations. [20]

It's no secret that the culinary tourism tourism has exploded throughout the nation, setting a precedent for growth and improvement in the field. The Indian Ministry of Tourism has increased its focus on developing culinary destinations, which have a significant and lasting effect on rural communities' socioeconomic status, in recent years. The government hopes that an increase in culinary tourism in rural regions would aid in the economic prosperity of the impacted towns. Because it makes use of the area's natural attractions and the community's beautiful traditions and cultures as attractive methods of luring visitors, culinary tourism has the potential to boost the tourism industry in the region. Positive responses to culinary tourism have led to its implementation as a strategy for combating poverty in rural areas. Vacationers who go to these less-traveled regions will get a more genuine feel for the country's rich cultural heritage. Despite its rising significance in India's domestic tourism industry and its relevance to community-based types of tourism more generally, culinary tourism is mostly ignored in the country. This essay aims to examine the consequences of actively participating in Madhya Pradesh homestay tourism. Improved management of culinary tourism may be achieved via tourism in tourism tourism, promotion of authentic cultural encounters, and enhancement of service quality. [21]

Madhya Pradesh, India, is gaining recognition as a top tourism for foodies and travellers alike because to the rise of "culinary tourism." The business fits within the experience sector, whereby the value of unique encounters is measured against the Fourth of July. They can expect the food sector to continue expanding so long as people continue to need to eat. The term "culinary tourism" refers to both traditional tourism fare and off-the-beaten-path food. Madhya Pradesh is expanding in all directions, which means there will be more visitors looking to try out the city's cuisine. This, in turn, will provide residents more chances to hone their skills in the kitchen and introduce them to new audiences. Indeed, this approach will open up a plethora of new opportunities for locals to earn money. The article highlights the idea that promoting regional food via tourism may help preserve culinary traditions while creating new job possibilities. A rise in culinary tourism to the Madhya Pradesh area is credited in the report as a key factor in the rise in popularity of local food. [22]

In the recent decade, an academic discussion has developed about the significance of local food's role in destination marketing and development, although little work been done to experimentally support the idea of local food. In order to fill this knowledge vacuum, this study interviews supply-side representatives in Yorkshire, England, to learn more about their thoughts and feelings about local food. The results imply that supply-side representatives have various conceptions and discursive constructs of what "local food" is. These are often broken down into three categories: physical, cultural, and societal. Future study directions are indicated, and practical consequences are provided for producers, service providers, and managers. [23]

While the tourism, ocean, and sun were formerly the main attractions for vacationers, postmodern tourism has expanded its product offerings to include a wider variety of experiences. More and more specialized industries are emerging as a means to prolong the tourism life cycle and attract more international visitors. Markets like this are growing, and one that has been getting a lot of attention lately is the food tourism industry. Given that eating is an integral part of every vacation, the last two decades have seen a rise in the popularity of "food tourism," which benefits both the local food and the visitors' wallets. Unlike most other resources, food may be produced all year round using locally sourced components. Destination managers must foster a sense of regional identity and preserve the cultural traditions of their areas if they want to attract and retain visitors. Food offers visitors an opportunity to try something new and immerse themselves in the culture of the place they are visiting. [24]

3. Findings

Trends in Culinary Tourism: Often referred to as "food tourism," culinary tourism is gaining popularity all over the world by providing travellers with a taste of local cuisine. Exploring local specialties, drinks, and cooking customs is becoming more and more popular, which is driving this trend. Dedicated Food Tours and Experiences: A lot of places now provide guided culinary tours that highlight their best restaurants, giving visitors firsthand access to genuine regional dining experiences. These trips frequently offer extra sightseeing options, enabling travellers to find hidden treasures outside of the usual tourist traps. Tourism Combined with Cooking Classes: A growing number of tourists are combining their love of cuisine with education by enrolling in cooking classes while on vacation. These classes, which are provided by eateries, resorts, and private chefs, give students practical experience in learning regional cuisines.

Popular Culinary Destinations: Due to their outstanding culinary offerings, nations such as France, Italy, Thailand, Mexico, Greece, and Japan have become well-known worldwide. Travellers swarm to these locations to partake in a variety of culinary customs and taste signature foods and drinks.

India's Culinary Tourism: With large investments and programmes aimed at promoting the nation's rich and varied food culture, India's culinary tourism industry is expanding quickly. This rise is being driven by factors including cooking courses, dining in local households, a shift towards localism, the availability of traditional foodstuffs, and government efforts.

Madhya Pradesh Culinary Tourism: With its varied cuisines and rich cultural legacy, Madhya Pradesh is quickly becoming one of India's top destinations for culinary tourism. Many traditional foods influenced by Mughal, Rajput, tribal, and other culinary traditions can be found in the area. Travellers looking for real food experiences are drawn to the region's delicacies, which include Murar Ke Kebabs, Kodu Til Ka Baat, Bhundeli Ghost, and Gwalia Kebabs.

https://doi.org/10.52202/074664-0035

Impact of Culinary Tourism: Culinary tourism fosters cultural interaction, employment creation, and the preservation of culinary traditions in addition to aiding in the economic development of travel destinations. The culinary tourism sector depends on both community involvement and government backing to grow and survive.

Opportunities and Challenges: Although culinary tourism offers chances for both cultural preservation and economic growth, it also faces obstacles including sustainability, environmentl effect, and authenticity maintenance. In order to overcome these obstacles, cooperation amongst interested parties is needed, as the promotion of regional foods and cultural heritage preservation.

4. Conclusion

Madhya Pradesh's central position in India makes it a popular tourism destination. For the last three years in a row-2017, 2016, and 2015Madhya Pradesh has been recognised as the best tourism state in the nation. For tourism stakeholders including tourism planners, destination marketers, and operators of culinary enterprises, the question of how happy international visitors were with their culinary experiences in the state is of crucial importance. Customers' happiness with a service or attraction has a significant role in determining whether they will return and if they will spread the word about it favourably (or the entire travel experience to a destination, to that matter).

Madhya Pradesh is good source of culinary and has a history of cultural foods for attracting more and more tourism. There is a gap for several research aiming at the improvement of economic benefits of state as well as the country in terms of culinary tourism. Once the visitor arrives in M P., a plan must be developed to increase the tourist's chances of spending money across a wide range of categories, such as retail, entertainment, and food. A visitor from another nation wants to buy some trinkets for his friends and family back home. In any case, he has plenty of chances to do so, notably at the entrance and departure points. Therefore, it's crucial to use more professional packaging and marketing techniques, and to persuade him that the identical products are cheaper in the point of origin.

References

1. Córdova-Buiza, F., Campos, E.G., Castaño-Prieto, L., García-García, L.: The Gastronomic Experience: Motivation and satisfaction of the gastronomic Tourist—The case of Puno City (Peru). Sustainability. 13, 9170 (2021). https://doi.org/10.3390/su13169170.
2. S. Redl, "Culinary Tourism for Young Adult Travellers and its connection to Destination Management," no. May, 2013.
3. Aslimoski, P., Gerasimoski, S.: Food and nutrition as tourist phenomenon. Procedia: Social & Behavioral Sciences. 44, 357–362 (2012). https://doi.org/10.1016/j.sbspro.2012.05.039.
4. Şahin, G.G.: Gastronomy Tourism as an Alternative tourism: An assessment on the gastronomy tourism potential of Turkey. International Journal of Academic Research in Business & Social Sciences. 5, (2015). https://doi.org/10.6007/ijarbss/v5-i9/1816.
5. P. Kumar, "A Critical Study on Market Demand of Culinary Professionals with Reference to Training Facilitation in National Level Hospitality Schools in Madhya Pradesh," no. February, (2021), doi: 10.13140/RG.2.2.29998.43840.
6. Sormaz, Ü., Akmeşe, H., Güneş, E., Aras, S.: Gastronomy in tourism. Procedia Economics and Finance. 39, 725–730 (2016). https://doi.org/10.1016/s2212-5671(16)30286-6.
7. Tovmasyan, G.: Exploring the role of gastronomy in tourism. Socioeconomic Challenges. 3, 30–39 (2019). https://doi.org/10.21272/sec.3(3).30-39.2019.
8. Khan, M.: The Opportunity of Tourism in Madhya Pradesh to develop the good Environment for Tourists'(Special Reference to Bhopal).
9. G. Shah and K. Shende, "A study on the importance of Food Tourism and its impact on Creating Career Opportunities amongst the Residents of Pune city," *Int. J. Res. IT Manag.,* 7 , 3 , 192-208, (2017), [Online]. Available: http ://aiss mschmct.in/wpcontent/uploads/2017/08/67.pdf.
10. Updhyay, Y., Sharma, D.: Culinary preferences of foreign tourists in India. Journal of Vacation Marketing. 20, 29–39 (2013). https://doi.org/10.1177/1356766713486143.
11. I. M. Darsana and P. H. Susanti, "Trends of Traditional Culinary Tourism Research in Tourism Sector Journals Around Indonesia," *Budapest Int. Res. Critics Inst. Humanit. Soc. Sci.,* 5, 1, pp. 6664-6674, (2022), [Online]. Available: https://bircujourn al .com/index .php/birci/article/view/4393.
12. J. D. Mulcahy, "Food (in) tourism is important, or is it?," *Voyag. between Fr. !rel. Cult. Tour. Sport,* no. January 2017, (2017).
13. Y. Perdomo, *Second Global Report on Gastronomy Tourism: Sustainability and Gastronomy.* 2017.

14. GoMP, 'Twenty Years Perspective Plan of Tourism for the State of Madhya Pradesh," *Final Rep. Twenty Years Perspect. Plan Tour. State Madhya Pradesh Final Rep., vol.* II, p. Annexures, (2003).

15. Boiko, V., Liubynskyi, O., Strikha, L., Zarakhovskyi, O.Y., Neilenko, S.: Development of culinary tourism in European countries, https://kiss.kstudy.com/DetailOa/Ar?key=54462212.
16. S. Duttagupta, "Foreign travellers' recommendation of culinary tourism in India based on cuisine image and satisfaction with experiences at culinary establishments: an exploratory study," 192, (2013).
17. A. K. Rai and P. Baredar, "Investigative and Impact analysis of tourism in the state of Madhya Pradesh," vol. 56, 13347-13349, (2013).
18. A. Soderstrom, "Attracting tourists with the use of local food : Using a multiple case study to investigate the role of local food in the marketing of tourist destinations in the UK," no. June, 2022, [Online]. Available: http ://urn. kb.se/re solve?u rn=u : nbn : se : uu:div a-478657 .
19. A. Sharma, "Gastronomic Tourism in India — A Transaction Cost," 1-2, doi: 10.1108/978-1-80262-937-820221005/full/html.
20. S. Basumatary, "THE ADVENT OF CULINARY TOURISM , ITS SCOPE & CHALLENGES," vol. 9, no. 5, 120-128, (2021).
21. C. S. Barua, "A Case Study on Homestay Tourism with Reference to Madhya Pradesh Region," vol. 3, no. 8, 873-878, (2022).
22. Kulkarni, V.N., Shah, G.: The Study of Culinary Tourism Management - A Tool for Revenue Generation and its Importance for Pune's Economic Status. International Journal of Trend in Scientific Research and Development. Volume-2, 1371–1385 (2018). https://doi.org/10.31142/ijtsrd10705.
23. Stalmirska, A.M.: Local Food in Tourism Destination Development: The Supply-Side Perspectives. Tourism Planning & Development. 21, 160–177 (2021). https://doi.org/10.1080/21568316.2021.1928739
24. Hsu, F.C.: Food tourism: consumer behaviour in relation to traditional food, https://doi.org/10.14264/uql.2015.373, (2015). https://doi.org/10.14264/uql.2015.373.

 https://doi.org/10.52202/074664-0035

Eco-Tourism - A reliable tool to maintain sustainability in the Eco-Sensitive Zone (ESZ) of North Sikkim

Bidisha Roy[1[0000-0001-6595-4073]] and Santosh Kumar Upadhyay[2[0000-0001-6871-2221]]

[1]Research Scholar, Department of Tourism & Hospitality, Assam Down Town University, Guwahati, Assam
[2]Associate Professor, Department of Tourism & Hospitality Assam Down Town, University, Guwahati, Assam

Abstract.

A significant approach to encourage sustainable development and the preservation of delicate ecosystems is eco-tourism. Ecologically Sensitive Zones (ESZ) are places, reserved to preserve certain species', natural habitats and safeguard biodiversity. Ministry of Environment, Forest and Climate Change (MOEF) on 27.8.2014, declared the following protected areas as Eco-Sensitive Zones (ESZs) in the State of Sikkim in violation of all legislative policies - Kitam Bird Sanctuary, Barsey Rhododendron Sanctuary, Fambong Lho Wildlife Sanctuary, Kyongnosla Alpine Sanctuary, Maenam Wildlife Sanctuary, Pangolakha Wildlife Sanctuary, Khangchendzonga National Park, Shingba Rhododendron Sanctuary. These Eco-Sensitive Zones usually serve as shock absorbers and have been notified keeping a mere 25-200 meters' width in violation of the Environment (Protection) Act, 1986 (EP Act), the Biological Diversity Act, 2002, the Precautionary Principle, the Principle of Sustainable Development, Intergenerational Equity and Principle of Eco-centrism. It seeks to lessen ecological imbalances and the conflict between humans and the natural world. The critical role of ecotourism as a tactical instrument for preserving sustainability in environmentally endangered regions is reviewed in this paper, with a special focus on North Sikkim, a region well-known for its untouched landscapes and rich biodiversity. The article addresses the ideas behind ecotourism as well as the obstacles that it confronts in North Sikkim. It emphasizes the necessity of an all- encompassing strategy that takes environmental preservation and community welfare into account. The article concludes with suggestions for legislators, locals, and tourists to encourage eco-friendly travel practices, beneficial for local socioeconomic well-being while protecting the region's natural assets.
Keywords: Ecotourism, Sustainability, Environmental Conservation

1. Introduction

A tourism approach termed "ecotourism," frequently referred to as “ethical tourism," or “sustainable tourism” intends to optimize the positive impacts of tourism while minimizing its detrimental effects on the environment, local communities, and cultures. It entails making judgments that advance the interests of the places visited, encouraging environmental preservation, honouring regional customs, and promoting the welfare of nearby populations. Formally entitled "Our Common Future," the Brundtland Report (1987) popularized the term "sustainable development." The World Commission on Environment and Development emphasized in its research the significance of forging a balance amid social justice, environmental preservation, and economic development. The report's fundamental principles paved the way for the incorporation of sustainability ideas into an array of business sectors, including tourism. Another comprehensive action plan for sustainable development, Agenda 21(1992) was endorsed by the United Nations Conference on Environment and Development (UNCED) in Rio de Janeiro. A special section on sustainable tourism was incorporated in Agenda 21's Chapter 10, which also reiterated the relevance of ecotourism approaches. A crucial turning point was the Cape Town Declaration (2002) on Ecotourism in Destinations. It brought the spotlight to the requirement of tourism to foster more desirable regions for people to migrate and settle in. In 2008, the Global Sustainable Tourism Criteria (GSTC) was established by a consortium of premier sustainable tourism organizations. These standards offer a framework that companies and tourist cultural respect, and locations can employ their operations to be socially, economically, and environmentally responsible. The connection between ecotourism and sustainable tourism is rooted in their mutual dedication to reducing adverse effects and fostering favourable results from tourism endeavours. In the tourism sector, environmental, social, and economic factors are acknowledged to be interconnected by both conceptions. Important facets of their partnership consist of environmental conservation, community involvement, and cultural benefit. While ecotourism strongly emphasizes the righteous attitude and responsibility of travellers and travel agencies, sustainable tourism analyses the destinations' longer-term economic and environmental viability. When combined, they offer a thorough strategy for developing a tourism sector that is socially and environmentally conscious.

Ecologically sensitive areas and striking a balance between development and conservation are central to the notion of Eco- Sensitive Zones (ESZ). Eco-sensitive zones are defined areas with certain limitations and regulations put in place to mitigate activity by humans that could result in an adverse impact on the environment. The concept and history of Eco-Sensitive Zones have changed in reaction to the realization of how crucial it is to preserve ecological processes, natural ecosystems, and biodiversity. International Conservation Movements, Legal Frameworks and International Agreements, National Legislation Policy, and Recognition of Fragile Ecosystems made these ESZs more noticeable all over the world. Conserving biodiversity, safeguarding wildlife habitats, maintaining ecological balance, and promoting sustainable resource use are the main goals of establishing Eco-Sensitive Zones. These zones seek to reconcile the necessity of protecting the environment with the requirements of nearby communities.

One of the four districts that make up the Indian state of Sikkim is North Sikkim. It is renowned for its breathtaking scenery, unusual cultural legacy, and varied flora and fauna. North Sikkim's salient features include- natural beauty & and landscapes, flora and fauna, cultural heritage, tourist destinations, accessibility, eco-sensitivity, tourism, and adventure activities. As per the National Tribunal Act (2010) under Section 14 & Section 18, some places are identified as Eco-Sensitive Zones (ECZ) - Barsey Rhododendron Sanctuary, Fambong Lho Wildlife Sanctuary,

https://doi.org/10.52202/074664-0036

Kitam Bird Sanctuary, Maenam Wildlife Sanctuary, Pangolakha Wildlife Sanctuary, Khangchendzonga National Park, Kyongnosla Alpine Sanctuary, Shingba Rhododendron Sanctuary. These Eco-sensitive Zones, which typically act as shock absorbers, have allegedly been notified that they are only 25 to 200 meters wide, in violation of the Biological Diversity Act of 2002, the Environment (Protection) Act of 1986 (also known as the EP Act), the Precautionary Principle, the Principle of Sustainable Development, Intergenerational Equity, and the Principle of Eco-centrism.

2. Review of Literature

Self-help groups are becoming more popular in developing countries for overall development. This research aims to conduct a thorough investigation of the elements that impede women's self-help groups' success. The report also suggests a path forward for resolving the issues for all institutions that support self-help, including banks, local, state, and federal governments, as well as non-governmental groups [13].

The study examines the correlation between the degree of engagement in tourism activities and the sense of community in Shiraz, Iran. The association between these variables has been ascertained through the application of Pearson correlation. The results obtained from the survey questionnaire indicated a strong correlation between the degree of participation in local tourism development and one's sense of community. The findings indicate that a strong sense of community can boost local participation and act as a catalyst for the growth of the tourism business [1].

This study reveals that gaining legal rights to land, opposing other extractive land uses, and cultural revival are the main factors influencing indigenous participation in ecotourism. Indigenous ecotourism will preserve and protect natural areas, uphold indigenous lifestyles, and benefit indigenous communities. It is made easier for indigenous people to participate in ecotourism in Sikkim because a significant portion of the tribal population owns landed property in the state [2].

The article reveals that from 1981 to 2011 Sikkim noticed domestic tourist growth at a compound annual rate of 11.87%, foreign tourist growth at an annual 6.98%, and domestic and foreign tourist growth at 11.50%. But with the rising rate of transport and communication, trade hotels and restaurants, banking and insurance the growth started declining. The paper aims to target the overall growth in tourism which can boost the economy the country as well [3].

The study's initial concept is that women's empowerment is the route to self-realization, and local tourism stakeholders' support serves as a sufficient foundation. Self-help groups (SHGs), founded by women in many rural communities, provide communication with a variety of stakeholders. This paper's goal is to look into how SHGs affect female self-employment and entrepreneurship in the tourism industry [14].

This paper's primary goal is to investigate how technology can be used to promote travel to India. The researcher chose the Andhra Pradesh districts of Krishna and Guntur in order to gather data from travel agents for this purpose. In order to limit the sample size of travel agents to 200 (100 from each district), a stratified random selection method has been used. Respondents are chosen using a random sampling technique. A research tool (questionnaire) is created based on the goal of the study in order to gather information from the respondents [4].

The tourism sector boosts the rate of economic growth and provides numerous opportunities for obtaining foreign exchange. The length of stay, amount of money spent by the sample respondents, and the effect of tourism on the local economy are all examined in this study. Since the analysis demonstrates that the tourism sector has no detrimental effects on the study area, the government should take the initiative to promote leisure, beach, religious, rural, and medical tourism, particularly in the underdeveloped areas, as these will directly help reduce the regional inequalities [9].

India's tourism sector is going through a transformation, and the government is launching new initiatives to digitalize it and boost productivity. The use of technologies like augmented reality and virtual reality will influence Indian tourism in the future. In this sense, the study will examine how technology innovations, such as artificial intelligence (AI) and the Internet of Things (IoT), have affected the travel and tourism sector and how they will help resolve operational problems there. The study offers insightful information that businesses in the Indian tourism industry can use to enhance and modify their technology-related strategies in response to market demands.

The study aims to attempt to carry out developmental projects through community-based organizations like self-help groups (SHGs). This article presents the findings of an evidence review on the effects of interventions delivered through SHGs on health, finance, agriculture, and empowerment outcomes in South Asia and sub-Saharan Africa. It also briefly summarizes theories regarding the efficacy of interventions utilizing SHGs. Even though SHG-based interventions typically have positive effects, the evidence base is small and doesn't often examine whether other delivery methods might be more efficient [6].

Sikkim is a well-liked travel destination for both domestic and foreign travelers. The South Sikkim village of Yuksam serves as the entry point to this mountain. According to trends, Yuksam is currently experiencing an increase in the number of tourists. This could have an impact on the village's infrastructure, environment, natural resources, culture, and ecotourism. Therefore, the purpose of this paper is to investigate ways to maintain tourism in Sikkim, specifically in Yuksam [8].

Examining the effectiveness of an integrative model for community involvement in tourism development was the aim of this study. The model, known as motivation—opportunity—ability (MOA), combined two research streams—"means" and "ends" oriented studies—to offer a comprehensive perspective on community participation research. The model received some support from the data, which indicates that motivation, opportunity, and ability to participate all play a role in how much a community member participates in the process of developing tourism. The general framework of community participation has been used to discuss the theoretical and practical implications [7].

This study evaluates how self-help groups (SHGs) can enhance community health interventions' efficacy and how that can affect SHG members' sense of empowerment. Analysis was done using qualitative information obtained in 2018 through in-depth interviews (IDI) with members of community-based Self-Help Groups (SHGs) who participated in the SWABHIMAAN program as Nutrition Friends (Poshan Sakhi-PS). The findings imply that in order to achieve better results, policies and programs pertaining to nutrition and health care should involve peer women from the community. In order to achieve the 2030 Sustainable Development Goals, women's empowerment and the closing of gender gaps in the workforce are essential [15].

The world over, there is widespread recognition of the significance of tourism as a tool for job creation and economic growth, especially in isolated and underdeveloped areas. In terms of both foreign exchange earnings and gross revenue, it is the largest service industry in the world [15].

https://doi.org/10.52202/074664-0036

According to the findings, there may be a long-term equilibrium between tourism and economic expansion. This supports the identification of the indicators that need to be given special attention when developing plans and strategies for the growth of the tourism industry. The long-run regression model's estimation indicates that the tourism industry's ability to drive economic growth is largely dependent on factors like foreign exchange earnings, foreign visitor spending, domestic tourism spending, and capital investment by all travel-related businesses [19].

The Sikkim Himalaya is a popular tourist destination because of its diverse ethnic groups, mountain peaks, holy lakes, and monasteries. It is also known for its high biodiversity and cultural heterogeneity. Over the course of five years (1988-1994), the number of visitors to Sikkim has increased by approximately 155% annually. In the Sikkim Himalaya, trekking, nature tourism, and leisure travel are all expanding quite quickly. This essay provides examples of how tourism can be developed and run sustainably to preserve local culture and flora while also benefiting the local population [20].

3. Objectives

- To sustain the biodiversity, ecological balance in Eco Sensitive areas of North Sikkim through ecotourism.
- To create awareness among community people regarding natural resources and community employment.

4. Research Methodology

4.1 Study Area

The epitome of Sikkim's beauty can be found in North Sikkim having N 27° 49' 25.0896", E 88° 33' 23.5116", longitude and latitude respectively, and spread over 4,226 square km^2 of entire Sikkim. The region of northern Sikkim is where the mountain peaks appear to reach the sky and consist of numerous passes, valleys, waterfalls, and mountain springs that enhance the area's natural splendor. Travel to North Sikkim offers guests a distinctly one-of-a-kind experience. Mostly a restricted area, most of this northernmost district of Sikkim requires special authorization to access. Khangchen zonga National Park and Shingba Rhododendron Sanctuary are the world famous tourist spots as well as Eco-Sensitive Zones of North Sikkim.

4.2 Data Source

Mainly based on secondary data sources gathered from different journals, articles, book chapters, and also from different websites.

4.3 Population Size

As the article is based on Eco-Sensitive Zones of North Sikkim, the footfall of tourists will be considered as the population.

4.4 Data Collection

Data Type – Secondary data (articles, journals, book chapters, websites)

5 Finding & Analysis

This article focused on PESTEL analysis to find out the external impact on Eco-Sensitive Zones(ESZ) of North Sikkim. As the ESZs are completely based on tourists' behaviour, external forces like political, economic, sociological, technological, environmental, and legal impact have been analyzed on tourists' behaviour to sustain the natural resources and positively promote ecotourism.

5.1 Political Impact

The government set up the Environmental Protection Act(EP) and Biological Diversity Act (BD) in the years 1986 and 2002 respectively to take precautionary measures and safeguard the natural resources for ecotourism. Political stability [11] and calmness in ESZs also create an impact on wildlife and enhance the natural environment. Additionally, the Precautionary Principle, the Principle of Sustainable Development [8], Intergenerational Equity, and the Principle of Eco-centrism are also strictly followed and maintained by the Government all over India. As per the Wildlife Protection Act (1972), seven sanctuaries have been notified by the state government of Sikkim as ESZs , among them, two are situated in North Sikkim only. There are substantial areas adjoining the parks, sanctuaries, and reserve forests that effectively and automatically act as buffer zones of protection.

5.2 Economic Impact

The more tourism takes place, the more national and international currency will be generated naturally. North Sikkim's economy [9] is directly proportionate to tourism, but tourism also forces economic development [3] to happen at the cost of destroying natural resources. Ecotourism safeguards natural resources by reducing footfalls to a limit that is not harmful to the environment. Only through ecotourism, the optimum economic growth [2] is possible without harming and disturbing the environment and wildlife.

5.3 Sociological Impact

Ecotourism encourages more greenery and discourages carbon emissions, and noise which are the cause of innumerable diseases. Society gets healthier surroundings without smoke, dust, and excessive sound of vehicles. Children grow up amongst lash-green, with a creative mind, full of oxygen, and natural playgrounds, and enjoy natural resources. Besides that, society people are habituated to living within the environment without degrading the resources [16]. This kind of living creates a significant balance between society people [19] and natural wildlife to strengthen each other. Society people also benefit from employment opportunities provided by different tourist spots development.

5.4 Technological Impact

Technology always destroys natural resources by applying different mechanisms. However, the adoption of environment- friendly technology promotes ecotourism and encourages sustainability. Jhum or Slash and Burn cultivation technique meets the rising population and also at the same time the demand for firewood and forest produce in the protected areas. Rainwater conservation technique, retreatment technique of sewage water, and using that water further for cultivation helps to maintain the underground water level and the over usage of water. More production of jute bags and recycled cloth bags ensures a ban on plastic in Eco-Sensitive Zones. Recycled items using techniques [4] also help in trash management in such areas that promote a healthy life for people as well as wild resources.

5.5 Environmental Impact

The environment always remains a big concern for ecotourism. Afforestation and reforestation can help to increase the wildlife of North Sikkim in different sanctuaries, natural parks, and protected areas, hence tourism and biodiversity conservation [20] will be possible. The rising issue of Global warming [12] can be automatically addressed at the same time. Future generations can be able to enjoy the same natural resources as present. Not only that, afforestation also reduces carbon levels in the air and gifts a healthy, pollution-free, harmful-disease-free society.

https://doi.org/10.52202/074664-0036

5.6 Legal Impact

Innumerable applications were submitted to the State government and Central Government's tourism department for safeguarding the Eco Sensitive Zones. Some instances are stated here as examples. Ministry of Environment, Forest and Climate Change (MOEF) declared Eco-Sensitive Zones(ECZ) in Sikkim on 27.08.2014. The legislative intent will be completely thwarted by the notifications issued by the MOEF by Section 3 of the EP Act regarding 8 ESZs that maintain a mere 25–200 meters, as a meager 25 meters would not act as buffer zones to protect the areas or as "shock absorbers." The application of respondents claims that the intent behind the State's declaration of eight ESZs has been a flagrant disrespect for all legal, judicial, and technical discussions and policies, especially about:

i) The guidelines of dated on 9.2.2011 published for the identification of ESZs
ii) Procedural excess vires during notice issuance;
iii) Disregarding public objections to draft notifications before final notification issuance.

Applicant contends that the MOEF has manifestly violated its policy, the National Board of Wildlife's policy, and the Apex Court's directive designating 10 kilometers as an ESZ perimeter around protected areas by demarcating arbitrary and inappropriate ESZ boundaries in the current instance.

In light of this, the applicant has requested that the Notifications dated August 27, 2014, be quashed, citing the Honourable NGT's sufficient judicial review powers as evidenced in the case of Wilfred J. & Anr vs Union of India (M.A. No. 182 of 2014 and MA No. 239 of 214 in Appeal No. 14 of 2014).

6. Suggestions:

Development of tourism, society, and balancing the natural resources can only be possible through ecotourism. Ecotourism boosts wildlife and attracts tourist footfalls, which directly reflects growth in the state economy as well as in the country's economic earnings. It also helps in community development by empowering community people, especially women of the society. Ecotourism can be boosted in different ways to make it a more familiar concept all over North Sikkim –

6.1 Self-Help Group

Self–help groups (SHGs) are small groups created for mutual help to meet financial crises [10], especially in the sustainable tourism sector [1] SHGs are becoming very popular due to community-driven development. These groups are extensive parts of the tourism industry by making different handicrafts, authentic food preparations of specific areas, and actively taking part in waste management of Eco-Sensitive Zones. Through SHGs, tourists can able to enjoy more organic foods and authentic crafts of North Sikkim. Strengthening the bond between SHGs and the tourism sector will boost income generation [17], personal benefits [7] empowerment [6], local control, self–improvement [14], and resource conservation [13], mutually and gradually without harming the natural resources of North Sikkim.

6.2 Awareness Creation & Women Involvement

In Bangladesh, a community-based cross-functional survey in 2018 revealed the fact that women are aware of their human rights [18] regarding material health which consists of higher education attainment, wealth status, and awareness. But only 20% of them are duty bearers while others stated their husband as duty bearers. Ecotourism can employ a strategy to involve women of the North Sikkim region in national awareness-creation paid activities and simultaneously level up their material wealth status. This strategy will help in creating awareness regarding natural resources, and rules of ECZs, among tourists and same time will involve the women empowerment [5] of those areas to contribute to their house-level-up activities.

7. Conclusion:

In summary, employing ecotourism provides an essential instrument to safeguard the unstable ecological balance in North Sikkim's Eco-Sensitive Zone (ESZ). To maintain the region's long-term sustainability, tourism must be approached with consideration and diligence due to its distinctive natural diversity. In addition to protecting the pristine environment, ecotourism promotes socioeconomic growth, cultural preservation, and community empowerment. Stakeholders (Community people, SHGs) can reduce the environmental effect of visitor activities, conserve biodiversity, and uphold the ecological integrity of North Sikkim's ESZ by using ecotourism practices. By encouraging low-impact activities, enforcing carrying capacity restrictions, and making investments in environmentally friendly infrastructure, sustainable tourism initiatives can help conserve natural resources and lower carbon footprints.

In addition, meaningful interaction with indigenous communities can be encouraged by ecotourism, which builds respect and understanding between the two parties. Through tourism-related projects, indigenous tribes in the area can be empowered and gain pride in maintaining their cultural heritage, better infrastructure, and economic prospects. Simply, ecotourism fosters a symbiotic relationship between tourists and the environment, preserving North Sikkim's ESZ's beauty and distinctiveness for coming generations. It is a comprehensive strategy that balances the interests of visitors, residents, and the environment, opening up the possibility of a resilient and sustainable future for this environmentally delicate area. Travelers, authorities, and industry stakeholders might collaborate to promote ecotourism, and North Sikkim could serve as a global example for ecotourism, demonstrating the benefits of reconciling human endeavours with the environment.

References

1. Aref, F., 2011. Sense of Community and Participation for Tourism Development Life Science Journal. Volume8. Issue1.
2. Binod Bhattarai, D. P., 2012. Eco-Tourism: An Engine of Economic Growth in Sikkim. Kurukshetra, November.pp. 43-46.
3. Chakrabarti, A., 2014. Economic Development and Tourism in Sikkim - A Critical Review. International Journal of Research in Commerce, Economics & Management, 4(3), pp. 2231-4245.
4. Darapureddy, B., 2022. Role of Technology in promoting Tourism (with reference to Travel Agents of Krishna & Guntur Districts of Andhra Pradesh). International Journal of Creative research Thoughts, 3 March, 10(3), pp. 2320-2882.
5. Dubey, N. and Thakur, M., 2023. Restoring Indian Tourism with Digital Innovations in New Normal: Tourism Business Perspective. Recent Advances in Commerce, Management, and Tourism, p.52.
6. Gugerty, M.K., Biscaye, P.E., Anderson, C.L.: Delivering development? Evidence on self-help groups as development intermediaries in South Asia and Africa. Development Policy Review. 37, 129–151 (2018).
7. Hung, K., Sirakaya-Turk, E., Ingram, L.J.: Testing the efficacy of an integrative model for community participation. Journal of Travel Research. 50, 276–288 (2010). https://doi.org/10.1177/0047287510362781.
8. Rubita, I.: Sustainability issue in tourism: A case study of Yuksam Village, Sikkim. Atna Journal of Tourism Studies. 7, 117–129 (2012). https://doi.org/10.12727/ajts.8.7.
9. Jansirani, S. and Priya, M.R., 2018. Tourism and Economic Development–A Case Study In Tamil Nadu. J. Eng. Res. Appl, 8(2), pp.01-08.
10. Deininger, K., Liu, Y.: Economic and social impacts of Self-Help groups in India. (2009). https://doi.org/10.1596/1813-9450-4884.

 https://doi.org/10.52202/074664-0036

11. Singha, K., Singh, M.: Political stability and its effect on economy: some lessons from Sikkim Himalaya. Journal of Social and Economic Development. 24, 340–354 (2022). https://doi.org/10.1007/s40847-022-00199-9.
12. Kumar, P., Sharma, M.C., Saini, R., Singh, G.K.: Climatic variability at Gangtok and Tadong weather observatories in Sikkim, India, during 1961–2017. Scientific Reports. 10, (2020). https://doi.org/10.1038/s41598-020-71163-y.
13. Nayak, A.K., Panigrahi, P.K., Swain, B.: Self-help groups in India: challenges and a roadmap for sustainability. Social Responsibility Journal. 16, 1013–1033 (2019). https://doi.org/10.1108/srj-02-2019-0054.
14. Vuković, D., Petrović, M.D., Maiti, M., Vujko, A.: Tourism development, entrepreneurship and women's empowerment – Focus on Serbian countryside. Journal of Tourism Futures. 9, 417–437 (2021). https://doi.org/10.1108/jtf-10-2020-0167.
15. Nikkhah, H.A., Redzuan, M.: The role of NGOs in promoting empowerment for sustainable community development. Journal of Human Ecology/Journal of Human Ecology. 30, 85–92 (2010). https://doi.org/10.1080/09709274.2010.11906276.
16. Ray, N., Das, D.K., Sengupta, P.P., Ghosh, S.: Rural Tourism and It's Impact on Socioeconomic Condition: Evidence from West Bengal, India. The Global Journal of Business Research. 6, 11–22 (2012).
17. Noorhayati, M.R., Hasmadi, I.M., Pakhriazad, H.Z., Wahidin, A.: Identifying motivation factors of the participation of local community in tourism industry in National Park, Pahang, Malaysia. In: CRC Press eBooks. pp. 107–112 (2016). https://doi.org/10.1201/b21184-20.
18. Perkins, J., Rahman, A.E., Siddique, A.B., Mazumder, T., Haider, M.R., Arifeen, S.E.: Awareness and perceptions of women regarding human rights related to maternal health in rural Bangladesh. Journal of Global Health. 9, (2019). https://doi.org/10.7189/jogh.09.010415.
19. Rout, H.B., Mishra, P.K., Pradhan, B.B.: Socio-Economic Impacts of Tourism in India: An Empirical analysis. Journal of Environmental Management and Tourism. 7, 762 (2017). https://doi.org/10.14505/jemt.v7.4(16).22.
20. Bhartwal, U., Monika, Tiwari, R.K., Sharma, N.K., Mathur, D., Saxena, S.: Dimensional role of the latest trends in sustainability at religious places in rural Rajasthan. Advances in Logistics, Operations, and Management Science Book Series. 135–145 (2023). https://doi.org/10.4018/979-8-3693-2065-5.ch010.

AUTHOR INDEX